BUILDING
CONSTRUCTION
TECHNOLOGY

U0731412

建筑
施工技术

中英版双语

主　编｜张孟阳　韦朝运　秦　康
副主编｜李伯阳　韦东奇　韦永丽

湖南大学出版社·长沙

图书在版编目（CIP）数据

建筑施工技术：汉、英/ 张孟阳，韦朝运，秦康主编.—长沙：湖南大学出版社，2024.10

ISBN 978-7-5667-2962-0

Ⅰ.①建…　Ⅱ.①张…　②韦…　③秦…　Ⅲ.①建筑施工—施工技术—教材—汉、英　Ⅳ.①TU74

中国国家版本馆CIP 数据核字（2023）第 075523号

建筑施工技术（中英双语版）

JIANZHU SHIGONG JISHU（ZHONG-YING SHUANGYU BAN）

主　　编：张孟阳　韦朝运　秦　康

策划编辑：贾志萍　胡建华

责任编辑：胡建华　尹鹏凯

印　　装：长沙市雅捷印务有限公司

开　　本：787 mm×1092 mm　1/16	印　　张：16.5	字　　数：454千字
版　　次：2024年10月第1版	印　　次：2024年10月第1次印刷	

书　　号：ISBN 978-7-5667-2962-0

定　　价：66.00元

出 版 人：李文邦

出版发行：湖南大学出版社

社　　址：湖南·长沙·岳麓山	邮　　编：410082	
电　　话：0731-88822559（营销部）	88821174（编辑部）	88821006（出版部）

传　　真：0731-88822264（总编室）

网　　址：http://press.hnu.edu.cn

目 录
Contents

1

土方工程

Earthworks

学习目标：

Learning objectives:

掌握土方工程施工的特点；能正确选用基坑支护形式；能正确选用降排水的方法；能正确选用土方施工机械；能正确选择回填的填土料及压实方法；能合理编制土方工程施工方案。

Be able to master the characteristics of earthwork construction; select the correct support for foundation pits; select the correct dewatering and drainage methods; select the correct earthwork construction machinery; select the correct filling materials and compaction methods for backfill; prepare earthwork construction schemes reasonably.

技能抽查要求：

Skill checking requirements:

能进行土方施工方案的编制。

Be able to prepare earthwork construction schemes.

建筑八大员岗位资格考试要求：

Qualification examination requirements for eight major posts of construction engineering:

能合理选择降排水方法、支护方法及开挖机械；能进行土方施工方案的编制，掌握土方工程施工质量及安全要求。

Be able to select the dewatering and drainage methods, support methods, and excavation machinery reasonably; prepare the earthwork construction schemes, master the construction quality and safety requirements of earthwork projects.

1.1 概述

1.1 Overview

1.1.1 土方工程的特点

1.1.1 Characteristics of Earthwork Construction

土方工程是建筑施工中的主要工程之一，是建筑施工过程中的第一道工序，包括一切土（石）方的开挖、填筑、运输，以及降水、排水等方面的活动。土方工程工程量大、施工周期长、劳动强度大、施工条件复杂，受地质、水文、气象等条件影响较大。因此，在组织土方工程施工时，应做好必要的工作，以确保工程质量。

Earthwork is a major project and the first process for building construction, including all earth(stone) excavation, filling, transportation, dewatering, and drainage works, featuring large quantities, long construction period, high labor intensity, and complex construction conditions, and greatly affected by geological, hydrological, and meteorological conditions. For this concern, necessary work shall be done when organizing earthwork construction to ensure the project quality.

1.1.2 土的工程分类

1.1.2 Engineering Classification of Soil

在建筑施工中，根据土的开挖难易程度，可以将土分为松软土、普通土、坚土、砂砾坚土、软石、次坚石、坚石、特坚石等八类（表1-1）。前四类为一般土，后四类为岩石。只有正确区分和鉴别土的种类，才能合理地选择施工方法、施工机械以及计算土方工程费用。

In construction, soil can be divided into eight categories, i.e., soft soil, ordinary soil, solid soil, gravelly solid soil, soft rock, sub-hard rock, hard rock, and extra-hard rock, according to the excavation difficulty(Table 1-1). The first four categories belong to general soil, and the last four are rocks. Only by correctly distinguishing and identifying soil categories, can we reasonably select the construction methods and machinery, and correctly calculate the earthwork cost.

表 1-1　土的工程分类及开挖方法

Table 1-1　Engineering Classification and Excavation Method of Soil

土的分类 Soil Category	土的名称 Soil Description	可松性系数 Loosening Coefficient		开挖方法 Excavation Method
		K_s	K_s'	
一类土 Cat-I soil （松软土） (soft soil)	砂；粉土；冲积砂土层；疏松的种植土；淤泥（泥炭） Sand; silt; alluvial sand layer; loose planting soil; sludge(peat)	1.08~1.17	1.01~1.03	用锹、锄头挖掘 Excavated with shovels and hoes
二类土 Cat-II soil （普通土） (ordinary soil)	粉质黏土；潮湿的黄土；夹有碎石、卵石的砂；粉质混卵（碎）石；种植土、填土 Silty clay; moist loess; sand and silty soil mixed with gravel and pebbles; planting soil, filling soil	1.14~1.28	1.02~1.05	用锹、锄头挖掘，少许用镐翻松 Excavated with shovels and hoes, and a few scarified with pickaxes

土的分类 Soil Category	土的名称 Soil Description	可松性系数 Loosening Coefficient		开挖方法 Excavation Method
		K_s	K'_s	
三类土 Cat-III soil （坚土） (solid soil)	软及中等密实黏土；重粉质黏土、砾石土；干黄土、含有碎石（卵石）的黄土；粉质黏土；压实的填土 Soft and medium dense clay; heavy silty clay; gravel soil; dry loess, loess with gravel and pebbles; silty clay; compacted filling soil	1.24~1.30	1.04~1.07	主要用镐，少许用锹、锄头，部分用撬棍 Mainly excavated with pickaxes, a few with shovels and hoes, and some with crowbars
四类土 Cat-IV soil （砂砾坚土） (gravelly solid soil)	坚硬密实的黏性土或黄土；中等密实的含碎石（卵石）黏性土；粗卵石；天然级配砾石；软泥灰岩 Hard and dense cohesive soil or loess; medium dense cohesive soil with gravel (pebbles); coarse pebbles; natural graded gravel; soft marl	1.26~1.32	1.06~1.09	用镐或撬棍，部分用楔子及大锤 Excavated with pickaxes or crowbars, some with wedges and sledgehammers
五类土 Cat-V soil （软石） (soft rock)	硬质黏土；中等密实的页岩、泥灰岩、白垩土；胶结不紧的砾岩；软石灰岩 Hard clay; medium dense shale, marl, chalk soil; loosely cemented conglomerate; soft limestone	1.30~1.45	1.10~1.20	用镐或撬棍、大锤，部分用爆破 Excavated with pickaxes or crowbars and sledgehammers, and some with blasting
六类土 Cat-VI soil （次坚石） (sub-hard rock)	泥岩；砂岩；砾岩；坚实的页岩、泥灰岩；密实的石灰岩；风化花岗岩、片麻岩及正长岩 Mudstone; sandstone; conglomerate; solid shale, marl; dense limestone; weathered granite, gneiss, and syenite	1.30~1.45	1.10~1.20	用爆破，部分用风镐 Blasting, some excavated with pneumatic picks
七类土 Cat-VII soil （坚石） (hard rock)	大理岩；辉绿岩；玢岩；粗、中粒花岗岩；坚实的白云岩、砂岩、砾岩、片麻岩、石灰岩；微风化安山岩、玄武岩 Marble, diabase; porphyrite; coarse-grained and medium-grained granite; solid dolomite, sandstone, conglomerate, gneiss, and limestone; slighty weathered andesite, basalt	1.30~1.45	1.10~1.20	爆破 Blasting
八类土 Cat-VIII soil （特坚石） (extra-hard rock)	安山岩；玄武岩；花岗片麻岩；坚实的细粒花岗岩、闪长岩、石英岩、辉长岩、辉绿岩、玢岩、角闪岩 Andesite; basalt; granitic gneiss; solid fine-grained granite, diorite, quartzite, gabbro, diabase, porphyrite, and amphibolite	1.45~1.50	1.20~1.30	爆破 Blasting

1.1.3 土的基本性质

1.1.3 Basic Properties of Soil

（1）土的组成

(1) Composition of soil

　　土一般由土颗粒（固相）、水（液相）和空气（气相）三部分组成，如图1-1所示。这三部分之间的比例关系随着周围条件的变化而变化，三者比例不同，反映出的物理状态也不同，如干燥、稍湿或很湿，密实、稍密或松散。这些指标是土最基本的物理性质，对评价土的工程性质，进行土的工程分类具有重要意义。

　　Generally speaking, soil is composed of three parts: soil particles(solid phase), water(liquid phase), and air(gas phase), as shown in Figure 1-1. The proportional relationship among the three parts varies with the change in surrounding conditions. Different proportions among the three parts reflect different physical states, such as dry, slightly wet or very wet; dense, slightly dense, or loose. These indexes, indicating the most basic physical properties, are of great significance for evaluating the engineering properties of soil and classifying soil for engineering purposes.

（2）土的物理性质

(2) Physical properties of soil

　　①土的含水量。

　　① Moisture content of soil.

　　土的含水量是指土中所含水的质量与土的固体颗粒的质量之比 w。

　　The moisture content of soil refers to the ratio of the mass of water contained in soil to the mass of solid particles in soil, the ratio is represented by w.

$$w = \frac{m_1 - m_2}{m_2} \times 100\% = \frac{m_w}{m_s} \times 100\% \qquad （1.1）$$

　　式(1.1)中 m_1 为含水状态时土的质量（kg），m_2 为烘干后土的质量（kg），m_w 为土中水的质量（kg），m_s 为土与水的质量（kg）。

　　Where, m_1 is the mass of soil in a water-bearing state (kg), m_2 is the mass of soil after drying

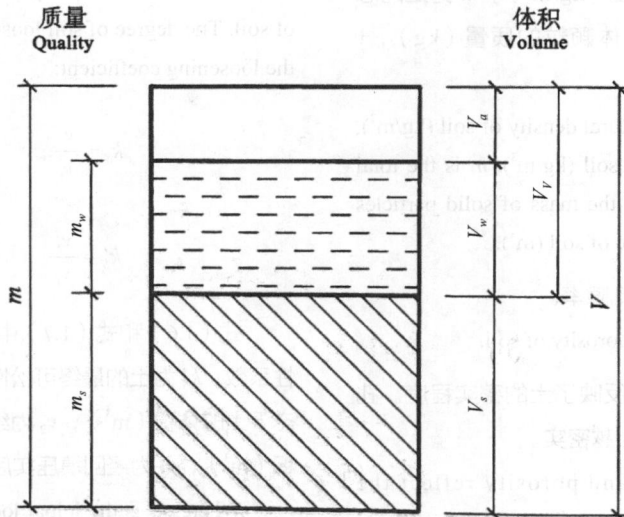

图1-1　土的三相示意图

Figure 1-1 Three-phase Schematic of Soil

(kg), m_w is the mass of water in soil (kg), and m_s is the mass of water and soil (kg).

②土的天然密度和干密度。

② Natural density and dry density of soil.

土的天然密度是指土在天然状态下单位体积的质量，可按下式计算：

The natural density of soil refers to the mass per unit volume of soil in the natural state, which can be calculated as per the following formula:

$$\rho = \frac{m}{v} \qquad (1.2)$$

土的干密度，指单位体积土中固体颗粒的质量，是填土压实质量的控制指标，土的干密度可以用下式表示：

The dry density of soil refers to the mass of solid particles per unit volume of soil, which is the index to control the fill compaction quality, it can be expressed by the following formula:

$$\rho_d = \frac{m_s}{v} \qquad (1.3)$$

式（1.2）（1.3）中 ρ 为土的天然密度（kg/m³），ρ_d 为土的干密度（kg/m³），m 为土的总质量（kg），m_s 为固体颗粒的质量（kg），v 为土的体积（m³）。

Where, ρ is the natural density of soil (kg/m³), ρ_d is the dry density of soil (kg/m³), m is the total mass of soil (kg), m_s is the mass of solid particles (kg), and v is the volume of soil (m³).

③土的孔隙比和孔隙率。

③ Void ratio and porosity of soil.

孔隙比和孔隙率反映了土的密实程度。孔隙比和孔隙率越小，土越密实。

The void ratio and porosity reflect the compactness of soil. The smaller the void ratio and porosity, the denser the soil.

孔隙比 e 是土的孔隙体积 v_v 与固体体积 v_s 的比值，用下式表示：

The void ratio e, is the ratio of the void volume v_v of soil to the volume of the solid v_s, it can be expressed by the following formula:

$$e = \frac{v_v}{v_s} \times 100\% \qquad (1.4)$$

孔隙率 n 是土的孔隙体积 v_v 与总体积 v 的比值，用百分率表示：

Porosity n is the ratio of the void volume v_v of soil to the total volume v, expressed in percentage:

$$e = \frac{v_v}{v} \times 100\% \qquad (1.5)$$

④土的可松性。

④ Looseness of soil.

天然状态下的土（原状土）经开挖后，其体积因松散而增加，即使经振动夯实，仍不能恢复到原来的体积，这种性质称为土的可松性。土的可松性程度用可松性系数表示：

After excavation, the volume of soil (undisturbed soil) in its natural state increases due to looseness, which cannot be restored to the original volume even after vibration and compaction. This property is called the looseness of soil. The degree of soil looseness is expressed by the loosening coefficient:

$$k_S = \frac{v_2}{v_1} \qquad (1.6)$$

$$k_S' = \frac{v_3}{v_1} \qquad (1.7)$$

式（1.6）和式（1.7）中 k_S 为土的最初可松性系数，k_S' 为土的最终可松性系数，v_1 为天然状态下土的体积（m³），v_2 为经开挖后土的松散体积（m³），v_3 为经回填压实后土的体积（m³）。

Where, k_S is the initial loosening coefficient of soil, k_S' is the final loosening coefficient of soil, v_1 is the volume of soil in natural state (m³), v_2 is the loose volume of soil after excavation (m³), and v_3 is the volume of soil after backfilling and compaction (m³).

可松性系数对土方的调配，计算土方运输量、填方量及选择运输工具都有影响，尤其是大型挖方工程，必须考虑土的可松性系数。

The loosening coefficient has an impact on earthwork allocation, calculation of earthwork transportation volume, filling volume, and selection of transportation tools. Especially for large excavation projects, the loosening coefficient of soil must be taken into account.

【例1-1】某工业厂房为钢筋混凝土条形基础，条形基础横截面面积为 3.0 m², 地基土为干黄土，基坑深 2.0 m，底宽 2.5 m。若需开挖 100 延米长基槽，请计算基槽挖土方量、填土量和弃土量（不考虑放坡，k_S=1.3, k'_S=1.05）。

[Example 1-1] An industrial workshop has a reinforced concrete strip foundation with a cross-sectional area of 3.0 m², dry loess as the foundation soil; the foundation pit is 2.0 m deep and 2.5 m wide at the bottom. If a 100 linear meter long foundation trench needs to be excavated, please figure out the cut, fill, and spoil volumes (Grading not considered, k_S=1.3, k'_S=1.05).

解：挖土量：V=2.0×2.5×100=500（m³）

Solution: Cut volume: V=2.0×2.5×100=500 (m³)

条形基础体积：V=3.0×100=300（m³）

Volume of strip foundation: V=3.0×100=300 (m³)

填土量：V=（500-300）/1.05×1.3≈247.6（m³）

Fill volume: V=(500-300)/1.05×1.3≈247.6 (m³)

弃土量：V=500×1.3-247.6=402.4（m³）

Spoil volume: V=500×1.3-247.6=402.4 (m³)

⑤土的渗透性。

⑤ Permeability of soil.

土的渗透性是指水在土体中渗流的性能，一般以渗透系数 k 表示。地下水在土中的渗流速度可按达西定律计算。

The permeability of soil refers to the permeability of water in the soil, usually expressed by the permeability coefficient k. The seepage velocity of groundwater in soil can be calculated according to Darcy's Law:

$$u = ki \qquad (1.8)$$

式中，u 为水在土中渗流速度；k 为土的渗透系数（m/d）；i 为水力坡度。

Where, u is the seepage velocity of water in soil; k is the permeability coefficient of soil (m/d); i is the hydraulic slope.

渗透系数反映出土透水性的强弱，它直接影响降水方案的选择和涌水量计算的准确性，可通过室内渗透试验或现场抽水试验确定，土的渗透系数见表 1-2。

The permeability coefficient, reflecting soil permeability, directly affects the selection of dewatering schemes and the accuracy of water inflow calculation. It can be determined by the indoor permeability test or field pumping test. See Table 1-2 for the permeability coefficient of general soil.

表 1-2 土的渗透系数参考

Table 1-2 Permeability Coefficients of Soil for Reference

土的名称 Soil Description	渗透系数k/（m/d） Permeability Coefficient k (m/d)	土的名称 Soil Description	渗透系数k/（m/d） Permeability Coefficient k (m/d)
黏土 Clay	<0.005	中砂 Medium sand	5.000~20.000

土的名称 Soil Description	渗透系数$k/$（m/d） Permeability Coefficient k (m/d)	土的名称 Soil Description	渗透系数$k/$（m/d） Permeability Coefficient k (m/d)
粉质黏土 Silty clay	0.015~0.100	均质中砂 Homogeneous medium sand	35.000~50.000
粉土 Silt	0.100~1.500	粗砂 Coarse sand	20.000~50.000
黄土 Loess	0.250~0.500	圆砾石 Round gravel	5.000~100.000
粉砂 Silty sand	0.500~1.000	卵石 Pebble	100.000~500.000
细砂 Fine sand	1.000~5.100		

1.2 土方工程量计算及土方调配

1.2 Earthwork Quantity Calculation and Earthwork Allocation

1.2.1 基坑、基槽

1.2.1 Foundation Pit and Trench

（1）基坑

(1) Foundation pit

基坑土方量的计算可按棱柱体（由两个平行的平面作为上下底的多面体）的体积公式来计算（图1-2）。

The earth volume of the foundation pit can be calculated approximately according to the volume formula of the prism(a polyhedron with two parallel planes as its topline and baseline) (Figure 1-2).

$$V = \frac{H}{6}(A_1 + 4A_0 + A_2) \qquad (1.9)$$

式中，H 为基坑深度（m）；A_1 为基坑上底面积（m²）；A_2 为基坑下底面积（m²）；A_0 为基坑中截面面积（m²）。

Where, H is the depth of the foundation pit (m); A_1 is the topline area of the foundation pit (m²); A_2 is the baseline area of the foundation pit (m²); A_0 is the cross-sectional area of the foundation pit (m²).

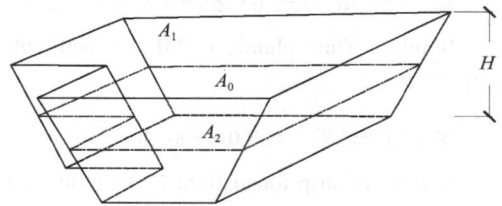

图1-2 基坑土方量计算

Figure 1-2 Calculation of Earth Volume of Foundation Pit

（2）基槽

(2) Foundation trench

基槽土方量可沿其长度方向分段后，按照上述方法计算（图1-3）。

The foundation trench, after being segmented along its length direction, can have the earth volume calculated by the same method as above (Figure 1-3).

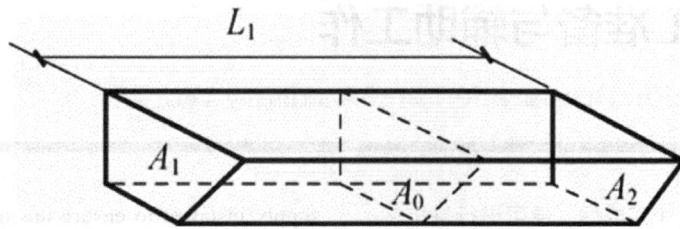

图1-3 基槽土方量计算

Figure 1-3 Calculation of Earth Volume of Foundation Trench

$$V_1 = \frac{L_1}{6}(A_1 + 4A_0 + A_2) \qquad (1.10)$$

式中，V_1 为第一段的土方量（m^3）；L_1 为第一段的长度（m）。

Where, V_1 is the earth volume of the first segment (m^3); L_1 is the length of the first segment (m).

然后将各段的土方量相加，即得总土方量 V：

Then, the earth volumes of each segment are added up to obtain the total earth volume V:

$$V = V_1 + V_2 + … + V_n \qquad (1.11)$$

式中，V_1、V_2、V_n 为各段的土方量（m^3）。

Where, V_1, V_2, and V_n represent the earth volumes of each segment (m^3).

1.2.2 土方调配

1.2.2 Earthwork Allocation

土方量计算完成后，即可着手土方的调配工作。土方调配，就是确定填、挖方区土方的调配方向和数量。好的土方调配方案，能使土方运输量或土方施工费最少，而且方便施工。

Upon completion of the earthwork calculation, earthwork allocation can be started. Earthwork allocation is to determine the allocation direction and quantity of earthwork in the filling and excavation area. A good earthwork allocation scheme should minimize the earth transportation volume or cost and facilitate construction.

土方调配的原则：

Principles of earthwork allocation:

①力求达到挖方与填方平衡和运距最短。

① Achieve cut-and-fill balance and minimize the haul distance.

②近期施工和后期利用。土方调配，必须依据现场具体情况、有关技术资料、工期要求、土方施工方法和运输方法来进行。

② Facilitate short-term construction and future utilization. Earthwork allocation must be properly carried out according to the specific site conditions, relevant technical data, construction period requirements, earthwork construction methods, and transportation methods.

施工者应综合上述原则，并经计算比较，选择经济合理的土方调配方案。

Based on the above principles and through calculation and comparison, an economical and reasonable allocation scheme shall be selected.

1.3 施工准备与辅助工作

1.3 Construction Preparation and Auxiliary Work

在场地平整工作完成后，便可进行基坑的开挖。基坑的开挖往往涉及一系列的问题，如边坡的稳定，深基坑的支护，降低地下水位以及基坑开挖方案的确定等。

Upon completion of site leveling, the excavation of the foundation pit can be launched. This work often involves a series of problems, such as slope stability, support of deep foundation pits, reduction of groundwater level, and the determination of foundation pit excavation scheme.

1.3.1 施工准备

1.3.1 Construction Preparation

（1）技术准备

(1) Technical preparation

施工者熟悉施工图纸，踏勘施工现场，掌握水文地质条件，从而确定合理施工方案及施工方法。

Get familiar with construction drawings, investigate the construction site, and master the hydrogeological conditions, so as to determine reasonable construction schemes and methods.

（2）现场准备

(2) Site preparation

场地需进行清理、平整，设置排水设施排除地面水，修筑场内道路及搭设临时建筑物，安装供水、供电等临时设施，从而确保土方施工顺利进行。

The site shall be cleaned and leveled, with drainage facilities set up to drain surface water, roads and temporary buildings built, and temporary facilities such as those for water supply and power

supply installed to ensure the smooth progress of earthwork construction.

1.3.2 土方边坡与土壁支撑

1.3.2 Earth Slope and Earth Wall Support

（1）土方边坡

(1) Earth slope

为保证土方边坡稳定，防止土壁塌方，确保施工安全，当挖方超过一定深度或填方超过一定高度时，应做一定形式的边坡或设置临时支撑。影响边坡稳定的因素很多，主要有土的种类、基坑开挖深度、水、坡顶荷载、振动等。

For stability of the earth slope, prevention of earth wall collapse, and construction safety, when cut exceeds a certain depth or fill exceeds a certain height, a certain form of slope or temporary support shall be applied. Many factors can affect slope stability, such as the soil category, excavation depth of the foundation pit, water, load on the slope top, and vibration.

土方边坡的坡度以开挖深度 H 与放坡的宽度 B 之比来表示，见图 1-4（a），即

The gradient of the earth slope is expressed by the ratio of excavation depth H to slope width B, as shown in Figure 1-4 (a)

土方边坡的坡度 $=H/B=1/(B/H)=1：m$

Gradient of earth slope $= H/B=1/(B/H)=1：m$

式中，$m=B/H$，称为边坡系数。

Where, $m=B/H$, called slope coefficient.

边坡坡度应根据土质、开挖深度、开挖方法、施工工期、地下水位、坡顶荷载及气候条

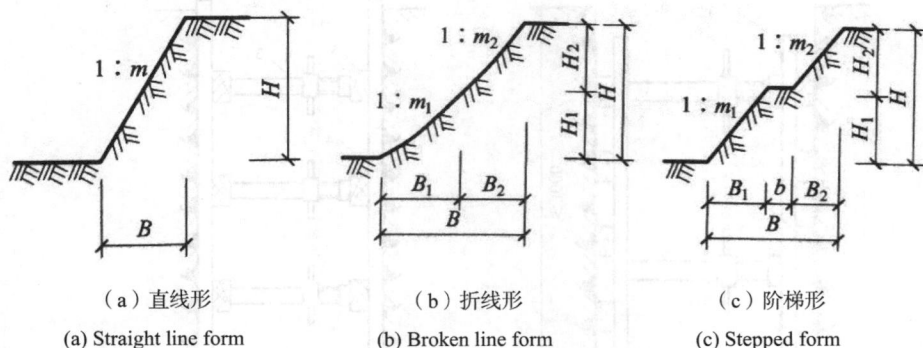

(a) 直线形 (b) 折线形 (c) 阶梯形
(a) Straight line form (b) Broken line form (c) Stepped form

图1-4　基坑边坡
Figure 1-4 Slope of Foundation Pit

件等因素确定，可做成直线形、折线形或阶梯形（图1-4）。

The slope gradient shall be determined according to the soil quality, excavation depth, excavation method, construction period, groundwater level, load on the slope top, and climatic conditions, and can be made into the straight line form, broken line form, or stepped form (Figure 1-4).

合适的边坡系数应满足安全与经济两方面的要求，既要保证边坡稳定，又不增加土方量。一般边坡系数 m 由设计文件规定，当设计文件未作规定时，应按照其他有关规定来选取。

Appropriate slope coefficient shall meet the requirements of safety and economy, so as to ensure slope stability without increasing the amount of earthwork. Generally, the slope coefficient m is specified in the design documents. If not specified in such a way, it shall be selected according to other specifications.

（2）土壁支撑

(2) Earth wall support

采用放坡开挖基坑（槽）法往往是比较经济的。但有时受场地的限制不能按要求放坡，可采用设置土壁支撑的施工方法。土壁支撑可分为横撑式支撑、锚锭式支撑及板桩式支撑等。

Sloping excavation of the foundation pits(trenches) is often economical. However, sometimes sloping may not be applied as required due to site restrictions. In this case, earth wall support can be adopted, including cross bracing, anchor bracing, and sheet pile bracing.

①横撑式支撑。

① Cross bracing.

开挖狭窄的基坑（槽）或管沟时，可采用横撑式支撑。横撑式支撑指贴附于土壁上的挡土板，可水平铺设或垂直铺设，可断续铺设或连续铺设（图1-5）。断续式水平挡土板支撑在湿度小的黏性土中及挖土深度小于 3 m 时采用。连续式水平挡土板支撑用于挖土深度不大于 5 m 的较潮湿或松散的土。连续式垂直挡土板支撑则常用于湿度很高和松散的土，挖土深度不限。

For the excavation of narrow foundation pits (trenches) or pipe ditches, cross bracing can be used. The cross bracing refers to the retaining plate attached to the earth wall which can be laid horizontally or vertically, intermittently or continuously (Figure 1-5). Bracing with intermittent horizontal retaining plates is used for cohesive soil with low moisture with an excavation of less

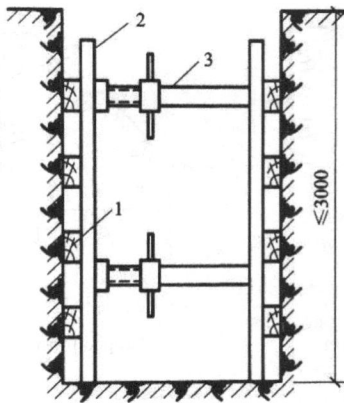

（a）断续式水平挡土板支撑（单位：mm）　　　（b）垂直挡土板支撑

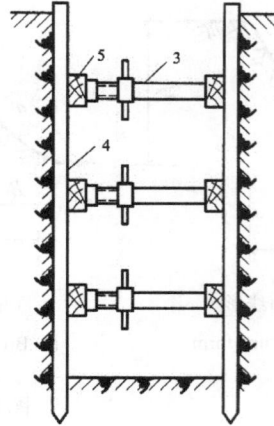

(a) Bracing with intermittent horizontal retaining plates　　(b) Bracing with vertical retaining plates

1.水平挡土板；2.竖楞木；3.工具式横撑；4.竖直挡土板；5.横楞木

1. horizontal retaining plate; 2. vertical joist; 3. tool-based cross bracing; 4. vertical retaining plate; 5. horizontal joist

图1-5　横撑式支撑

Figure 1-5 Cross Bracing

than 3 m. Bracing with continuous horizontal retaining plates is used for damp or loose soil with an excavation depth of not more than 5 m. Bracing of continuous vertical retaining plates is often used for loose soil and soil with high moisture, without limiting the excavation depth.

②锚锭式支撑。

② Anchor bracing.

锚锭式支撑指水平挡土板支在柱桩的内侧，柱桩一端打入土中，另一端用拉杆与锚桩拉紧，在挡土板内侧回填土。这种方式可在开挖较大型、深度不大的基坑或使用机械挖土，不能安设横撑时使用，锚锭式支撑形式见图1-6。

The anchor bracing refers to the horizontal retaining plate that is erected on the inner side of the pile, which has one end driven into the soil and the other end tightened with the anchor pile via a tie bar, with soil backfilled on the inner side of the retaining plate. This bracing is suitable for excavating large foundation pits with small depths

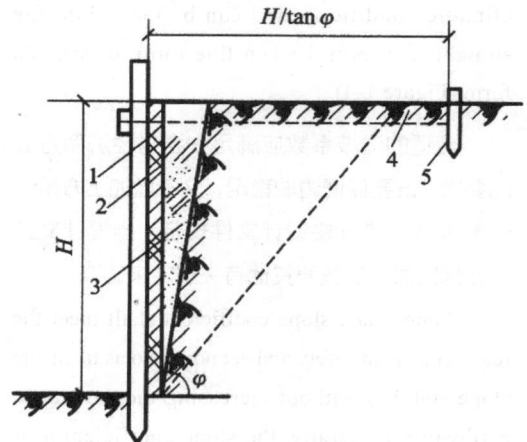

1.柱桩；2.挡土板；3.回填土；4.拉杆；5.锚桩

1. pile; 2. retaining plate; 3. backfill; 4. tie bar;

5. anchor pile

图1-6　锚锭式支撑

Figure 1-6 Anchor Bracing

or for mechanical excavation where crossing bracing cannot be used. See Figure 1-6 for the form of anchor bracing.

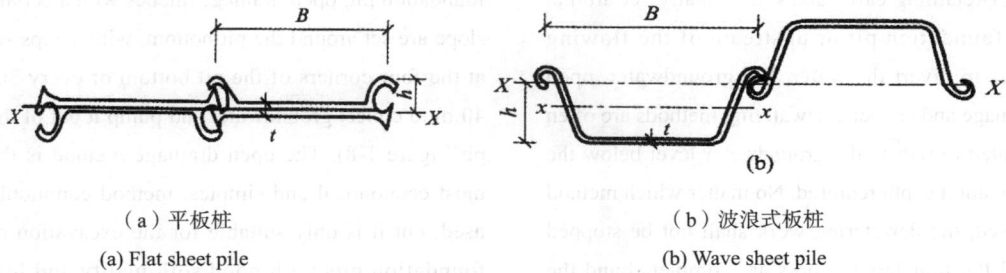

| （a）平板桩 | （b）波浪式板桩 |
| (a) Flat sheet pile | (b) Wave sheet pile |

图1-7 钢板桩示意图

Figure 1-7 Schematic of Steel Sheet Piles

③板桩式支撑。

③ Sheet pile bracing.

板桩式支撑特别适用于地下水位较高且土质为细颗粒、松散饱和的土的支护，可防止流砂现象的发生。板桩种类很多，有木板桩、钢板桩及钢筋混凝土板桩等，其中钢板桩应用最广。

Sheet pile bracing is especially suitable for supporting loose, fine-grained, saturated soil with a high groundwater level, to prevent the occurrence of quicksand. There are many types of sheet piles, including wooden sheet piles, steel sheet piles, and reinforced concrete sheet piles, among which steel sheet piles are the most widely used.

钢板桩又可分平板桩和波浪式板桩两类。平板桩防水和承受轴向压力性能良好，易打入地下，但长轴方向抗弯强度较小，如图1-7（a）所示。波浪式板桩的防水和抗弯性能都较好，施工中多采用该类板桩，如图1-7（b）所示。

Steel sheet piles further include flat sheet piles and wave sheet piles. Flat sheet piles, with good waterproof performance and axial compression resistance, are easy to be driven into the ground but have bad bending resistance in the long axis direction, as shown in Figure 1-7 (a). Wave sheet piles, with good waterproof and bending resistance, are widely used in construction, as shown in Figure 1-7 (b).

1.3.3 排水与降低地下水位

1.3.3 Drainage and Reduction of Groundwater Level

当基坑底面低于地下水位时，开挖基坑的过程将会切断土壤的含水层，使得地下水不断渗入基坑，导致地基承载力不断下降，同时也容易造成基坑边坡塌方。因此，为了保证工程质量和施工安全，在基坑开挖前或开挖过程中，必须采取措施降低地下水位，使基坑在开挖过程中坑底始终保持干燥。对于地面水，一般采取在基坑四周或流水的上游设排水沟、截水沟或挡水土堤等办法解决。对于地下水，则常采用明排水法和人工降低地下水位法，使地下水位降至所需开挖的深度以下。无论采用何种方法，排水与降水工作都应持续到基础工程施工完毕并回填土后才可停止。

When the bottom of the foundation pit is lower than the groundwater level, excavation will cut off the aquifer of the soil and let groundwater seep into the foundation pit, which will keep degrading the foundation pit's bearing capacity and result in slope collapse. Therefore, for project quality and construction safety, measures must be taken to reduce the groundwater level before or during the excavation of the foundation pit, so as to keep the pit bottom dry all the time. For surface water, drainage ditches, intercepting ditches, or water-retaining earth dikes are usually set around the foundation pit or upstream of the flowing water to divert the water. For groundwater, open drainage and artificial dewatering methods are often adopted to reduce the groundwater level below the excavation depth required. No matter which method is used, the dewatering work shall not be stopped until the foundation works are completed and the soil is backfilled.

（1）明排水法

(1) Open drainage method

在逐层开挖基坑的过程中，在坑底周围设置具有一定坡度的排水明沟，并在坑底四角或每 30~40 m 设置集水井，使地下水流入集水井内，然后用水泵将水抽出坑外（图 1-8），这种方法为明排水法。明排水法是一种常用的最经济、最简单的方法，但仅适用于土质较好且地下水位不高的基坑。当土为细砂或粉砂时，易发生流砂现象，此时可采用人工降低地下水位的方法。

foundation pit, open drainage ditches with a certain slope are set around the pit bottom, with sumps set at the four corners of the pit bottom or every 30-40 m to collect groundwater and pump it out of the pit(Figure 1-8). The open drainage method is the most economical and simplest method commonly used, but it is only suitable for the excavation of foundation pits with good soil quality and low groundwater level. When soil is composed of fine sand or silty sand, quicksand is likely to happen. In this case, the artificial dewatering method can be used.

①集水井与排水明沟的设置。

① Setting of sumps and open drainage ditches.

基坑四周的排水沟及集水井应随基坑开挖逐层设置，并设置于基础轮廓 0.3 m 以外、地下

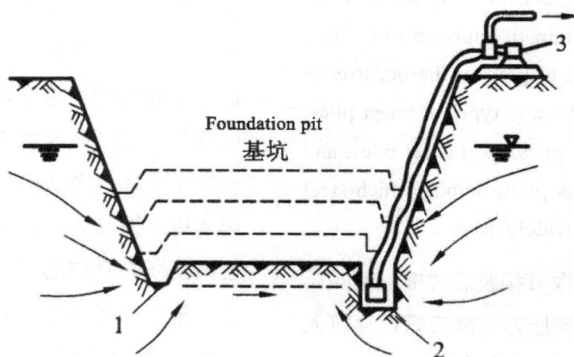

1.排水沟；2.集水井；3.水泵

1. drainage ditch; 2. sump; 3. water pump

图 1-8　集水井排水

Figure 1-8 Dewatering of Sump

During the layered excavation of the

水流的上游；排水沟的坡度为 1‰ ~5‰，断面尺寸一般不小于 0.5 m × 0.5 m；集水井的直径或宽度一般为 0.7~1.0 m，其深度宜比排水沟的深度低 0.5~1.0 m。

Drainage ditches and sumps around the

foundation pit shall be set layer by layer along with the pit excavation, and shall be set 0.3 m away from the foundation contour, upstream of the groundwater flow. The drainage ditch shall have a gradient of 1‰-5‰ and a cross-sectional size of usually not less than 0.5 m×0.5 m. The sump shall have a diameter or width of 0.7-1.0 m and a depth of 0.5-1.0 m lower than the drainage ditch.

②水泵的选用。

② Selection of water pump.

明排水用水泵从集水井中抽水，常用的水泵有潜水泵、离心水泵和泥浆泵。一般所选用水泵的抽水量为基坑涌水量的 1.5~2.0 倍。

Open drainage pumps water from the sump. Common pumps include the submersible pump, centrifugal pump, and mud pump. In most cases, the pumping capacity of the pump selected is 1.5-2.0 times the water inflow of the foundation pit.

③流砂的发生与防治。

③ Occurrence and prevention of quicksand.

采用明排水法开挖基坑时，当基坑挖至地下水位以下，如果坑底、坑壁的土粒形成流动状态并随地下水渗流不断涌入基坑，即称为流砂现象。

With the open drainage method, when the foundation pit is excavated below the groundwater level, if the soil particles at the pit bottom and wall keep flowing into the foundation pit with the seepage of groundwater, then quicksand occurs.

a. 发生流砂现象的原因。

a. Causes of quicksand.

如图 1-9（a）所示，由于高水位的左端（水头为 h_1）与低水位的右端（水头为 h_2）之间存在压力差，水经过长度为 l，断面积为 F 的土体由左端向右端渗流。

As shown in Figure 1-9(a), due to a pressure difference between the high water level at the left end(head: h_1) and the low water level at the right end(head: h_2), the water seeps from left to right into the soil mass with a length of l and a cross-sectional area of F.

如图 1-9（b）所示，动水压力 G_D 的大小与水力坡度成正比，即水位差 h_1-h_2 越大，则 G_D 越大；而渗透路程越长，则 G_D 越小。动水压力的作用方向与水流的方向相同。当水流在水位差的作用下对土颗粒产生向上的压力时，土颗粒受到了向上的浮力。如果动水压力等于或大于土的饱和密度时，则土颗粒处于悬浮状态，土的抗剪强度等于零，土颗粒能随着渗流的水一起流动，这便会导致流砂现象。

As shown in Figure 1-9(b), the magnitude of

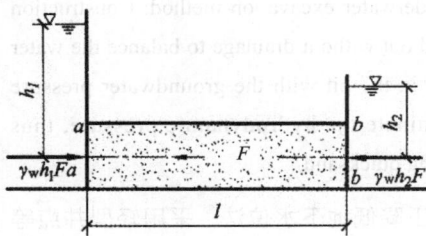

（a）动水压力对地基土的影响

(a) Impact of hydrodynamic pressure on foundation soil

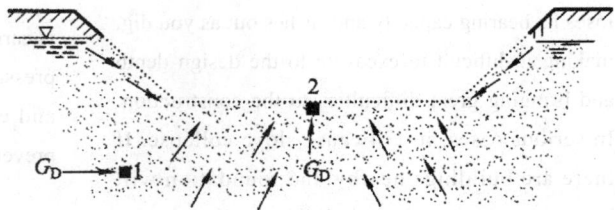

（b）水在土中渗流时的力学现象

(b) Mechanical phenomenon of water seepage in soil

图1-9 动水压力原理图

Figure 1-9 Schematic of Hydrodynamic Pressure

the hydrodynamic pressure G_D is proportional to the hydraulic slope, namely, the larger the water head h_1-h_2, the greater the G_D, while the longer the infiltration path, the smaller the G_D. The direction of action of the hydrodynamic pressure is the same as that of the water flow. When the water flow exerts upward pressure on soil particles under the action of the water head, soil particles are subject to upward buoyancy. If the hydrodynamic pressure is equal to or greater than the saturated density of the soil, soil particles are in a suspended state, and the shear strength of the soil is equal to zero. In this case, soil particles can flow with the seepage water, which causes the so-called "quicksand phenomenon".

b. 流砂的防治办法。

b. Prevention and control methods of quicksand.

当发生流砂现象时，土完全丧失承载力，土边挖边冒，很难挖到设计深度，给施工带来极大困难，严重时会引起边坡塌方，如果附近有建筑物，还会引起地基被掏空而使建筑物下沉、倾斜，甚至倒塌。消除流砂现象的关键在于控制动水压力的大小与方向。所以，在基坑开挖中，防治流砂的原则是"治流砂必治水"，主要途径是消除、减小或平衡动水压力或者改变动水压力的方向。其具体措施有：

When quicksand occurs, the soil completely loses its bearing capacity and rushes out as you dig, making it difficult to excavate to the design depth and bringing great difficulties to the construction. In serious cases, it will cause slope collapse. If there are buildings nearby, the foundation will be hollowed out and the buildings will sink, tilt, or even collapse. The key to eliminate quicksand is controlling the magnitude and direction of hydrodynamic pressure. Therefore, in the excavation of foundation pits, the principle is "preventing and controlling quicksand is to control water", and the main way is to eliminate, reduce, or balance the hydrodynamic pressure or change the direction of hydrodynamic pressure. The specific measures include:

抢挖法：组织分段开挖，使挖土速度超过冒砂速度，挖到标高后立即铺竹筏或芦席，并抛大石块以平衡动水压力，压住流砂，此法可解决轻微流砂现象。

Urgent excavation method: Excavate in sections to increase the excavation speed over the sand inrush speed. Lay bamboo rafts or reed mats immediately after reaching the elevation, and fill in large stones to balance the hydrodynamic pressure and press the quicksand. This method can solve slight quicksand.

打板桩法：将板桩打入坑底下面一定深度，增加地下水从坑外流入坑内的渗流长度，减小水力坡度，从而减小动水压力，防止流砂产生。

Sheet pile driving method: Sheet piles are driven into a certain depth below the pit bottom to increase the seepage length of groundwater flowing into the pit from outside and reduce the hydraulic slope, thus reducing the hydrodynamic pressure and preventing quicksand.

水下挖土法：不排水施工，使坑内水压力与地下水压力平衡，消除动水压力，从而防止流砂产生。

Underwater excavation method: Construction is carried out without drainage to balance the water pressure in the pit with the groundwater pressure and eliminate the hydrodynamic pressure, thus preventing quicksand.

人工降低地下水位法：采用轻型井点等降水，使地下水降到坑底以下，水不致流入坑内，从而防止流砂产生。

Artificial dewatering method: Light well points are adopted to lower groundwater below the pit bottom and prevent water from flowing into the pit, thus preventing quicksand.

地下连续墙法：在基坑周围浇筑一道混凝土连续墙，以支承土壁、截水并防止流砂产生。此外，还可以选择在枯水期施工等方法。

Underground diaphragm wall method: A concrete diaphragm wall is poured around the foundation pit to support the earth wall, intercept water, and prevent quicksand. In addition, construction in the dry season is also an option.

（2）人工降低地下水位法

(2) Artificial dewatering method

当土层是软土层或者含有淤泥层、细砂时，不宜采用明排水法，因为在基坑中直接排水，地下水将产生自下而上或从边坡向基坑方向流动的动水压力，容易导致边坡塌方和产生流砂现象，并使基底土结构遭受破坏，这种情况应考虑采用人工降低地下水位法。

For a soft layer or when the soil layer contains silt and fine sand, the open drainage method is not recommended, because direct discharge in the pit will generate a hydrodynamic pressure from the groundwater as it flows from the bottom up or from the slope to the foundation pit, which is easy to cause slope collapse and quicksand and damage the base soil structure. In this case, artificially dewatering method should be considered.

人工降低地下水位就是在基坑开挖前，预先在基坑周围埋设一定数量的滤水管（井），利用抽水设备不断抽出地下水，使地下水位降低到坑底以下，直至基础工程施工完毕。人工降低地下水位法改善了工作条件，防止了流砂现象的发生。同时，由于地下水位在降低过程中动水压力向下作用与土体自重作用，使基底土层压密，提高了地基土的承载能力。

Artificial dewatering is to, before the excavation of the foundation pit, bury a proper number of filter pipes(wells) around the foundation pit and continuously pump out the groundwater to lower the groundwater level below the pit bottom until the foundation project is completed. Artificial dewatering improves working conditions and prevents the occurrence of quicksand. Also, due to the downward action of the hydrodynamic pressure during groundwater fall and the self-weight of the soil, the base soil layer is compacted, improving the bearing capacity of the foundation soil.

人工降低地下水位法按其系统的设置、吸水原理和方法的不同，可分为轻型井点、喷射井点、电渗井点、管井井点和深井井点，其中轻型井点应用最广泛。

According to the different system settings, water absorption principles and methods, artificial dewatering method can be divided into light well points, jet well points, electro-osmotic well points, tube well points, and deep well points, among which light well points are most widely used.

①轻型井点降低地下水位。

① Dewatering by light well points.

轻型井点降低地下水位是指沿基坑四周每隔一定距离将若干直径较小的井点管埋入蓄水层内，井点管上端伸出地面，通过弯联管与总管相连并引向水泵房，利用抽水设备将地下水从井点管内不断抽出，使地下水位降至坑底以下的过程，如图1-10所示。

Dewatering by light well points is to bury a number of well pipes with small diameters into the aquifer at intervals around the foundation pit. The upper end of the well pipe extends out of the ground, connected to the header pipe via an elbow pipe, and led to the water pump house. Groundwater is continuously pumped out from well pipes to lower the groundwater level below the pit bottom, as shown in Figure 1-10.

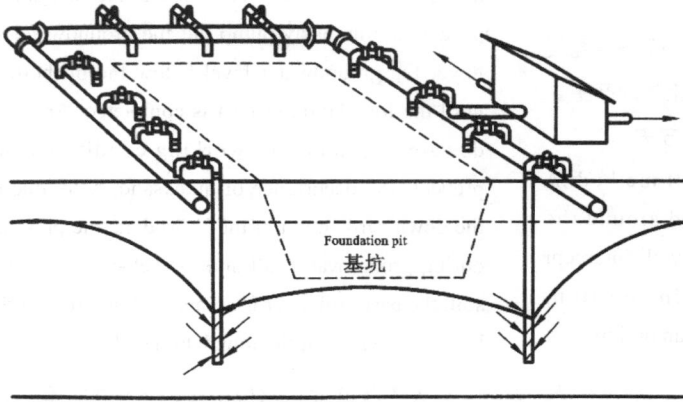

图1-10 轻型井点降低地下水位示意图

Figure 1-10 Schematic of Dewatering by Light Well Points

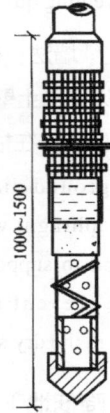

图1-11 滤管构造（单位：mm）

Figure 1-11 Structure of Filter Pipe

轻型井点由管路系统和抽水设备两部分组成。

A light well point consists of two parts: the pipeline system and pumping equipment.

抽水设备由真空泵、离心泵和集水箱等组成。

Pumping equipment consists of vacuum pump, centrifugal pump, and water tank.

②轻型井点的布置。

② Layout of light well points.

轻型井点的布置，应根据基坑的大小和深度、土质、地下水位的高低与流向、降水深度要求等因素确定。设计时主要考虑平面和高程两个方面。轻型井点管路系统的滤管构造如图1-11所示。

The layout of light well points shall be determined according to the size and depth of the foundation pit, soil quality, height and flow direction of groundwater, and dewatering depth requirements. The plane and elevation are mainly considered in the design. The structrue of filter pipe of the light well point is shown in Figure 1-11.

a. 平面布置。

a. Plane layout.

单排线状井点布置（图1-12）：基坑或沟槽宽度小于6 m，且降水深度不超过5 m。布置在地下水流的上游一侧，两端延伸长度一般不小于沟底宽度。

Layout of single-row linear well points (Figure 1-12): the foundation pit or trench has a width of less than 6 m, and the dewatering depth is limited to 5 m. They are arranged upstream of the groundwater flow, with the extension length at both ends not less than the width of the ditch bottom.

双排线状井点布置（图1-13）：基坑或沟槽宽度大于6 m或土质不良。

Layout of double-row linear well points (Figure 1-13): The width of the foundation pit or trench exceeds 6 m or the soil quality is poor.

环形井点布置（图1-14）：基坑开挖面积较大。

Layout of annular well points (Figure 1-14): The excavation area of the foundation pit is large.

b. 高程布置。

b. Elevation layout.

轻型井点的降水深度从理论上讲可达10 m左右，但由于抽水设备的水头损失，实际降水深度一般不大于6 m。井点管的埋设深度 H（不包括滤管）可按式（1.12）计算：

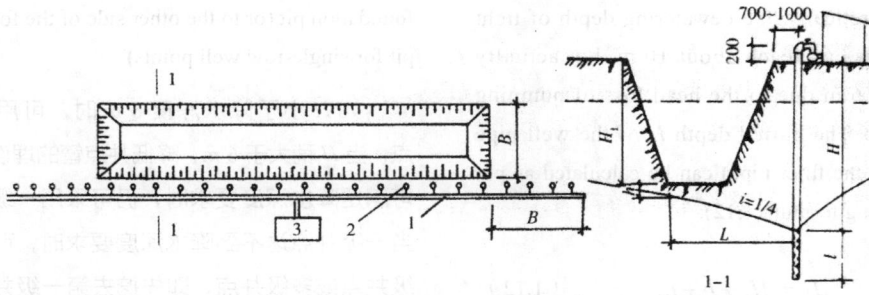

（a）平面布置

(a) Plane layout

（a）平面布置

(a) Plane layout

（b）高程布置

(b) Elevation layout

1.总管；2.井点管；3.抽水设备

1. header pipe; 2. well pipe; 3. pumping equipment

图1-12　单排线状井点布置图（单位：mm）

Figure 1-12 Layout of Single-row Linear Well Points

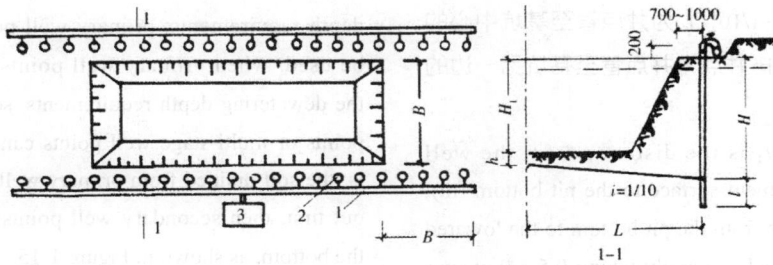

（a）平面布置

(a) Plane layout

（b）高程布置

(b) Elevation layout

1.总管；2.井点管；3.抽水设备

1. header pipe; 2. well pipe; 3. pumping equipment

图1-13　双排线状井点布置图（单位：mm）

Figure 1-13 Layout of Double-row Linear Well Points

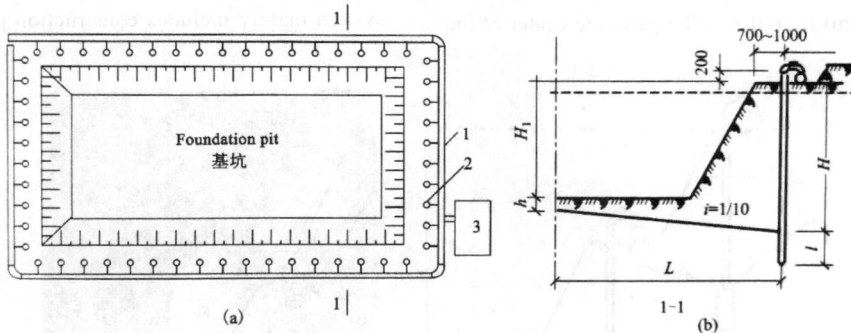

Foundation pit
基坑

（a）平面布置

(a) Plane layout

（b）高程布置

(b) Elevation layout

1.总管；2.井点管；3.抽水设备

1. header pipe; 2. well pipe; 3. pumping equipment

图1-14　环形井点布置图（单位：mm）

Figure 1-14 Layout of Annular Well Points

Theoretically, the dewatering depth of light well points can reach about 10 m, but actually limited to 6 m due to the head loss of pumping equipment. The buried depth H of the well pipe (excluding the filter pipe) can be calculated as per the following formula(1.12):

$$H = H_1 + h + iL \qquad (1.12)$$

式中，H_1 为井点管理设面到基坑底面的距离（m）；h 为基坑底面至降低后的地下水位线的距离，一般取 0.5~1.0 m（人工开挖取下限，机械开挖取上限）；i 为降水曲线坡度，可取实测值或按经验，单排井点取 1/4，双排井点、环形井点取 1/15~1/10；L 为井点管至基坑中心的水平距离，单排井点为井点管至基坑另一边的距离（m）。

Where, H_1 is the distance from the well point's management surface to the pit bottom (m); h is the distance from the pit bottom to the lowered groundwater level, generally taking 0.5-1.0 m (use the lower limit for manual excavation and upper limit for mechanical excavation); i is the gradient of the dewatering curve, taking the measured value or according to experience, using 1/4 for single-row well points and 1/15-1/10 for double-row well points and annular well points; L is the horizontal distance (m) from the well pipe to the center of the foundation pit (or to the other side of the foundation pit for single-row well points).

如 H 小于降水深度 6 m 时，可用一级井点；当 H 稍大于 6 m，降低井点管的埋设面后，可满足降水深度要求时，仍可采用一级井点；当一级井点达不到降水深度要求时，可采用二级井点或多级井点，即先挖去第一级井点所疏干的土，然后在其底部埋设第二级井点，见图1-15。

If H is less than the dewatering depth of 6 m, primary well points can be used; when H is slightly more than 6 m and the buried surface of the well pipe can be lowered to meet the dewatering depth requirements, primary well points can still be used; when primary well points cannot meet the dewatering depth requirements, secondary well points or multi-stage well points can be used, that is, the soil drained from primary well points is dug out first, then secondary well points are buried at the bottom, as shown in Figure 1-15.

③轻型井点的施工。

③ Construction of light well points.

轻型井点的施工，主要包括施工准备、井点系统的安装、使用及拆除。

The construction of the light well point system mainly includes construction preparation,

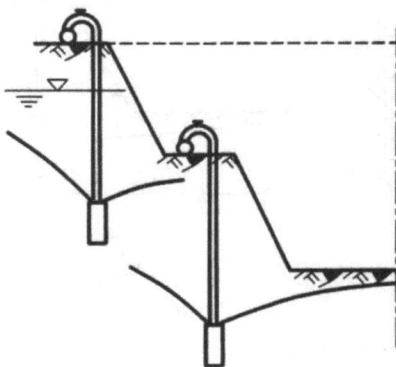

图1-15　二级轻型井点示意图

Figure 1-15 Schematic of Secondary Light Well Points

installation, use, and removal of the well point system.

在井点系统安装时，先根据降水方案埋设总管，再冲孔、埋设井点管，然后用弯联管将井点管与总管连接，最后安装抽水设备。

During the installation of the well point system, the header pipe is buried first according to the dewatering scheme, then punched, with well pipes buried. After that, well point pipes are connected with the header pipe via elbow pipes, and finally, the pumping equipment is installed.

井点管的埋设一般用水冲法，水冲法分为冲孔与埋管两个过程。冲孔时，利用起重设备将冲管吊起，并插在井点位置上，利用高压水泵冲松土体使其下沉，冲孔应垂直，直径一般为300 mm，以保证井管壁有一定厚度砂滤层，冲孔深度要比滤管底深0.5 m左右，以防冲管拔出时部分土颗粒沉于底部而触及滤管。

The burial of well point pipes is generally done by the water jetting method, including punching and burial. During punching, the jet pipe is hoisted by lifting equipment and inserted into the well point. The high-pressure water pump is used to loosen the soil when sinking the pipe. The punched hole shall be vertical, with a diameter of 300 mm, so as to keep a sand filter layer of proper thickness for the well pipe wall. The punching depth shall be about 0.5 m deeper than the filter pipe's bottom, in case some soil particles sink to the bottom and touch the filter pipe when the jet pipe is pulled out.

轻型井点系统全部安装完毕后，应进行抽水试验，以检查井点管有无淤塞或漏气、漏水现象。在井点系统的使用过程中，应连续抽水，时抽时停会抽出大量泥沙，使滤管淤塞，并可能造成附近建筑物因土粒流失而沉降开裂。

After the light well point system is completely installed, a pumping test shall be carried out to check whether the well point pipe is blocked or leaked. During the use of the well point system, pumping shall be continuous instead of intermittent. Otherwise, it may extract a large amount of mud and sand to block the filter tube and may cause settlement and cracking of nearby buildings due to the loss of soil particles.

1.4　土方的回填与压实
1.4 Backfill and Compaction of Earthwork

土方回填主要有地基填土、基坑（槽）或管沟回填等。土方回填是一项很重要的工作，要求回填土应有一定的密实性，使回填土土层不致产生较大的沉陷。在实际施工中，一些建筑物沉降过大，室内地坪和散水出现大面积严重开裂，主要原因之一就是回填压实的密实度没有达到设计规范要求。

Earthwork backfill mainly includes foundation fill, foundation pit(groove) or pipe ditch backfill. As an important job, the backfill must be properly compacted to avoid large settlement for the backfill layer. In actual construction, the large settlement of some buildings causes wide and serious cracks in indoor floors and aprons. One major reason is that the compactness of backfilling does not meet the design specifications.

1.4.1 回填土料的选择

1.4.1 Selection of Backfill

回填土料应符合设计要求，以保证填方的强度与稳定性。凡含水量过大的黏土，含有 8% 以上的有机物（腐烂物）的土，含有 5% 以上的水溶性硫酸盐的土以及淤泥、垃圾土、冻土、膨胀土等均不能作为回填土。

The backfilling soil shall meet the design requirements to ensure the strength and stability of the fill. Clay with excessive water content, soil containing more than 8% organic matter(rotten), soil containing more than 5% water-soluble sulfate, silt, garbage soil, frozen soil, and expansive soil cannot be used as backfill.

同一回填工程应尽量采用同类土回填；如采用不同土回填时，必须按土类不同分层夯填，并将透水性大的置于透水性小的土层之下，以防填土内形成水囊。

The same type of soil shall be used for the same backfilling project as far as possible. If different types of soil are used for backfilling, they must be tamped and filled in layers according to their types, with those of high water permeability placed under those of low water permeability to avoid water pockets in the filled soil.

1.4.2 填土压实方法

1.4.2 Fill Compaction Method

填土压实的方法一般有碾压、夯实、振动压实，如图 1-16 所示。

Fill compaction methods include rolling, tamping, and vibrating compaction, as shown in Figure 1-16.

（1）碾压法

(1) Rolling

碾压原理是利用沉重的滚轮碾压土壤表面，使土壤在压力作用下压实。碾压法适用于大面积填土工程。

Rolling is to roll the soil surface with heavy rollers so that the soil can be compacted under pressure. This method is suitable for large-area filling works.

碾压机械有光面碾（压路机）、气胎碾和羊足碾。光面碾是一种以内燃机为动力的自行式压路机，砂土和黏性土均可压实，应用最普遍。气胎碾在工作时是弹性体，其压力均匀，填土质量好。羊足碾靠拖拉机牵引，由于它与土接触面小，单位面积压力大，故压实效果好，主要用于黏性土的压实。因在砂土中使用羊足碾会使土颗粒受到羊足碾较大单位压力后向四周移动，从而使土的结构遭到破坏，碾压机械压实填方时，行驶速度不宜过快，一般光面碾控制在 2 km/h，羊足碾控制在 3 km/h，否则会

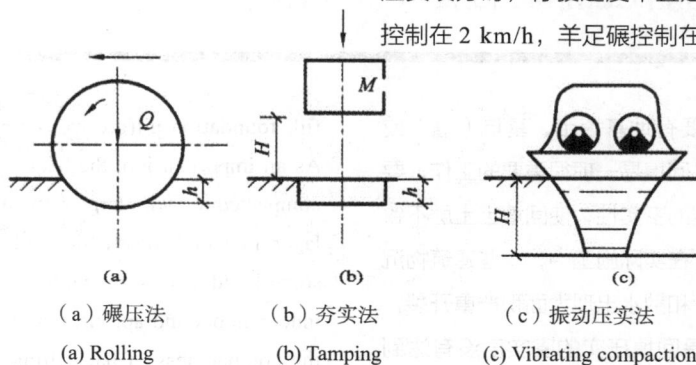

(a)	(b)	(c)
（a）碾压法	（b）夯实法	（c）振动压实法
(a) Rolling	(b) Tamping	(c) Vibrating compaction

图1-16 填土压实方法

Figure 1-16 Fill Compaction Method

影响压实效果。

Rolling machines include the smooth roller (road roller), pneumatic tired roller, and tamping roller. As the most widely used type, the smooth roller is a self-propelled roller powered by the internal combustion engine, which can be used to compact sandy soil and cohesive soil. The pneumatic tired roller is an elastomer during work, demonstrating uniform pressure and good filling quality. The tamping roller is pulled by a tractor. Due to its small contact surface with soil and large pressure per unit area, it shows a good compaction effect, mainly used for the compaction of cohesive soil. However, the use of a tamping roller in sandy soil will force soil particles away under the large unit pressure of the roller, thus damaging the soil structure. When compacting the fill, the roller shall drive at a proper speed, not too fast, normally at 2 km/h for smooth rollers and 3 km/h for tamping rollers. Otherwise, the compaction effect will be affected.

（2）夯实法
(2) Tamping

夯实法是利用夯锤自由下落的冲击力来夯实土壤，主要用于小面积回填。夯实法分人工夯实和机械夯实两种。

Tamping is to compact the soil with the impact force of a free-falling rammer. This method is mainly used for small-area backfilling. Tamping includes manual tamping and mechanical tamping.

夯实机械有夯锤、内燃夯土机和蛙式打夯机。人工夯土用的工具有木夯、石夯等。夯锤是借助起重机悬挂一重锤进行夯土的夯实机械，适用于夯实砂性土、湿陷性黄土、杂填土以及含有石块的填土等。

Tamping machines include rammers, internal combustion rammers, and frog rammers, and tools for manual tamping include wooden rammers and stone rammers. A rammer is a ramming machine that hangs a hammer with the help of a crane to ram soil, suitable for ramming sandy soil, collapsible loess, miscellaneous fill, and fill containing stones.

（3）振动压实法
(3) Vibrating compaction

振动压实法是将振动压实机放在土层表面，借助振动机械使压实机机械振动，使土颗粒在振动力的作用下发生相对位移而达到紧密状态的方法。这种方法振实非黏性土的效果较好。

Vibrating compaction is to place a vibrating compactor on the soil surface and make the compactor vibrate with the help of a vibrator so that soil particles will be displaced under the action of vibrating force to reach a tight state. This method is recommended for compacting non-cohesive soil.

使用振动碾进行碾压，可使土体受振动和碾压两种作用，碾压效率高，适用于大面积填方工程。

The vibratory roller, when used for rolling, can apply both vibration and rolling effects on the soil, achieving high rolling efficiency, and is suitable for large-area filling works.

1.4.3 影响填土压实的因素
1.4.3 Factors Affecting Fill Compaction

影响填土压实的因素很多，主要有压实功、土的含水量，以及每层铺土厚度。

Many factors may affect the compaction of fill, mainly including compaction effort, soil moisture content, and the soil-paving thickness of each layer.

（1）压实功
(1) Compaction effort

填土压实后的密度与压实机械在其上所施加的功有一定的关系。土的密度与所耗的功的关系如图 1-17 所示。当土的含水量一定，在开始压实时，土的密度急剧增加，待到接近土的最大密度时，压实功虽然增加许多，但土的密度则变化甚小。在实际施工中，对于砂土只需碾压或夯击 2~3 遍，对粉土只需 3~4 遍，对粉质黏土或黏土只需 5~6 遍。此外，松土不宜用重型碾压机械直接滚压，否则土层有强烈起伏现象，压实效率不高。如果先用轻碾压实，再用重碾压实就会取得较好的效果。

The density of the filled soil after compaction is related to the work applied by the compaction machine. The relationship between the soil density and the compaction effort is shown in Figure 1-17. When the moisture content of the soil is constant, the soil density rises sharply at the beginning of compaction. When the density approaches the maximum density of the soil, although the compaction effort increases greatly, the soil density changes little. In actual construction, sand only needs to be rolled or tamped 2-3 times, silt only needs to be rolled 3-4 times, and silty clay or clay only needs to be tamped 5-6 times. In addition, loose soil should not be directly rolled by heavy rolling machines; otherwise, the soil layer will show strong fluctuations, and the efficiency is not high. Light rolling for compaction before heavy rolling will yield better results.

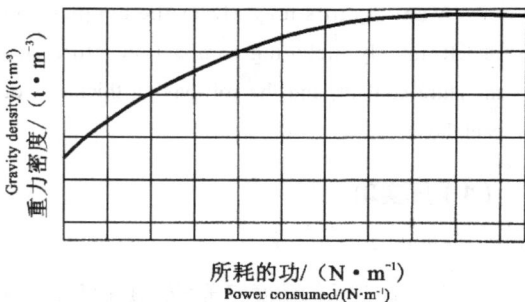

图1-17　土的密度与压实功的关系示意

Figure 1-17 Relation between Soil Density and Compaction Effort

（2）土的含水量

(2) Soil moisture content

在同一压实功条件下，填土的含水量对压实质量有直接影响。较为干燥的土颗粒之间的摩阻力较大，因而不易压实。当含水量超过一定限度时，土颗粒之间孔隙由水填充而呈饱和状态，也不能压实。当土的含水量适当时，水起了润滑作用，土颗粒之间的摩阻力减小，压实效果好。每种土都有其最佳含水量，土在这种含水量的条件下，使用同样的压实功进行压实，所得到的密度最大（图 1-18）。工地现场简单检验黏性土最佳含水量的方法一般是"手握成团落地开花"。为了保证填土在压实过程中处于最佳含水量状态，当土过湿时，应予以翻松晾干，也可掺入同类干土或吸水性土料；当土过干时，则应预先洒水润湿。

Under the same compaction effort conditions, the moisture content of the fill has a direct impact on the compaction quality. Dry soil particles have relatively large friction, thus not easy to compact. When the moisture content exceeds a certain limit, pores between soil particles are saturated by water and cannot be compacted. When the soil has an appropriate moisture content, water can facilitate lubrication to reduce the frictional resistance between soil particles and achieve a good compaction effect. Each type of soil has its optimal moisture content. Under the optimal moisture condition, the soil is compacted with the same compaction effort to obtain the maximum density (Figure 1-18). Generally, the way to simply test the optimal moisture of cohesive soil on the construction site is to see whether the soil lumps when held in hands and scatters when hitting the ground. To keep the fill in the optimal moisture state during compaction, the soil, if too wet, shall be scarified and dried, or added with the same kind of dry soil or water-absorbing soil; or if too dry, shall be watered in advance.

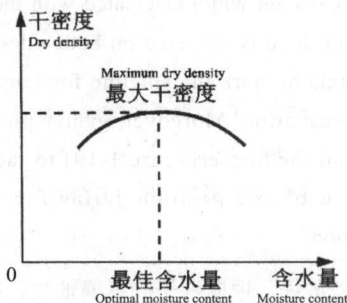

干密度
Dry density

Maximum dry density
最大干密度

0　　最佳含水量　　含水量
　　Optimal moisture content　　Moisture content

1-18　土的干密度与含水量的关系

Figure 1-18 Relation between Soil Density and Moisture Content

（3）每层铺土厚度

(3) Soil-paving thickness of each layer

土在压实功的作用下，其压力随深度增加而逐渐减小，其影响深度与压实机械、土的性质和含水量等有关。铺土厚度应小于压实机械压土时的作用深度，但其中还有最优土层厚度的问题。铺得过厚，要压很多遍才能达到规定的密实度；铺得过薄，也要增加机械的总压实次数。

Under the action of compaction effort, the stress of soil gradually decreases with the increase of depth, and its influence depth depends on compaction machinery, soil properties, and moisture content. The soil-paving thickness shall be smaller than the action depth of the compaction machine, but it also involves the problem of the optimal soil-layer thickness. If the pavement is too thick, it needs to be compacted many times to achieve the specified compactness; if the pavement is too thin, it requires more times of compaction by the machine.

压实功、土的含水量，以及每层铺土厚度之间是相互影响的。为了保证压实质量，提高压实机械的压实效率，重要工程中，应根据土质和所选用的压实机械在施工现场进行压实试验，以确定达到规定密实度所需的压实次数、铺土厚度及最优含水量。

The compaction effort, soil moisture content, and thickness of each soil-paving layer are mutually affected. For satisfactory compaction quality and better production efficiency of compaction machinery, compaction tests shall be carried out on the construction site for important works according to the soil quality and the selected compaction machinery to determine the compaction times, soil-paving thickness, and optimal moisture content required to achieve the specified compactness.

1.5　基坑（槽）施工

1.5 Foundation Pit(Trench) Construction

基坑（槽）的施工，首先应进行房屋定位和标高引测，然后根据基础的底面尺寸、埋置深度、土质好坏、地下水位的高低及季节性变化等不同情况，考虑施工需要，确定是否需要留工作面、边坡，增加排水设施及设置支撑，从而定出挖土边线和进行放灰线等工作。

For foundation pit(trench) construction, operators shall first conduct the house positioning and elevation survey and consider construction needs(whether to reserve the working face, slope; add drainage facilities; or set up supports) based on such factors as the bottom sizes of the foundation, buried depth, soil quality, groundwater level, and seasonal changes, so as to set out excavation sideline and gray line.

1.5.1 放线
1.5.1 Setting-out

基槽放线：根据房屋主轴线控制点，首先将外墙轴线的交点用木桩测设在地面上，并在桩顶钉上铁钉作为标志。房屋外墙轴线测定以后，再根据建筑物平面图，将内部开间所有轴线都一一测出。最后根据边坡系数计算的开挖宽度在中心轴线两侧用石灰在地面上撒出基槽开挖边线。同时在房屋四周设置龙门板（图1-19），以便于基础施工时复核轴线位置。

to the excavation width calculated with the slope coefficient, lime is scattered on both sides of the central axis to mark the sideline for foundation trench excavation. Moreover, gantry plates are set around the house(Figure 1-19) to facilitate the review of axis position during foundation construction.

柱基放线：根据柱子的纵横轴线，用经纬仪在矩形控制网上测定基础定位桩的端点。同时在每个柱基中心线上，测定基础定位桩，每个基础的中心线上设置四个定位木桩，其位与基础开挖线的距离为0.5~1.0 m。

Setting-out of column base: According to the vertical and horizontal axes of the column, measure

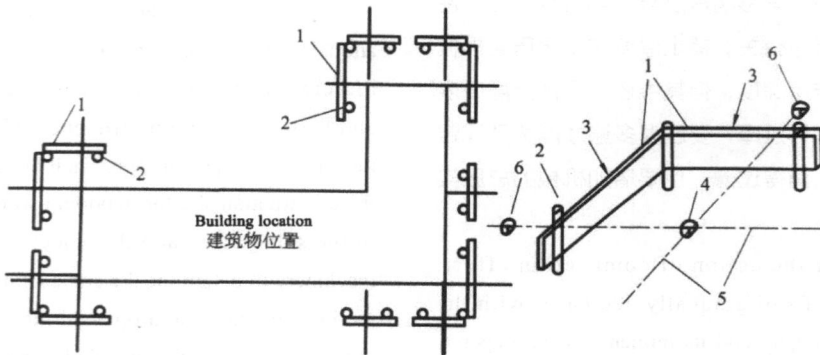

1.龙门板；2.龙门桩；3.轴线钉；4.角桩；5.轴线；6.控制桩
1.gantry plate; 2. gantry pile; 3. axis nail; 4. corner pile; 5. axis; 6. control pile

图1-19　建筑定位
Figure 1-19 Building Positioning

Setting-out of foundation trench: According to the control point of the house's main axis, the intersection point of the exterior wall axes is measured and set on the ground with a wooden pile, with nails fixed on the pile top as marks. After the exterior wall axes of the house are determined, measure all the axes of internal bays one by one according to the building plan. Finally, according

the endpoint of the foundation postioning pile on the rectangular control network with a theodolite. Meanwhile, measure the foundation positioning piles on the centerline of each column foundation. Four wooden positioning piles are set on the centerline of each foundation, and the distance from the pile position to the foundation excavation line is 0.5-1.0 m.

1.5.2 基坑（槽）开挖

1.5.2 Foundation Pit(Trench) Excavation

　　土方开挖应遵循"开槽支撑、先撑后挖、分层开挖、严禁越挖"的原则。

　　Earth excavation shall follow the principles of "supporting trenches during excavation, supporting before excavation, layered excavation, and no excessive excavation".

　　开挖基坑（槽）应按规定的尺寸合理安排开挖顺序和分层开挖高度，连续地进行施工，尽快地完成。土方开挖施工要求标高、断面准确，土体应有足够的强度和稳定性，所以在开挖过程中要随时注意检查。挖出的土除预留一部分用作回填外，不得在场地内任意堆放，应把多余的土运到弃土地区，以免妨碍施工。为防止坑壁滑坡，根据土质情况及坑（槽）高度，在坑顶两边一定距离（一般为 1.0 m）内不得堆放弃土，在此距离外堆土高度不得超过 1.5 m，否则，应验算边坡的稳定性。在桩基周围，墙基或围墙一侧，不得堆土过高。在坑边放置有动载的机械设备时，也应根据验算结果，在距坑边较远处放置，如地质条件不好，还应采取加固措施。

　　The excavation sequence and layered excavation height of the foundation pit(trench) shall be reasonably arranged according to the specified sizes, for continuous construction and scheduled completion. Since the earthwork excavation requires accurate elevations and sections, and the soil mass shall have sufficient strength and stability, regular inspection is required during excavation. The excavated soil, except some reserved for backfilling, shall be transported to spoil ground instead of being stacked within the site for not interfering with construction. For the avoidance of pit wall collapse, according to the soil quality and pit(trench) depth, no soil shall be piled up within a certain distance(normally 1.0 m) on both sides of the pit top, and the height of the piled soil outside this distance shall not exceed 1.5 m. Otherwise, the slope stability shall be checked. The soil must not be piled up too high around the pile foundation and at one side of the fence or wall foundation. Mechanical equipments with a dynamic load, when placed at the pit edge, shall be kept far away from the edge by referring to the checking calculation results, and take reinforcement measures if the geological conditions are unsatisfactory.

　　为了防止基底土（特别是软土）受到水或其他因素的扰动，基坑（槽）挖好后，应立即做垫层或浇筑基础，否则，挖土时应在基底标高以上保留 150~300 mm 厚的土层，待基础施工时再行挖去。如用机械挖土，为防止基底土被扰动、结构被破坏，不应直接挖到坑（槽）底，应根据机械种类，在基底标高以上留出 200~400 mm。待基础施工前用人工铲平修整。挖土不得挖至基坑（槽）的设计标高以下，如有个别处超挖，应用与基土相同的土料填补，并夯实到要求的密实度，如图 1-20 所示。

　　To prevent the base soil(especially soft soil) from disturbance due to immersion or other reasons, upon finishing excavation of the foundation pit(trench), operators shall make cushions or pour the foundation immediately. Otherwise, a 150-300 mm thick soil layer shall be reserved above the base elevation during excavation, and this layer will be removed during foundation construction. For mechanical excavation, to prevent the base soil from disturbance and protect the structure, operators shall not directly excavate to the pit(trench) bottom, instead reserving 200-400 mm above the base elevation according to the type of machinery. This part shall be manually trimmed and leveled before foundation construction. Excavation shall not reach below the design elevation of the foundation pit(trench). Any part over-excavated shall be made up with the same soil as the foundation soil and tamped to the required compactness, as shown in Figure 1-20.

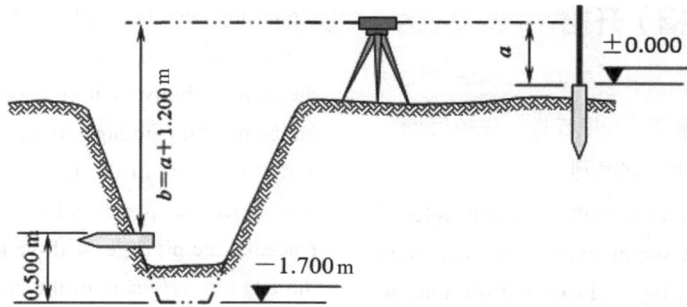

图1-20 基坑（槽）开挖

Figure 1-20 Foundation Pit(Trench) Excavation

在软土地区开挖基坑（槽）时，应符合下列规定：

The excavation of foundation pits(trenches) in the mollisol area shall meet the following requirements:

①施工前必须做好地面排水和降低地下水位工作，地下水位应降低至基坑底以下0.5~1.0 m后，方可开挖。降水工作应持续到回填完毕。

① Ground drainage and dewatering measures must be completed before construction. Excavation must not be started until the groundwater level is lowered to 0.5-1.0 m below the bottom of the foundation pit. Dewatering shall last until backfilling is completed.

②施工机械行驶道路应填筑适当厚度的碎石或砾石，必要时应铺设工具式路基箱（板）或梢排等。

②The access road for construction machinery shall be paved with crushed stones or gravels of appropriate thickness, and if necessary, with tool-type subgrade boxes(boards) or top rows.

③相邻基坑（槽）开挖时，应遵循先深后浅或同时进行的施工顺序，并应及时做好基础。

③ During the excavation of adjacent foundation pits(trenches), the sequence from deep to shallow or concurrent construction shall be followed, and the foundation shall be completed in time.

④在密集群桩上或其附近开挖基坑（槽）时，应在打桩完成后间隔一段时间，再对称挖土。在密集群桩附近开挖基坑（槽）时，还应采取措施防止桩基位移。

④ When excavating the foundation pit(trench) in or near dense group piles area, operators shall wait for some time upon completion of the piling to start symmetrical excavation. When the foundation pit(trench) is excavated near dense group piles area, measures shall be taken to prevent the pile foundation from displacement.

⑤挖出的土不允许放在坡顶或建筑物（构筑物）附近。

⑤ The excavated soil must not be piled up on the slope crest or near buildings(structures).

基坑（槽）开挖有人工开挖和机械开挖之分，对于大型基坑应优先考虑选用机械化施工，以加快施工进度。

The foundation pit(trench) excavation includes manual excavation and mechanical excavation. For large foundation pits, mechanized construction is preferred to speed up the construction progress.

深基坑开挖过程中，随着土的挖出，下层土因逐渐卸载而有可能回弹，尤其在基坑挖至设计标高后，如搁置时间过久，回弹现象更为明显。如弹性隆起，在基坑开挖和基础工程初期发展很快，将加大建筑物后期的沉降。因此，对深基坑开挖后的土体回弹应有适当的估计。

如在勘察阶段，土样的压缩实验中应补充卸荷弹性实验等，还可以采取结构措施，例如，在基底设置桩基，或事先对结构下部土质进行深层地基加固。施工中减少基坑弹性隆起的一个具体方法是加速建造主体结构或逐步利用基础的质量来代替被挖去土体的质量。

During the excavation of deep foundation pits, with the removal of soil, the subsoil may rebound due to gradual unloading, especially after the foundation pit is excavated to the design elevation. If it is left unused for a long time, the rebound effect will be more significant. For example, elastic uplift develops rapidly during foundation pit excavation and in the initial stage of foundation works, which will increase building settlement in the later stage. Therefore, the soil rebound after deep foundation pit excavation should be properly estimated. In the survey stage, the unloading elasticity test shall be supplemented for the compression test of soil samples. Also, structural measures can be taken, such as laying pile foundations on the base or strengthening the deep foundation for substructure soil in advance. During construction, an effective way to reduce the elastic uplift of the foundation pit is to accelerate the construction of the main structure or gradually replace the mass of the excavated soil with the mass of the foundation.

1.6　土方工程的质量标准和安全技术

1.6 Quality Standards and Safety Technology of Earthwork

1.6.1 质量标准

1.6.1 Quality Standards

①柱基、基坑、基槽和管沟基底的土质必须符合设计要求，并严禁扰动基底土层。

① The quality of soil at the column foundation, foundation pit, foundation trench, and pipe ditch base must meet the design requirements, and the base soil layer must not be disturbed.

②填方的基底处理，必须符合设计要求或施工规范。

② The base treatment of fill must meet the design requirements or construction specifications.

③填方和柱基、基坑、基槽、管沟的回填的土料必须符合设计要求和施工规范。

③ Soil materials for the fill and backfill of the column foundation, foundation pit, foundation trench, and pipe ditch must meet the design requirements and construction specifications.

④填方和柱基、基坑、基槽、管沟的回填必须按规定分层夯压密实。取样测定的压实土的干密度90%以上应符合设计要求，其余10%的最低值与设计值的差不应大于 0.08 g/cm³，且不应集中。土的实际干密度可用"环刀法"测定。其取样组数：柱基回填取样不少于柱基总数的 10%，且不少于 5 个；基槽、管沟回填每层按长度 20~50 m 取样一组；基坑和室内填土每层按 100~500 m 取样一组；场地平整填土每层按 400~900 m 取样一组，取样部位应在每层压实后的下半部。

④ The backfill of the column foundation, foundation pit, foundation trench, and pipe ditch must be compacted in layers as required. The soil is sampled to test its dry density after compaction, 90% shall meet the design requirements, and the rest 10% shall have its minimum value differing less than 0.08 g/cm^3 from the design value and shall not be concentrated. The actual dry density of soil can be tested with the cutting-ring method. Regarding the number of sampling groups, for column foundation backfilling, not less than 10% of the total column foundations and at least 5 samples shall be collected; for backfilling of the foundation trench and pipe ditch, a group of samples is collected from each layer at the length of 20-50 m; for the foundation pit and indoor fill, a group of samples is collected from each layer at the length of 100-500 m; for fill used for site leveling, a group of samples is collected from each layer at the length of 400-900 m, from the lower half of each layer compacted.

⑤土方工程的允许偏差和质量检验标准，应符合表1-3、表1-4的规定。

⑤ The allowable deviation and quality inspection standard of earthwork shall comply with the provisions of Table 1-3 and Table 1-4.

表1-3　土方开挖工程质量检验标准

Table 1-3 Quality Inspection Standards for Earthwork Excavation Works

项目 Works	序 S/N	项目 Item	允许偏差和允许值/mm Allowable deviation and allowable value/mm						检验方法 Inspection method
			桩基、基坑、基槽 Pile foundation, foundation pit, and foundation trench	挖方场地平整 Leveling of the cut site		管沟 Pipe ditch	地（路）面基层 Ground (pavement) base course		
				人工 Manual	机械 Machinery				
主控项目 Control works	1	标高 Elevation	−50	±30	±50	−50	−50		用水准仪检查 Check with the level gauge
	2	长度、宽度（由设计中心线向两边量） Length and width (measured from the design centerline to both sides)	+200 −50	+300 −100	+500 −150	+100			用经纬仪和钢尺检查 Check with the theodolite and steel ruler
	3	边坡坡度 Slope gradient	按设计要求 As per design requirements						
一般项目 Ordinary works	1	表面平整度 Surface roughness	20	20	50	20	20		用2 m靠尺和楔形塞尺检查 Check with the 2 m guiding rule and wedge plug ruler

项目 Works	序 S/N	项目 Item	允许偏差和允许值/mm Allowable deviation and allowable value/mm					检验方法 Inspection method
			桩基、基坑、基槽 Pile foundation, foundation pit, and foundation trench	挖方场地平整 Leveling of the cut site		管沟 Pipe ditch	地（路）面基层 Ground (pavement) base course	
				人工 Manual	机械 Machinery			
	2	基本土性 Basic soil properties	按设计要求 As per design requirements					观察或图样分析 Observation or pattern analysis

注：地（路）面基层的偏差只适用于直接在挖、填方上做地（路）面的基层。

Note: Due to the deviation of the ground(pavement) base course, it is only applicable to make the ground(pavement) base course directly on cut and fill.

表1-4 填土工程质量检验标准

Table 1-4 Quality Inspection Standards for Filling Works

项目 Works	序 S/N	项目 Item	允许偏差和允许值/mm Allowable deviation and allowable value/mm					检验方法 Inspection method
			桩基、基坑、基槽 Pile foundation, foundation pit, and foundation trench	挖方场地平整 Leveling of the cut site		管沟 Pipe ditch	地（路）面基层 Ground (pavement) base course	
				人工 Manual	机械 Machinery			
主控项目 Control works	1	标高 Elevation	−50	±30	±50	−50	−50	用水准仪检查 Check with the level gauge
	2	分层压实系数 Coefficient of layered compaction	按规定方法 As specified					用水准仪检查 Check with the level gauge
一般项目 Ordinary works	1	表面平整度 Surface roughness	20	20	30	20	20	用2 m靠尺和楔形塞尺检查 Check with the 2 m guiding rule and wedge plug ruler
	2	回填土料 Earth backfill	按设计要求 As per design requirements					取样检查或直观鉴别 Sampling inspection or visual inspection
	3	分层厚度及含水量 Layer thickness and moisture content	按设计要求 As per design requirements					用水准仪检查或抽样检查 Check with the level gauge or collect samples

1.6.2 安全技术

1.6.2 Safety Technology

①基坑开挖时，两人操作间距应大于2.5 m，多台挖掘机在同作业面开挖的机间距应大于10 m。挖土应由上而下，分层分段按顺序进行，严禁先挖坡脚或逆坡挖土，或采用底部掏空塌土法挖土。

① During foundation pit excavation, the distance between two operators shall exceed 2.5 m, and the distance between multiple excavators operating on the same working face shall exceed 10 m. Excavation shall proceed in layers and sections from the top down. It is strictly prohibited to excavate from the slope toe or against the slope or excavate by hollowing out the soil at the bottom.

②基坑开挖应按要求放坡。操作时应随时注意土壁变动情况，如发现有裂纹或部分坍塌现象，应及时进行支撑或放坡，并注意支撑是否稳固，关注土壁的变化。

② Sloping is required for foundation pit excavation. During operation, attention shall be paid to the change of soil wall at any time. If cracks or partial collapse are found, support or sloping shall be applied in time, and attention shall be paid to the support stability and the change of soil wall.

③基坑挖土使用吊装设备吊土时，起吊后，坑内操作人员应立即离开吊点的垂直下方，坑内人员应戴安全帽。起吊设备距坑边至少1.5 m，以防止造成坑壁塌方。

③ When lifting equipment is used for foundation pit excavation, operators in the pit must not stand directly under the lifting point, and people in the pit must wear safety helmets. The lifting equipment shall be at least 1.5 m away from the pit edge to prevent the collapse of the pit wall.

④用手推车运土，应先铺好道路。卸土回填，不得放手让车自动翻转。用翻斗汽车运土时，运输道路的坡度、转弯半径应符合有关安全规定。

④ When soil is transported by a trolley, the road shall be paved first. When the unloaded earth is backfilled, operators shall hold the trolley instead of releasing it for self-overturn. If the earth is transported by dump trucks, the slope and turning radius of the road shall comply with relevant safety regulations.

⑤在深基坑中上下应先挖好阶梯或设置靠梯，或开斜坡道，采用防滑措施，禁止踩踏支撑上下。坑四周应设安全栏杆或悬挂危险标志。

⑤ Steps or ladders shall be set up, or slopes shall be made (with anti-skid measures) for accessing the deep foundation pit. It is forbidden to step on the support for access. Safety railings or danger signs shall be available around the pit.

⑥应经常检查基坑（槽）设置的支撑是否有松动变形等不安全迹象，特别是雨后更应加强检查。

⑥ The support set in the foundation pit(trench) shall be checked frequently, especially after the rain, for unsafe signs such as looseness and deformation.

⑦坑（槽）沟边1 m以内不得堆土、堆料和停放机具，1 m以外堆土，其高度不宜超过1.5 m。坑（槽）沟边与附近建筑物的距离不得小于1.5 m，必要时必须对坑（槽）进行加固。

⑦ No soil, materials, and machines are allowed within 1 m from the pit(trench) edge, and for the range outside 1 m, the stacking height shall not exceed 1.5 m. The pit(trench) edge shall be kept at least 1.5 m from nearby buildings and shall be reinforced under dangerous conditions.

⑧基坑（槽）和管沟回填前，应检查坑（槽）壁有无塌方迹象，填土夯实过程中，应随时注意边坡土的变化，必要时需采取适当的支护措施。基坑回填应分层进行，基础或管道、地沟回填应避免造成两侧压力不平衡，使基础或墙体位移或倾倒。

⑧ Before backfilling the foundation pit(trench) and pipe ditch, check the pit(trench) wall for signs of collapse. During filling and compaction, pay attention to the change of slope soil and take appropriate supporting measures when necessary.

The foundation pit shall be backfilled in layers. The foundation, pipeline, or trench shall be backfilled in such a way as to prevent pressure imbalance on both sides, thus protecting the foundation or wall from displacement or dumping.

1.7 土方工程冬期和雨期施工

1.7 Earthwork Construction in Winter and Rainy Seasons

1.7.1 土方工程冬期施工

1.7.1 Earthwork Construction in Winter

在冬季进行施工的过程称为冬期施工。冬季气温下降，不少地区温度在 0℃ 之下（即负温），土壤、混凝土、砂浆等所含的水分冻结，建筑材料容易脆裂，给建筑施工带来许多困难。

The construction undergone in winter is called winter construction. In winter, the temperature in many areas drops below 0°C (i.e. negative temperature), the moisture in soil, concrete, and mortar is frozen, and building materials are prone to brittle fracture, bringing many difficulties to building construction.

当室外连续 5 日平均气温低于 5℃ 或日最低气温低于 −5℃ 时，就要采取冬期施工措施，以保证工程质量。由于冬季施工需保温覆盖，消耗较多热能，增加工程造价，因此如场地平整、地基处理、室外装饰、屋面防水及高空灌筑混凝土等工程项目要尽量避免在冬季施工。对于不得不在冬季施工的项目，则须因时因地制宜，制定冬期施工措施，并及时掌握气温变化。

When the outdoor daily average temperature remains below 5°C for 5 consecutive days, or the minimum daily temperature drops below −5°C,

winter construction measures shall be taken to ensure project quality. Since winter construction requires thermal insulation and consumes more heat energy, the construction cost will grow. Therefore, some works shall not be carried out in winter, such as site leveling, foundation treatment, outdoor decoration, roof waterproofing, and high-altitude concrete pouring. For works that have to be done in winter, winter construction measures shall be prepared according to local conditions and considering the temperature change.

土方工程应尽量安排在入冬之前施工较为合理。对冬期开挖的工程，要随挖、随砌、随回填，严防地基受冻。对跨年度工程及冻前不能交付正常使用的工程，应对地基采取相应的过冬保温措施。

If conditions permit, earthwork should be done before winter. For works excavated in winter, the excavated parts shall be built and backfilled in time to protect the foundation from freezing. For works spanning over a year or failing to be delivered for normal use before freezing, proper winter insulation measures shall be taken for the foundation.

在反复冻融地区，连续 10 日昼夜平均温度在 −3℃ 以下时，工程施工应按冬期施工办理。

昼夜平均温度虽然上升到 -3℃ 以上，但冻土未完全融化时，亦应按冬期施工办理。

In areas with repeated freezing and thawing, winter construction starts when the average temperature at day and night remains below -3℃ for over 10 consecutive days. Even if the average day and night temperature rises above -3℃, the frozen soil is not completely melted, winter construction shall also be considered.

1.7.2 土方工程雨期施工

1.7.2 Earthwork Construction in Rainy Seasons

雨期施工条件较差，容易发生伤亡事故，故在施工中更应注意安全。如有可能，应避免在雨期施工。如无法避开雨期施工，则要按照相关规定保证土方工程质量和施工进度，避免雨期施工带来的损失。

The construction conditions in rainy seasons are inferior, and likely to cause casualties. Therefore, more attention shall be paid during construction. If conditions permit, construction should be launched in other seasons than in rainy seasons. If construction has to be done in rainy seasons, the earthwork quality and construction progress shall be ensured according to relevant provisions to avoid losses as working in rainy seasons.

2

地基处理与基础工程
Foundation Treatment and Foundation Works

学习目标：

Learning objectives:

掌握地基的加固处理方法、适用范围、施工工艺及质量检查要点；掌握浅埋式基础的施工要点；熟悉桩基础施工工艺、质量要求；掌握桩基础的质量验收标准及检测方法，能编制常见的基础工程施工方案。

Be able to master the foundation reinforcement methods, scope of application, construction technology, and quality inspection points; master the construction points of shallow buried foundations; get familiar with the construction technologies and quality requirements of pile foundations; master the quality acceptance standards and testing methods of pile foundations, and prepare construction schemes of common foundation works.

技能抽查要求：

Skill checking requirements:

能对常见基础的施工质量进行检测，能编制常见基础工程的施工方案。

Be able to test the construction quality of common foundations and prepare construction schemes of common foundation works.

建筑八大员岗位资格考试要求：

Qualification examination requirements for eight major posts of construction engineering:

掌握砌体和钢筋混凝土基础施工工艺；掌握钢筋混凝土预制桩和灌注桩的施工工艺和施工方法；掌握基础工程施工的质量与安全要求。

Be able to master the fundamental construction technology of masonry and reinforced concrete; master the construction technologies and methods of precast reinforced concrete piles and cast-in-situ piles; master the quality and safety requirements of foundation engineering construction.

2.1 地基处理及加固

2.1 Foundation Treatment and Reinforcement

任何建筑物都必须有可靠的地基和基础，建筑物的全部重量（包括各种荷载）最终将通过基础传递给地基，所以，对某些地基的处理与加固就成为基础工程施工中的一项重要内容。在施工过程中如发现地基土质过软或过硬，不符合设计要求时，应本着使建筑物各部位沉降尽量趋于一致，以减小地基不均匀沉降的原则对地基进行处理。

Any building must have a reliable foundation and base, and all the weight of the building(including various loads) will eventually be transferred to the foundation through the base. Therefore, the treatment and reinforcement of some foundations have become an important part of foundation engineering construction. During construction, if the foundation soil is too soft or too hard to meet the design requirements, the foundation shall be treated following the principle of making the settlement of each building part consistent to reduce the uneven settlement of the foundation.

当建筑物直接建造在未经人工处理的天然土层上时，这种地基称为天然地基。若天然地基不能满足强度和变形要求时，则必须进行地基处理。地基处理就是按照上部结构对地基的要求，对地基进行必要的加固或改良，提高地基土的承载力，保证地基稳定，减少房屋的沉降或避免产生不均匀沉降，消除湿陷性黄土的湿陷性及提高其抗液化能力等。

When buildings are built directly on natural soil layers without artificial treatment, such foundations are called natural foundations. If failing to meet the strength and deformation requirements, the natural foundation must be treated. As

for foundation treatment, it is to reinforce or improve the foundation to meet the superstructure requirements, so as to improve the bearing capacity of the foundation soil, ensure foundation stability, reduce the settlement or avoid uneven settlement of houses, eliminate the collapsibility of collapsible loess, and raise its liquefaction resistance.

常见的地基处理方法主要有：换土、重锤夯实、振冲、化学固结、砂桩挤密、深层搅拌、堆载预压等。

Ordinary foundation treatment methods mainly include soil replacement, heavy tamping, vibroflotation, chemical consolidation, sand pile compaction, deep mixing, and surcharge preloading.

2.1.1 换土地基

2.1.1 Soil Replacement Foundation

当建筑物基础下的持力层比较软弱，不能满足上部荷载对地基的要求时，常采用换土垫层法来处理软弱地基。换土垫层法是先将基础底面以下一定范围内的软弱土层挖去；然后回填强度较高、压缩性较低、没有腐蚀性的材料，如中粗砂、碎石或卵石、灰土、素土、石屑、矿渣、素混凝土等，最后分层夯实后作为地基的持力层。实践证明，换土垫层可以有效地处理某些荷载不大的建筑地基问题。例如，一般的三、四层房屋，路堤，油罐和水闸等的地基。换土地基按其回填的材料可分为砂地基、碎（砂）石地基、灰土地基等。

When the bearing stratum under the building foundation is too weak to bear the upper loads acting on the foundation, the soil cushion is usually replaced to treat the soft foundation. For soil cushion replacement, operators shall first remove the soft soil layer within a certain range below the foundation bottom; then backfill materials with high strength, low compressibility, and no corrosivity, such as medium-coarse sand, gravel or pebbles, lime soil, plain soil, aggregate chips, slag, and plain concrete, which are compacted in layers to serve as the bearing stratum of the foundation. The practice has proved that soil cushion replacement can effectively solve foundation problems of some buildings with limited loads, such as the foundations of three-floor or four-floor houses, embankments, oil tanks, and sluices. According to the backfill materials, the soil replacement foundation can be divided into the sand foundation, gravel(sand) foundation, and lime soil foundation.

砂地基和砂石地基是将基础下一定范围内的土层挖去，然后用强度较大的砂或碎石等回填，并经分层夯实至密实，以提高地基承载力、减少沉降、加速软弱土层的排水固结、防止冻胀和消除膨胀土的胀缩。该地基具有施工工艺简单、工期短、造价低等优点。适用于处理透水性强的软弱黏性土地基，但不宜用于湿陷性黄土地基和不适水的黏土地基，以免聚水而引起地基下沉和降低地基的承载力（图 2-1）。

For sand foundation and sand-gravel foundation, the soil layer within a certain range under the foundation is excavated, then backfilled with sand or gravel with higher strength, and tamped in layers to a dense state, so as to improve the foundation's bearing capacity, reduce settlement, accelerate the drainage consolidation of the soft soil layer, avoid frost heave, and eliminate the expansion and contraction of expansive soil. This foundation features simple construction, short construction period, and low cost. It is suitable for treating soft cohesive soil foundations with strong water permeability, but not suitable for collapsible loess foundations and impervious clay foundations, so as to avoid foundation settlement and reduction

of bearing capacity due to water accumulation (Figure 2-1).

图2-1　换填地基施工现场

Figure 2-1 Construction Site of Foundation Replacement

2.1.2 重锤夯实地基

2.1.2 Heavy Tamping Foundation

重锤夯实（图 2-2）是利用起重机将夯锤（1.5~3 t）提升到一定高度，然后自由落下，重复夯击基土表面，使地基表面形成一层较为均匀的硬壳层，从而使地基得到加固。该做法施工简单，费用较低；但布点较密，夯击遍数多，施工工期会相对较长；且夯击能量小，孔隙水难以消散，加固深度有限；当土的含水量稍高，易形成橡皮土，处理起来较为困难。该方法适用于地下水位以上、稍湿的黏性土、砂土、饱和度 Sr ≤ 60% 的湿陷性黄土、杂填土以及分层填土地基的加固处理。但当夯击振动对邻近的建筑物、设备及施工中的砌筑工程或混凝土浇筑工程等产生有害影响时，或地下水位高于有效夯实深度以及在有效深度内存在软黏土层时，不宜采用。

For heavy tamping (Figure 2-2), a crane lifts the hammer ram (1.5-3 t) to a certain height and releases it for free fall to repeatedly tamp the surface of the foundation soil, forming a relatively uniform hard shell on the foundation surface, so that the foundation can be reinforced. Although featuring simple construction and low cost, this practice requires dense distribution of points, numerous hits, and a long construction period, and yields small tamping energy, limits reinforcement depth, difficult to dissipate pore water. When the soil moisture content is slightly high, it is easy to form rubber soil, which is difficult to treat. It is suitable for the reinforcement of foundations above the groundwater level, with slightly wet cohesive soil, sandy soil, collapsible loess with saturation $Sr \leq 60\%$, miscellaneous fill, and layered fill. However, when tamping vibration has harmful effects on adjacent buildings, equipments, and undergoing masonry works or concrete pouring, or when the groundwater level is higher than the effective compaction depth and a soft clay layer exists within the effective depth, this method should not be used.

图2-2　重锤夯实地基施工现场

Figure 2-2 Construction Site of Heavy Tamping Foundation

2.1.3 振冲地基

2.1.3 Vibroflotation Foundation

振冲地基，又称振冲复合地基，是以起重机吊起振动器，启动潜水电机带动偏心块，使振冲器产生高频振动，同时开动水泵，通过喷嘴喷射高压水流成孔，然后分批填以砂石骨料形成一根根桩体，桩体与原地基构成复合地基，从而提高地基的承载力，减少地基的沉降和避免不均匀沉降的一种快速、经济有效的加固方法。该方法具有技术可靠、机具简单、操作技术易于掌握、施工简便、三材（钢材、水泥、木材）用料少、加固速度快等特点。

Vibroflotation foundation, also known as vibroflotation composite foundation, is to lift the vibrator with a crane and start the submersible motor to drive the eccentric block so that the vibroflotation device generates high-frequency vibration. Meanwhile, start the water pump and jet high-pressure water through the nozzle to form holes. Then, fill gravel aggregate in batches to form piles. The pile body and the original foundation constitute a composite foundation, improving its bearing capacity and reducing the settlement and uneven settlement of the foundation. It is a fast, economical, and effective reinforcement method. The method features reliable technology, simple machines and tools, easy operation, simple construction, less material consumption(steel, cement, wood), fast reinforcement speed.

振冲地基法按加固机理和效果的不同，可分为振冲置换法和振冲密实法两类。前者适用于处理不排水、抗剪强度不小于 20 kPa 的黏性土、粉土、饱和黄土及人工填土等地基。后者适用于处理砂土和粉土等地基，其中，不加填料的振冲密实法仅适用于处理黏土粒含量小于10% 的粗砂、中砂地基。

According to the different reinforcement mechanisms and effects, the vibroflotation foundation method can be divided into two methods: vibroflotation replacement and vibroflotation compaction. The former is suitable for treating foundations with undrained cohesive soil, silt, saturated loess, and artificial fill with a shear strength of not less than 20 kPa. The latter is suitable for treating foundations with sandy soil and silt. The vibroflotation compaction method without filler is only suitable for treating coarse sand and medium sand foundations with a clay particle content of less than 10%.

（1）施工要点

(1) Key points of construction

①施工前应在现场进行振冲试验，以确定成孔合适的水压、水量、速度、填料方法、达到土体密实时的密实电流值、填料量和留振时间。

① Before construction, a vibroflotation test shall be carried out at the site to determine the appropriate water pressure, water volume, speed, and filling method for hole forming, as well as the compaction current, filling amount, and vibration time required for dense soil.

②振冲前，应按设计图定出冲孔中心位置并编号。

② Before vibroflotation, the punching center shall be determined and numbered according to the design drawing.

③启动水泵和振冲器，水压可用400~600 kPa，水量可用200~400 L/min，使振冲器以1~2 m/min的速度徐徐沉入。当下沉达到设计深度时，振冲器应在孔底适当停留并减小射水压力，以便排除泥浆进行清孔；也可以1~2 m/min的速度将振冲器连续沉至设计深度以上0.3~0.5 m时，将其往上提到孔口，再用同法沉至孔底来成孔。如此往复1~2次，使孔内泥浆变稀，排泥1~2 min后，再将振冲器提出孔口。

③ Start the water pump and vibroflot, set the water pressure to 400-600 kPa and water volume to 200-400 L/min, and slowly sink the vibroflot at 1-2 m/min. When reaching the design depth, the vibroflot shall stay at the hole bottom for some time while reducing the water injection pressure, so as to remove the slurry for hole cleaning. Alternately, the vibroflot can sink to a place of 0.3-0.5 m above the design depth without stop at a speed of 1-2 m/min, then lifted up to the mouth, and down to the hole bottom in the same way. Repeat the above steps 1-2 times to dilute the mud in the hole. Discharge the sludge for 1-2 min and lift the vibroflot out of the mouth.

④填料和振密方法：成孔后，将振冲器提出孔口，从孔口往下填料，然后再下降振冲器至填料中进行振密（图2-3），待密实电流达到规定的数值，再将振冲器提出孔口。如此自下而上反复进行直至孔口，成桩操作即宣告完成。

④ For filling and compaction, the normal method is to lift the vibroflot out of the mouth, fill the hole through the mouth, and lower the vibroflot to compact the fill (Figure 2-3). When the compaction current reaches the specified value, lift the vibroflot out of the mouth. Repeat the above steps from the bottom up until the fill reaches the mouth. Then, the pile-forming operation is completed.

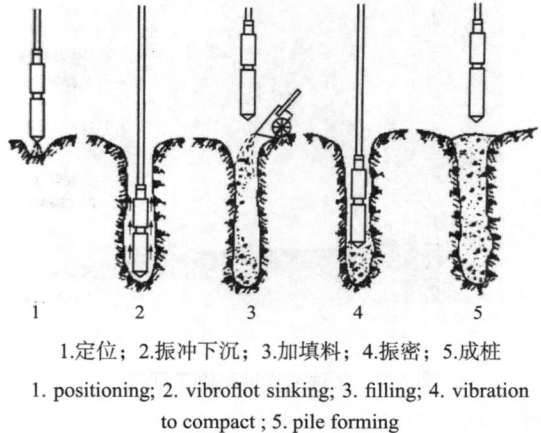

1.定位；2.振冲下沉；3.加填料；4.振密；5.成桩

1. positioning; 2. vibroflot sinking; 3. filling; 4. vibration to compact ; 5. pile forming

图2-3　振冲法成桩施工工艺

Figure 2-3 Construction Process of Pile Forming by Vibroflotation Method

⑤振冲桩施工时桩顶部约1 m范围内的桩体密实度难以保证，一般应予以挖除，另做地基，或用振动碾压使之压实。

⑤ During the construction of vibrosinking piles, it is difficult to maintain the compactness of the pile body within about 1 m of the pile top. Generally, this part should be removed, with another foundation made, or compacted via vibratory rolling.

⑥冬期施工时应将表层冻土破碎后成孔。每班施工完毕后应将供水管和振冲器水管内积水排净，以免水冻结影响施工。

⑥ In winter construction, the surface frozen soil shall be crushed for hole forming. After the work of each shift, the accumulated water in the water supply pipes and the vibroflot's water pipes shall be drained in case it freezes to affect construction.

（2）振冲地基质量检验标准及方法

(2) Quality inspection standard and method for vibroflotation foundation

①振冲地基的质量检验标准：振冲地基的质量检验标准应符合表2-1的规定。

① The quality inspection standard of the vibroflotation foundation shall comply with the provisions of Table 2-1.

表2-1　振冲地基质量检验标准及方法
Table 2-1 Quality Inspection Standards and Methods for Vibroflotation Foundation

项目 Works	序 S/N	检查项目 Check item	允许偏差或允许值 Allowable deviation or value	检测方法 Test method
主控项目 Control works	1	填料粒径 Size of filler	设计要求 Design requirements	抽样检查 Sampling inspection
	2	密实电流（黏性土） Compaction current (cohesive soil) 密实电流（砂性土或粉土）（以上为功率30 kW振冲器）密实电流（其他类型振冲器） Compaction current (sandy soil or silt) (30 kW vibroflot used), compaction current (other types of vibroflots)	50~55 A 40~50 A （1.5~2.0）A_0	电流表读数 Ammeter reading 电流表读数，A_0为空振电流 Ammeter reading, A_0 is the zero-load vibroflotation current
	3	地基承载力 Foundation bearing capacity	设计要求 Design requirements	按规定方法 As specified
一般项目 Ordinary works	1	填料含泥量 Silt content of filler	<5%	抽样检查 Sampling inspection
	2	振冲器喷水中心与孔径中心偏差 Deviation of the vibroflot's jetting center from the aperture center	≤50 mm	用钢尺量 Measured with the steel ruler
	3	成孔中心与设计孔位中心偏差 Deviation of the hole-forming center from the design hole center	≤100 mm	用钢尺量 Measured with the steel ruler
	4	桩体直径 Pile diameter	<50 mm	用钢尺量 Measured with the steel ruler
	5	孔深 Borehole depth	±200 mm	用量钻杆或重锤测 Measured with drill rod or heavy hammer

②振冲地基的质量检验方法。

② Quality inspection method of vibroflotation foundation.

施工前应检验振冲器的性能，电流表、电压表的准确度及填料的性能；施工中应检查密实电流、供水压力、供水量、填料量、孔底留振时间、振冲点位置、振冲器施工参数等；施工结束后，应在有代表性的地段做地基强度或地基承载力检验。

Before construction, the vibroflot performance, accuracy of ammeters and voltmeters, and filler performance shall be checked. During construction, the compaction current, water supply pressure and volume, filling volume, vibration time at the hole bottom, position of vibroflotation points, and construction parameters of the vibroflot shall be checked. Upon completion of construction, tests shall be carried out in representative sections to check the foundation's strength or bearing capacity.

2.2 桩基础工程

2.2 Pile Foundation Works

一般建筑物都应充分利用地基土层的承载能力，尽量采用浅基础。但若浅层土质不良，无法满足建筑物对地基变形和强度方面的要求时，可利用下部坚实土层或岩层作为持力层，这就要采取有效的施工方法建造深基础。深基础主要有桩基础、墩基础、沉井和地下连续墙等几种，其中桩基最为常用。

Generally, the bearing capacity of the foundation soil layer shall be fully utilized to hold buildings, and a shallow foundation shall be adopted if conditions permit. However, if the shallow soil is inferior and cannot meet the foundation deformation and strength requirements for buildings, the lower solid soil or rock layer can be used as the bearing stratum. This requires effective construction methods to build a deep foundation. Deep foundations mainly include pile foundations, pier foundations, open caisson, and underground diaphragm walls, among which the pile foundation is the most commonly used.

2.2.1 桩基的作用和分类

2.2.1 Function and Classification of Pile Foundations

（1）作用

(1) Function

桩基一般由设置于土中的桩和承接上部结构的承台组成（图2-4）。

A pile foundation is usually composed of a pile set in soil and a pile cap holding the superstructure (Figure 2-4).

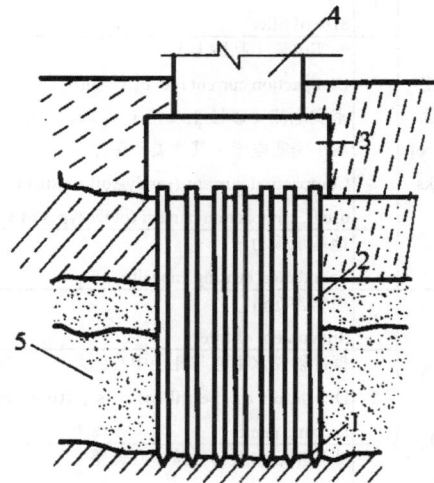

1.持力层；2.桩；3.桩基承台；
4.上部建筑；5.软弱土层

1. bearing stratum; 2. pile; 3. pile cap;
4. superstructure; 5. weak soil layer

图2-4 桩基示意图

Figure 2-4 Schematic of a Pile Foundation

桩的作用是将上部建筑物的荷载传递到深处承载力较大的持力层上；或使软弱土层受到挤压，以提高土壤的承载力和密实度，从而保证建筑物的稳定性和减少地基沉降。

The pile can transfer the load of the superstructure to the deep bearing stratum with a large bearing capacity, or squeeze the weak soil layer to improve the bearing capacity and compactness of the soil, thus maintaining the stability of the building and reducing the foundation settlement.

承台的作用是将桩基中的各根桩连成一个整体，共同承受上部结构的荷载。根据承台与地面的相对位置不同，一般有低承台和高承台桩基之分。前者承台底面位于地面以下，后者则高出地面，位于地面以上。一般来说，采用高承台主要是为了减少水下施工作业和节省基础材料，常用于桥梁和港口工程中。而低承台承受荷载的条件比高承台好，特别是在水平荷载作用下，承台周围的土体可以发挥一定的作用。一般的房屋和构筑物中，大都采用低承台桩基。

The pile cap can connect all piles in the pile foundation into a whole to share the load of the superstructure. According to its position against the ground, the pile cap can be high or low type. A low pile cap has its bottom below the ground, while a high pile cap has its bottom over the ground. Generally speaking, the high pile cap is mainly used to reduce underwater construction and save foundation materials, often used in bridge and port engineering. The low pile cap outperforms the high pile cap in load bearing, especially under the horizontal load, when the soil around the pile cap can help. For ordinary buildings and structures, pile foundations with low pile caps are mostly used.

（2）分类

(2) Classification

①按承载性质分。

① By bearing property.

a. 摩擦型桩。

a. Friction pile.

摩擦型桩又可分为摩擦桩和端承摩擦桩。摩擦桩是指在极限承载力作用下，柱顶荷载由桩侧阻力承受的桩；端承摩擦桩是指在极限承载力作用下，桩顶荷载由桩侧阻力及柱端阻力共同承受的桩。

Friction piles are further divided into friction piles and end-bearing friction piles. Friction pile refers to the pile whose top load is borne by pile side resistance under the ultimate bearing capacity; end-bearing friction pile refers to the pile whose top load is shared by pile side resistance and column end resistance under the ultimate bearing capacity.

b. 端承型桩。

b. End-bearing piles.

端承型桩又可分为端承桩和摩擦端承桩。端承桩是指在极限承载力作用下，桩顶荷载由桩端阻力承受的桩；摩擦端承桩是指在极限承载力作用下，桩顶荷载主要由柱端阻力承受的桩。

End-bearing piles are further divided into end-bearing piles and friction end-bearing piles. End-bearing pile refers to the pile whose top load is borne by pile end resistance under the ultimate bearing capacity; friction end-bearing pile refers to the pile whose top load is mainly borne by the pile end resistance under the ultimate bearing capacity.

②按使用功能分。

② By use functions.

根据使用功能的不同，可分为竖向抗压桩、竖向抗拔桩、水平受荷桩、复合受荷桩。

Piles can be divided into vertical compressive piles, vertical uplift piles, horizontally loaded piles, and composite loaded piles by use functions.

③按桩身材料分。

③ By pile materials.

根据桩身材料的不同，可分为混凝土桩、钢桩、组合材料桩。

Piles can be divided into concrete piles, steel piles, and composite piles by pile materials.

④按成桩方法分。

④ By pile-forming methods.

依据成桩方法的不同，可分为非挤土桩（如干作业法桩、泥浆护壁桩、套筒护壁桩）、部分挤土桩（如部分挤土灌注桩、预钻孔打入式预制桩等）、挤土桩（如挤土灌注桩、挤土预制桩等）。

Piles can be divided into non-soil-displaced piles(such as dry operation piles, slurry-supported bored piles, and sleeved piles), partial displacement piles(such as partial displacement cast-in-situ piles and pre-bored driven precast piles), and displacement piles(such as cast-in-situ displacement piles and precast displacement piles) by pile-forming methods.

⑤按桩制作工艺分。

⑤ By pile fabrication process.

依据制作工艺的不同，可分为预制桩和现场灌注桩，现在使用较多的是现场灌注桩。

Piles can be divided into precast piles and cast-in-situ piles by pile fabrication process, the latter are now widely used.

2.2.2 钢筋混凝土预制桩施工工艺

2.2.2 Construction Process of Reinforced Concrete Precast Piles

（1）锤击沉桩施工工艺

(1) Pile sinking by hammer

①特点及原理。

① Characteristics and principle.

锤击沉桩是利用桩锤下落时的瞬时冲击机械能，克服土体对桩的阻力，使其静力平衡状态遭到破坏，导致桩体下沉、达到新的静力平衡状态，如此反复地锤击桩头，桩身也就不断地下沉。锤击沉桩是预制桩最常用的沉桩方法。

该法施工速度快、机械化程度高、适应范围广且现场文明程度高，但施工时有噪声等公害，在城市中心和夜间施工时有所限制。

Pile sinking by hammer is to overcome the resistance of the soil mass against the pile by the mechanical energy generated by the instantaneous impact from the falling pile hammer, to destroy its static equilibrium state, thus sinking the pile and reaching a new static equilibrium state. Repeated hammering to the pile head in this way can keep driving the pile body into the ground. Pile sinking by hammer is the most commonly used pile-sinking method for precast piles. This method features fast construction speed, high degree of mechanization, wide application range, and high degree of HSE-compliant construction. However, it involves noise during construction, limiting its use in downtown and at night.

②沉桩机械设备。

② Mechanical equipment for pile sinking.

打桩所用的机具设备，主要包括桩锤、桩架及动力装置三部分。

The machines and equipment used for driving piles mainly include the pile hammer, pile frame, and power unit.

桩锤包括落锤、单动汽锤、双动汽锤、柴油打桩锤和液压锤等。

Pile hammers include the drop hammer, single-acting steam hammer, double-acting steam hammer, diesel pile hammer, and hydraulic hammer.

常用的桩架形式有滚筒式桩架、多功能桩架、履带式桩架三种。选择桩架时应考虑桩锤的类型、桩的长度和施工条件等因素。

Common pile frames include the roller pile frame, multi-functional pile frame, and crawler pile frame. For the selection of pile frames, the type of pile hammer, pile length, and construction conditions shall be considered.

③沉桩工艺方法。

③ Pile sinking process.

a. 锤击沉桩施工工艺流程。

a. Construction process of pile sinking by hammer.

确定桩位和沉桩顺序→桩机就位→吊桩喂桩→校正→锤击沉桩→接桩→再锤击沉桩→送桩→收锤→切割桩头。

Determination of pile position and pile sinking sequence→pile driver in place→lift the piles and place them in the designated holes→correction→hammering to sink piles→pile extension→hammering again→pile delivery→end of hammering→pile head cutting.

b. 沉桩顺序。

b. Pile sinking sequence.

沉桩顺序直接影响打桩速度和打桩质量，应综合桩距、桩机性能、工程特点和工期要求考虑确定。常见的打桩顺序见图2-5所示。

The pile sinking sequence directly affects the pile driving speed and quality, and shall be determined by a comprehensive consideration of pile spacing, pile driver performance, engineering characteristics, and project duration requirements. The common pile-driving sequence is shown in Figure 2-5.

当桩较稀时（桩中心距大于4倍桩径或桩边长时），土壤的挤压影响可忽略不计，可采

用由一侧向单一方向打（逐排打设），此法桩的就位和起吊方便，打桩效率高，但土壤向一个方向挤压，如图2-5（a）所示。

When piles are sparsely laid (the center distance of piles over 4 times the pile diameter or the pile's side length), the extrusion effect of the soil can be ignored, and the pile can be driven from one side along a single direction (row by row). This method is convenient for pile positioning and lifting, yielding a high piling efficiency, but the soil is extruded in one direction, as shown in Figure 2-5(a).

当桩较密时（桩中心距小于等于4倍桩径或桩边长时），由于打桩对土体的挤密作用，使先打的桩因受水平推挤而造成偏移和变位，或被垂直推挤拔起造成浮桩，此时可采用由中间向四周打设，或由中间向两侧对称施打的方法，如图2-5（b）和图2-5（c）所示。

When piles are densely laid (the center distance of piles less than or equal to 4 times the pile diameter or the pile's side length), due to the compaction effect of pile driving on the soil mass, the previously driven pile is offset and displaced due to horizontal squeezing or is vertically pushed/lifted to cause a floating pile. Therefore, pile driving shall proceed from the middle to the periphery or from the middle to both sides, as shown in Figure 2-5 (b) and 2-5 (c).

打设标高不一的桩，应遵"先深后浅"的

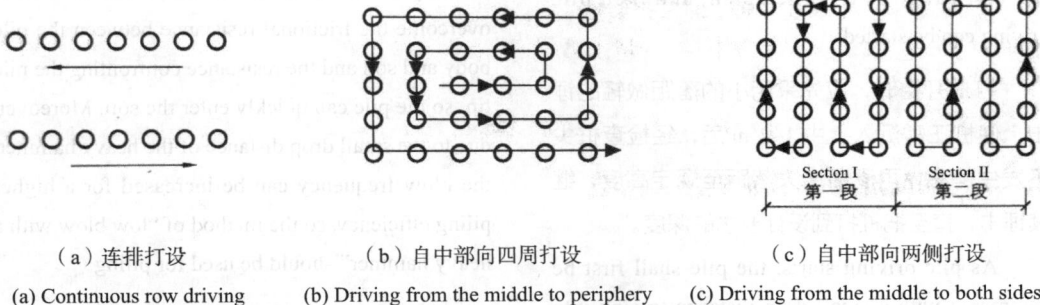

（a）连排打设　　　　　（b）自中部向四周打设　　　　　（c）自中部向两侧打设

(a) Continuous row driving　　(b) Driving from the middle to periphery　　(c) Driving from the middle to both sides

图2-5　打桩顺序

Figure 2-5 Piling Sequence

原则；对不同规格的桩，应遵循"先大后小、先长后短"的原则。

For piles with different elevations, the principle of "deep piles before shallow ones" shall be followed. For piles of different specifications, the principle of "large piles before small ones and long piles before short ones" shall be followed.

④打桩。

④ Piling.

在桩架就位后即可吊桩，将桩垂直对准桩位中心，缓缓放下插入土中。桩插入时垂直度偏差不得超过 0.5%，桩就位后在桩顶安上桩帽，然后放下桩锤轻轻压住桩帽。桩锤、桩帽和桩身中心线应在同一垂直线上。在桩的自重和锤重作用之下，桩沉入土中一定深度而达到稳定的状态。这时再校正一次桩的垂直度，即可进行打桩。

After the pile frame is put in place, the pile can be lifted by aiming the pile at the center of the pile position and slowly lowering it into the soil. When the pile is inserted, the verticality deviation shall not exceed 0.5%. After the pile is put in place, the pile cap is installed on the pile top, and then the pile hammer is lowered to gently touch the pile cap. The centerlines of the pile hammer, pile cap, and pile body shall coincide in the same vertical line. Under the dead weight of the pile and hammer weight, the pile is sunk into the soil to a certain depth to reach a stable state. At this time, correct the verticality of the pile again, and then pile driving can be started.

打桩开始时，应先采用小的落距做轻的锤击，使桩正常沉入土中 1~2 m 后，经检查桩尖不发生偏移，再逐渐增大落距至规定高度，继续锤击，直至把桩打到设计要求的深度。

As pile driving starts, the pile shall first be hammered with a small drop distance to sink into the soil for 1-2 m. Then, check and make sure the pile tip is not offset and gradually increase the drop distance to the specified height. After that, keep hammering until the pile reaches the designed depth.

打桩有"轻锤高击"和"重锤低击"两种方式。这两种方式，即使所做的功相同，所得到的效果却不相同。轻锤高击，所得的动量小，而桩锤对桩头的冲击力大，因而回弹也大，桩头容易损坏，大部分能量均消耗在桩锤的回弹上，故桩难以入土。相反，重锤低击，所得的动量大，而桩锤对桩头的冲击力小，因而回弹也小，桩头不易被打碎，大部分能量都可以用来克服桩身与土壤的摩阻力和桩尖的阻力，故桩很快入土。此外，重锤低击的落距小，因而可提高锤击频率，打桩效率也高，所以打桩宜采用"重锤低击"方式。

There are two driving methods: "high blow with a light hammer" and "low blow with a heavy hammer". These two methods yield different results even with the same work. When the light hammer strikes high, the momentum obtained is small, but the impact of the pile hammer acting on the pile head is large, resulting in a large rebound, likely to destroy the pile head. In this case, most energy is consumed by the rebound of the pile hammer, so it is difficult to drive the pile into the soil. On the contrary, when the heavy hammer strikes low, the momentum obtained is large, while the impact of the pile hammer acting on the pile head is small, resulting in a small rebound, so the pile head is not easy to be broken. Most energy can be used to overcome the frictional resistance between the pile body and soil and the resistance confronting the pile tip, so the pile can quickly enter the soil. Moreover, due to the small drop distance of the heavy hammer, the blow frequency can be increased for a higher piling efficiency, so the method of "low blow with a heavy hammer" should be used for piling.

打桩系隐蔽工程施工，应做好打桩记录。用落锤、单动汽锤或柴油锤打桩时，从开始即需记录桩身每沉入 1 m 所需要的锤击数。当桩下沉接近设计标高时，应在规定落距下，测定每一阵（每 10 击为一阵）的贯入度，使其达到设计承载力所要求的最小贯入度。

As concealed works, piling shall be recorded in detail. When the pile is driven with a drop hammer, single-acting steam hammer, or diesel hammer, the number of blows required for every 1 m of sinking shall be recorded from the beginning. When the sinking pile approaches the design elevation, the penetration of each array(every 10 blows) shall be measured under the specified drop distance to achieve the minimum penetration required by the designed bearing capacity.

⑤质量要求。

⑤ Quality requirements.

打桩质量包括两个方面的内容：一是能否满足设计规定的贯入度或标高的设计要求；二是桩打入后的偏差是否在施工规范允许的范围内。

Piling quality includes two aspects: Whether it can meet the design requirements of the penetration degree or elevation specified in the design; whether the deviation of the pile after driving is within the scope allowed by the construction code.

打桩的控制原则是：

The control principles of piling are as follows:

a. 桩尖达到坚硬的黏性土、碎石土，中密以上的砂土或风化岩等土层时，应以贯入度控制为主，桩尖进入持力层深度或桩尖标高可作参考；若贯入度已达到而桩尖标高未达到时，应继续锤击 3 阵，其每阵 10 击的平均贯入度不应大于规定的数值。

a. When the pile tip reaches the hard cohesive soil, gravelly soil, medium-dense or above sandy soil, or weathered rock, the penetration control shall be the main focus, with reference to the pile tip's depth into the bearing stratum or its elevation. If the penetration is achieved but the pile tip elevation has not been met, hammering shall continue for another 3 arrays, with the average penetration of 10 blows per array not greater than the specified value.

b. 桩端位于其他软土层时，以桩端设计标高控制为主，贯入度可作参考。

b. When the pile end resides in other soft soil layers, the design elevation control of the pile end shall be the main focus, with reference to the penetration.

c. 桩打入后的垂直度偏差和平面位置偏差在国家施工规范允许的范围内。

c. The verticality deviation and plane position deviation after pile driving fall into the allowable range specified by national construction specifications.

（2）静力压桩施工工艺

(2) Static pile pressing

①特点及原理。

① Characteristics and principle.

静力压桩施工是在软土地基上，利用静力压桩机或液压压桩机用无振动的静压力（自重和配重）将预制桩压入土中的一种沉桩工艺，在我国沿海软土地上广泛应用。与锤击沉桩相比，具有施工无噪声、无振动、节约材料、降低成本、提高施工质量、沉桩速度快等特点，特别适用于城市内桩基工程施工。其工作原理是：通过安置在压桩机上的卷扬机的牵引，由钢丝绳、滑轮及压梁，将整个桩机的自重反压在桩顶上，以克服桩身下沉时与土的摩擦力，迫使预制桩下沉。

Static pile pressing is a pile-sinking technology applied on weak soil foundations that presses precast piles into the soil with non-vibration static pressure(self-weight and counterweight) from the static pile press or hydraulic pile press. It is widely used for soft ground in coastal regions of China. Compared with pile sinking by hammer, this method features no noise, no vibration, less material consumption, less cost, higher construction quality, and faster pile-sinking speed. It is especially suitable for pile foundation works in cities. Working principle: Under the traction of the winch placed on the pile driver, the dead weight of the whole pile driver is pressed against the pile top via the wire rope, pulleys, and pressing beams, so as to overcome the friction with the soil when the pile body sinks and force the precast pile to sink.

②压桩机械设备。

② Mechanical equipment for pile pressing.

压桩机有两种类型：一种是机械静力压桩机（图2-6），它由压桩架（桩架与底盘）、传动设备（卷扬机、滑轮组、钢丝绳）、平衡设备（铁块）、量测装置（测力计、油压表）及辅助设备（起重设备、送桩）等组成；另一种是液压静力压桩机，它由液压吊装机构、液压夹持机构、压桩机构（千斤顶）、行走及回转机构、液压及配电系统、配重铁块等部分组成。

Two types of pile presses are available: One is the mechanical static pile press(Figure 2-6), which consists of the pile-pressing frame(pile frame and chassis), transmission equipment(winch, pulley block, and wire rope), balance equipment (iron blocks), measuring devices(dynamometer and oil pressure gauge), and auxiliary equipment (lifting equipment and pile follower). The other is the hydraulic static pile press, which consists of a hydraulic lifting mechanism, hydraulic clamping mechanism, pile-pressing mechanism(jack), traveling and slewing mechanism, hydraulic and power distribution system, and counterweight iron blocks.

③压桩工艺方法。

③ Pile pressing process methods.

a. 静力压桩施工工艺流程。

a. Construction process of static pile pressing.

测量定位→桩机就位→吊桩插桩→桩身对中调直→静压沉桩→接桩→再次静压沉桩→终止压桩→切割桩头。

Measurement and positioning → pile press

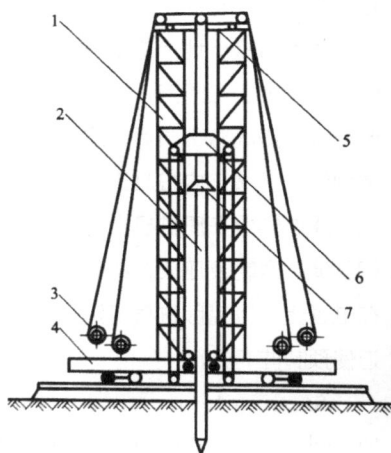

1.桩架；2.柱；3.卷扬机；4.底盘；5.顶梁；6.压梁；7.柱帽
1. pile frame; 2. column; 3. winch; 4. chassis; 5. top beam; 6. pressing beam; 7. column cap

图2-6 机械静力压桩机

Figure 2-6 Mechanical Static Pile Press

in place → pile lifting and insertion → centering and straightening of pile body → pile sinking under static pressure → pile extension → pile sinking under static pressure again → end of pile pressing → pile head cutting.

b. 静力压桩施工方法。

b. Construction method of static pile pressing.

用起重机将预制桩吊运或用汽车运至桩机附近，再利用桩机自带的起重装置将桩吊入夹持器中，夹持油缸将桩从侧面夹紧，压桩油缸做伸程动作，把桩压入土层中。伸程完毕，夹持油缸回程松夹，压桩油缸回程，重复上述动作，可实现连续压桩操作，直至将桩压入预定深度土层。

The precast pile is moved by a crane or truck to the vicinity of the pile press and lifted into the gripper with the lifting device of the pile press. Then, the clamping cylinder tightly clamps the pile from the side, and the pile-pressing cylinder extends to press the pile into the soil. After the extension is completed, the clamping cylinder returns to loosen the clamp, and the pile-pressing cylinder returns. The above steps are repeated for continuous pile pressing until the pile is pressed into the soil to a predetermined depth.

c. 桩拼接方法。

c. Pile splicing method.

钢筋混凝土预制长桩在起吊、运输时受力极为不利，因而一般先将长桩分段预制，再在沉桩施工时将桩接长。常用的接头连接方法有以下两种：

Since long reinforced concrete precast piles are stressed during lifting and transportation, they are often prefabricated in sections first and spliced during pile sinking. Two splicing methods are commonly used:

浆锚接头：用硫磺水泥或环氧树脂配制成的黏结剂，把上段桩的预留插筋黏结于下段桩

的预留孔内。

Mortar-anchor splice : The reserved dowel bars of the upper pile are inserted into the reserved holes of the lower pile and bonded with an adhesive made of sulfur cement or epoxy resin.

焊接接头：在每段桩的端部预埋角钢或钢板，施工时使上下两段桩的桩端紧密接触，用扁钢贴焊成整体。

Welded splice : Angle steel or steel plate is embedded at the end of each pile section. During construction, the pile ends of the upper and lower pile sections are put in close contact and welded into a whole with flat steel.

d. 压桩施工要点。

d. Key points of pile pressing construction.

压桩应连续进行，即便因故停歇，时间也不宜过长，否则压桩阻力将大幅增长导致桩压不下去或桩机被抬起。

Pile pressing shall proceed continuously without downtime(or with limited downtime if it has to be stopped); otherwise, the pile pressing resistance will surge, resulting in the failure of pile pressing or lifting of the pile press.

压桩的终压控制很重要。一般来说，对于纯摩擦桩，终压时以设计桩长为控制条件；对于长度大于 21 m 的端承摩擦桩，应以设计桩长控制为主，终压力值作为对照；对于一些设计承载力较高的桩基，终压力值宜尽量接近压桩机的满载值；对于长 14~21 m 的静压桩，应以终压力达满载值为终压控制条件；对于桩周围土质较差且设计承载力较高的桩，宜复压 1~2 次为佳；对长度小于 14 m 的桩，宜连续多次复压，特别对长度小于 8 m 的短桩，连续复压的次数应适当增加。

The final pressing control is crucial for pile pressing. Generally speaking, for pure friction piles, the design pile length is referred to as the control condition for final pressing. For end-bearing friction piles longer than 21 m, the design pile length shall be referred to as the major control factor, with reference to the final pressure. For some pile foundations with higher design bearing capacities, the final pressure should approach the full load value of the pile press. For 14-21 m long static pressure piles, the final pressure reaching the full load value shall be taken as the control condition for final pressing. For piles with a high design bearing capacity and poor soil around, it is better to press them another 1-2 times. Piles shorter than 14 m should be repeatedly pressed, and especially for piles shorter than 8 m, more repeated presses are required.

静力压桩单桩竖向承载力，可通过桩的最终压力值进行大致判断。如判断的终止压力值不能满足设计要求，应立即采取送桩加深处理或补桩，以保证桩基的施工质量。

For static pile pressing, the vertical bearing capacity of a single pile can be roughly judged by the final pressure value of the pile. If the determined end pressure value fails to meet the design requirements, pile following and deepening or pile filling shall be carried out immediately to ensure the construction quality of the pile foundation.

2.2.3 现浇混凝土桩施工工艺

2.2.3 Construction Process of Cast-in-situ Concrete Piles

现浇混凝土桩是一种直接在现场桩位上使用机械或人工等方法就地成孔，然后在孔内浇筑混凝土或安放钢筋笼再浇筑混凝土而成的桩。按其成孔方法不同，可分为钻孔灌注桩、沉管灌注桩、人工挖孔灌注桩等。

Cast-in-situ concrete pile is a pile formed by directly forming a hole at the pile position by mechanical or manual methods, and then pouring concrete in the hole, or placing a reinforcement cage before concreting. According to the different hole-forming methods, it can be divided into cast-in-situ bored piles, tube-sinking cast-in-situ piles, and hand-dug cast-in-situ piles.

（1）钻孔灌注桩

(1) Cast-in-situ bored pile

钻孔灌注桩是指利用钻孔机械钻出桩孔，并在孔中浇筑混凝土（或先在孔中吊放钢筋笼）而成的桩。根据钻孔机械的钻头是否在土壤的含水层中施工，又分为泥浆护壁成孔和干作业成孔两种施工方法。

Cast-in-situ bored pile refers to the pile formed by drilling a pile hole with a drilling machine and pouring concrete in the hole(or lowering a reinforcement cage into the hole first). It is further divided into slurry-supported bored piles and dry-forming piles by judging whether the drill bit operates in the soil aquifer.

①泥浆护壁钻孔灌注桩。

① Slurry-supported cast-in-situ bored pile.

泥浆护壁钻孔灌注桩适用于地下水位较高的地质条件。按钻孔设备可分为冲击、冲抓钻成孔灌注桩，回转钻成孔灌注桩，潜水钻成孔灌注桩。前三种适用于碎石土、砂土、黏性土及风化岩地基，后一种则适用于黏性土、淤泥、淤泥质土及砂土，如图2-7所示。

Slurry-supported cast-in-situ bored pile is suitable for geological conditions with a higher groundwater level. According to the drilling equipment, it is further divided into percussion drilling bored piles, percussion-grab drilling bored piles, rotary drilling bored piles, and submersible drilling bored piles. The first three types are suitable for gravel, sandy soil, cohesive soil, and weathered rock foundations, and the last one is suitable for cohesive soil, silt, mucky soil, and sandy soil, as shown in Figure 2-7.

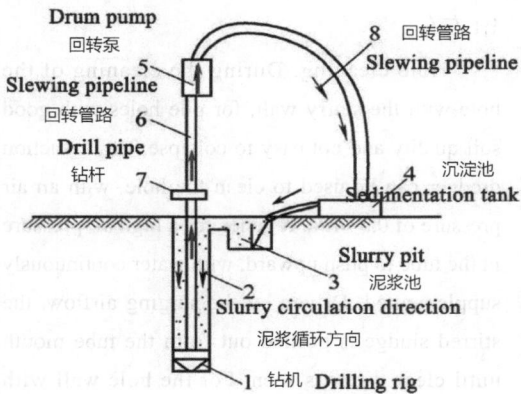

图2-7 泥浆护壁工作原理图

Figure 2-7 Working Principle of Slurry-supported Bored Pile

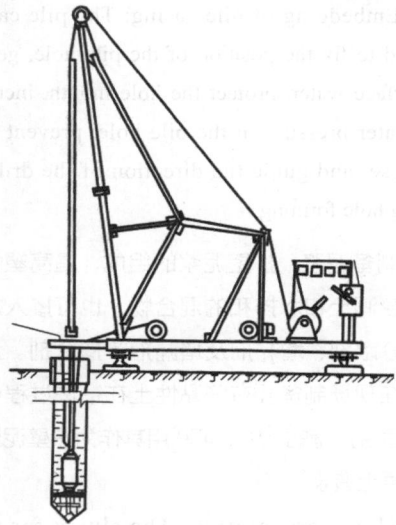

2-8 潜水钻机钻孔示意图

Figure 2-8 Schematic of Drilling with Submersible Drilling Rig

a. 施工设备。

a. Construction equipment.

主要有冲击、冲抓、回转钻机及潜水钻机。在此主要介绍潜水钻机（图2-8）。

Construction equipment mainly includes percussion, percussion-grab, rotary drilling rigs, and submersible drilling rigs. Here, we introduce the submersible drilling rig(Figure 2-8).

潜水钻机由防水电机、减速机构和钻头等组成。电机和减速机构设在绝缘和密封装置的电钻外壳内，且与钻头紧密连接在一起，因而能共同潜入水下作业。目前常用的潜水钻机钻孔直径为400~800 mm，最大钻孔深度50 m。既适用于水下钻孔，也可用于地下水位较低的干土层中钻孔。

The submersible drilling rig consists of a waterproof motor, reduction mechanism, and drill bit. The motor and reduction mechanism are enclosed in the electric drill shell of the insulation and sealing device and are closely connected with the drill bit so that they can submerge together for underwater operation. Currently, the ordinary submersible drilling rig has a drilling diameter of 400~800 mm and a maximum drilling depth of 50 m. It is suitable for both underwater drilling and drilling in dry soil layers with low groundwater levels.

b. 施工流程与工艺。

b. Construction process and technology.

场地平整→桩位放线→开挖浆池、浆沟→护筒埋设→钻机就位→钻孔、泥浆循环、清除泥渣→清孔→下钢筋笼→水下浇筑混凝土→成桩。

Site leveling → setting out of pile position → excavation of slurry pool and ditch → embedding of pile casing → drilling rig in place → drilling, slurry circulation, and desludging → hole cleaning → lowering of reinforcement cage → underwater concreting → pile forming.

护筒埋设：护筒的作用是固定桩孔位置，防止地面水流入，保护孔口，增高桩孔内水压力，防止塌孔以及成孔时引导钻头方向。

Embedding of pile casing: The pile casing is used to fix the position of the pile hole, get rid of surface water, protect the hole mouth, increase the water pressure in the pile hole, prevent hole collapse, and guide the direction of the drill bit during hole forming.

制备泥浆：护壁泥浆的组成，是高塑性黏土或膨润土和水拌和的混合物，也可掺入加重剂、分散剂、增黏剂及堵漏剂等掺合剂。泥浆一般在现场制备，有些黏性土在钻进过程中可形成适合护壁的浆液，可利用其作为护壁泥浆，即"原土造浆"。

Slurry preparation: The slurry for wall protection is composed of high-plasticity clay or a mixture of bentonite and water, and can also be mixed with such admixtures as the weighting agent, dispersant, tackifier, and plugging agent. Slurry is often prepared on site. Some cohesive soil can form slurry suitable for wall protection during drilling, so it can be used as wall protection slurry, that is, "making slurry with original soil".

泥浆具有保护孔壁、防止塌孔、排出土渣、冷却与润滑钻头、减少钻进阻力等作用。钻进过程中，护壁泥浆与钻孔的土屑混合，边钻边排出携带土屑的泥浆，当钻孔达到规定深度后，应运用泥浆循环进行孔底清渣。

The slurry can protect the hole wall, prevent hole collapse, facilitate mucking, cool and lubricate the drill bit, and reduce drilling resistance. During drilling, the slurry is mixed with soil cuttings, with the mixture discharged as drilling progresses. When the borehole reaches the specified depth, the slag at the hole bottom is removed by slurry circulation.

清孔：清孔时，对于土质较好不易坍塌的桩孔，可用空气吸泥机清孔，气压为 0.5 MPa，使管内形成强大的高气压，使气流向上涌，同时不断地补足清水，被搅动的泥渣随气流上涌从孔口被推出，直至喷出清水为止。对于稳定

性较差的孔壁应采用泥浆循环法清孔或抽筒排渣，清孔后的泥浆相对密度应控制在 1.15~1.25；原土造浆的孔，清孔后泥浆相对密度应控制在 1.1 左右。

Hole cleaning: During the cleaning of the hole with the slurry wall, for pile holes with good soil quality and not easy to collapse, an air suction dredger can be used to clean the hole, with an air pressure of 0.5 MPa. It generates a high air pressure in the tube to push upward, with water continuously supplemented. Driven by the surging airflow, the stirred sludge is pushed out from the tube mouth until clear water is seen. For the hole wall with poor stability, the hole can be cleaned by the slurry circulation method or deslagged by the slag extractor. After hole cleaning, the relative density of the slurry shall be controlled at 1.15-1.25. For the hole with slurry made from the original soil, the relative density of slurry after hole cleaning shall be controlled at about 1.1.

孔底沉渣必须设法清除，端承桩的沉渣厚度不得大于 50 mm，摩擦桩沉渣厚度不得大于 150 mm。

The sediment at the hole bottom must be removed, and the sediment thickness of end-bearing piles must not exceed 50 mm. The sediment thickness of friction piles must not exceed 150 mm.

水下浇筑混凝土：泥浆护壁钻孔灌注桩的水下混凝土浇筑常用导管法，混凝土强度等级不低于 C25。商品混凝土的坍落度值一般为 180~220 mm。导管一般用无缝钢管制作，直径为 200~300 mm，每节长度为 2~3 m，最下一节为脚管，长度不小于 4 m，各节管用法兰盘和螺栓连接。浇筑混凝土时，导管应始终埋入混凝土中 0.8~1.3 m，但最大埋入深度也不宜超过 5 m。

Underwater concreting: The tremie method is usually used for underwater concreting of slurry-supported cast-in-situ piles, with C25 or higher concrete. The slump value of commercial concrete is usually 180-220 mm. The tremie is normally made of seamless steel pipes, with a diameter of 200-300 mm and a length of 2-3 m for each section. The bottom section is the foot pipe, at least 4 m long. Pipe sections are connected by flanges and bolts. During concreting, the tremie shall always be buried in the concrete by 0.8-1.3 m, but the maximum buried depth should not exceed 5 m.

②干作业成孔灌注桩。

② Dry-forming cast-in-situ pile.

适用于成孔深度内无地下水的一般黏性土、砂土及人工填土，无须护壁，成孔深度 8~20 m、成孔直径 300~600 mm，不宜用于地下水位以下的各类土及淤泥质土。

It is suitable for general cohesive soil, sandy soil, and artificial fill without groundwater within the hole depth, with no need for wall protection. The hole depth is 8-20 m and the hole diameter is 300-600 mm. It is not suitable for various soils and mucky soil below the groundwater level.

③施工中常见问题及处理。

③ Common problems in construction and treatment.

a. 孔壁坍塌。

a. Hole wall collapse.

在钻孔过程中，如发现排出的泥浆中不断出现气泡，或泥浆突然漏失，这表示有孔壁坍塌的现象。导致孔壁坍塌的主要原因是土质松散，泥浆护壁不好，护筒周围未用黏土紧密填封及护筒内水位不高。钻进时如出现孔壁坍塌，首先应保持孔内水位并加大泥浆相对密度以稳定钻孔的护壁。如坍塌现象严重，应立即回填黏土，待孔壁稳定后再钻。

During drilling, if bubbles keep popping up from the discharged mud, or the mud suddenly sinks, the hole wall may collapse. Such collapse is mainly caused by loose soil, poor wall support by slurry, untight sealing of the casing with clay, and low water level in the casing. If the hole wall collapses during drilling, the water level in the hole shall be maintained, and the relative density of the slurry shall be increased to stabilize the borehole wall. For a serious collapse, clay shall be backfilled immediately, and drilling must not resume until the hole wall becomes stable.

b. 钻孔偏斜。

b. Hole deflection.

钻杆不垂直，钻头导向部分短、导向性差、土质软硬不一，或者遇上大孤石等，都会引起钻孔偏斜。因此，除钻头加工精确、钻杆安装垂直外，操作时还要注意经常观察。当钻孔偏斜时，可提起钻头，上下反复扫钻几次，以便削去硬土，如纠正无效，应在孔中部回填黏土至偏孔处 0.5 m 以上再重新钻进。

Drilling deflection may occur if the drill pipe is not vertical, the guiding part of the drill bit is short, guiding performance is poor, the soil hardness differs, or boulders are encountered. Preventive measures include accurate processing of the drill bit and vertical installation of the drill pipe. Also, frequent inspection is required. When the hole is deflected, operators can lift the drill bit and sweep it up and down repeatedly several times to cut off the hard soil. If the correction is ineffective, clay shall be backfilled in the middle of the hole to a place 0.5 m above the deflected part before re-drilling can be started.

c. 孔底虚土。

c. Loose soil at hole bottom.

在干作业施工中，由于钻孔机械结构所限，孔底常残存一些虚土，它来自扰动残存土、孔壁坍落土以及孔口落土。施工时，孔底虚土量超出规定范围时必须清除，防止虚土影响桩承载力。目前常用的治理虚土的方法是用 20 kg 重的铁饼人工辅助夯实，或采用孔底压力灌浆法。

In dry-forming construction, due to the structural limitation of drilling machines, some loose soil often remains at the hole bottom. It comes from disturbed residual soil, soil collapsed from the hole wall, and soil falling from the hole mouth. During construction, the loose soil at the hole bottom, if in excess of the volume required in the specification, must be removed to avoid interference with the bearing capacity of the pile. Currently, the common method to treat loose soil is using 20 kg heavy discus for manual compaction or applying the hole bottom pressure-grouting method.

d. 断桩。

d. Pile breaking.

水下灌注混凝土桩的质量除混凝土本身质量外，是否断桩是鉴定其质量好坏的关键。预防时应注意三方面的问题：一是力争首批混凝土浇灌一次成功；二是浇筑混凝土过程中导管要埋在混凝土中；三是严格控制现场混凝土配合比。

For underwater cast-in-situ concrete piles, besides the quality of the concrete, pile breaking is also a key factor to identify the pile quality. Notice three problems to prevent pile breaking: First, the first batch of concrete shall be poured successfully at one time; second, the tremie shall be buried in the concrete during pouring; third, the field concrete mix proportion shall be strictly controlled.

（2）沉管灌注桩

(2) Tube-sinking cast-in-situ pile

沉管灌注桩是指利用锤击打桩法或振动打桩法，将带有活瓣式桩尖或预制钢筋混凝土桩靴的钢套管沉入土中，然后边浇筑混凝土边锤击或振动套管将混凝土捣实而成的桩。前者称为锤击沉管灌注桩，后者称为振动沉管灌注桩。

Tube-sinking cast-in-situ pile refers to a pile formed by sinking a steel casing with a flap-type pile tip or a precast reinforced concrete pile shoe into the soil by hammering or vibrating the casing while concreting to compact the concrete. The former is called the hammer-driven tube-sinking cast-in-situ pile, and the latter is called the vibro-driven tube-sinking cast-in-situ pile.

①锤击沉管灌注桩。

① Hammer-driven tube-sinking cast-in-situ pile.

锤击沉管灌注桩是采用落锤、蒸汽锤或柴油锤将钢套管沉入土中成孔，然后灌注混凝土或钢筋混凝土，再拔出钢套管成桩。

The hammer-driven tube-sinking cast-in-situ pile is formed by sinking the steel casing into the soil with a drop hammer, steam hammer, or diesel hammer to form the hole, pouring concrete or reinforced concrete, and pulling out the steel casing.

a. 施工工艺。

a. Construction process.

场地平整→桩机就位→立管→对准桩位、套入桩靴、压桩入土→检查→桩锤轻击→检查有无偏移→正常施打至设计标高→第一次浇灌混凝土→边拔管，边锤击，边继续浇灌混凝土→安放钢筋笼继续浇灌混凝土至桩顶设计标高，如图2-9所示。

Site leveling → pile driver in place → tube erection → align with pile position, sleeve the pile shoe and press it into the soil → check → hammer the pile slightly → check for deviation → drive it to the design elevation → first concreting → pull out the tube while hammering and concreting → lower the reinforcement cage and continue concreting until the design elevation of the pile top, as shown in Figure 2-9.

桩靴与桩管。桩靴可分为混凝土预制桩靴和活瓣式桩靴两种，其作用是阻止地下水及泥砂进入桩管。桩管一般采用无缝钢管，直径为

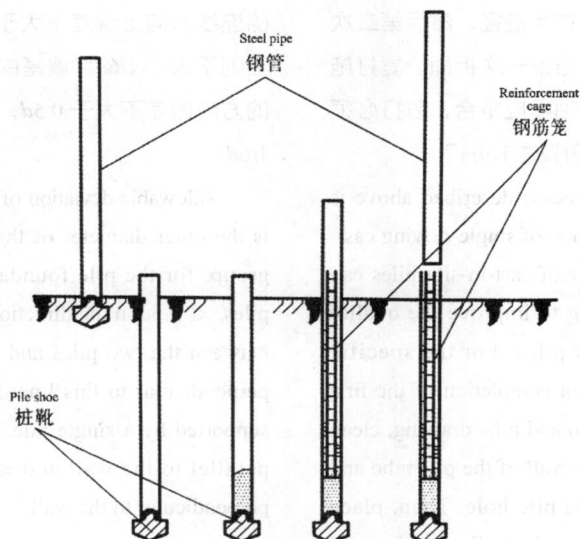

图2-9 锤击沉管灌注桩施工过程

Figure 2-9 Construction Process of Hammer-driven Tube-sinking Cast-in-situ Pile

270~600 mm，其作用是形成桩孔。

Pile shoe and pile tube. Pile shoe includes concrete precast pile shoe and flap-type pile shoe, used to prevent groundwater and silt from entering the pile tube. The pile tube is usually made of 270-600 mm seamless steel pipe and is used to form the pile hole.

成孔。由于锤击沉管灌注桩成孔时不排土，而沉管时会把土挤压密实，所以群桩基础或桩中心距小于3~3.5倍的桩径，应制定合理的施工顺序，以免影响相邻桩的质量。

Hole forming. For hole forming of the hammer-driven tube-sinking cast-in-situ pile, the soil is not discharged but squeezed and compacted during tube sinking. Therefore, for the pile group foundation or when the pile center distance is 3-3.5 times smaller than the pile diameter, a reasonable construction sequence shall be prepared to avoid interference with adjacent piles.

混凝土浇筑与拔管。浇筑混凝土和拔起桩管是保证质量的重要环节。当桩管沉到设计标高后，应停止锤击，检查管内无泥浆或水进入

后，即放入钢筋笼，边浇筑混凝土边拔管，拔管时必须边振（打）边拔，以确保混凝土振捣密实。拔管速度必须严格控制，对于一般土层，以不大于1 m/min为宜；在软土及硬土交界处，应控制在0.8 m/min以内。

Concreting and tube drawing. Concreting and tube drawing are important links to ensure quality. When the pile tube sinks to the design elevation, hammering shall be stopped. Verify that no mud or water exists in the tube and lower the reinforcement cage. Draw out the tube while pouring concrete. Keep vibration(driving) during tube drawing to ensure that the concrete is compacted. The tube-drawing speed must be strictly controlled. For normal soil layers, the speed should not exceed 1 m/min; at the boundary of soft soil and hard soil, the speed should be controlled within 0.8 m/min.

前文所述的这种施工工艺称为单打灌注桩的施工。为了提高桩的质量和承载能力，可采用复打扩大灌注桩的直径，其施工方法是在第一次单打施工完毕并拔出桩管后，清除桩管外壁上和桩孔周围地面上的污泥，立即在原桩位上再次安放桩尖，再进行第二次沉管，使未凝

固的混凝土向四周挤压扩大桩径，然后第二次灌注混凝土，拔管方法与第一次相同。复打施工要注意前后两次沉管轴线应重合，复打必须在第一次灌注的混凝土初凝之前进行。

This construction process described above is designed for the construction of single-driving cast-in-situ piles. The diameter of cast-in-situ piles can be enlarged by re-driving to improve the quality and bearing capacity of piles. For the specific construction method: Upon completion of the first single-driving construction and tube drawing, clean up the sludge on the outer wall of the pile tube and on the ground around the pile hole. Then, place the pile tip again on the original pile position and sink the tube the second time(which squeezes unset concrete to expand the pile diameter). After that, pour concrete the second time. The tube drawing method is the same as the first time. During the re-driving construction, note that two tube-sinking axes shall coincide, and re-driving must be done before the initial setting of the concrete poured previously.

b. 质量要求。

b. Quality requirements.

锤击沉管灌注桩混凝土强度等级应不低于C25；混凝土坍落度，在有筋时宜为 80~100 mm，无筋时宜为 60~80 mm；碎石粒径，有筋时不大于 25 mm，无筋时不大于 40 mm；桩尖混凝土强度等级不得低于 C30。

For hammer-driven tube-sinking cast-in-situ piles, the concrete shall have a strength of at least C25; a concrete slump of 80-100 mm (with reinforcement) or 60-80 mm(without reinforcement); a partial gravel size of not larger than 25 mm(with reinforcement) or 40 mm(without reinforcement); and a strength of not lower than C30 for the pile tip.

桩位允许偏差：群桩不大于 0.5d（d 为桩管外径）；对于两根桩组成的桩基，在两根桩的连线方向上偏差不大于 0.5d，垂直此线方向上则不大于 1/6d；墙基由单桩支承的，平行墙的方向偏差不大于 0.5d，垂直墙的方向不大于 1/6d。

Allowable deviation of pile position: ≤ 0.5d (d is the outer diameter of the pile tube) for the pile group; for the pile foundation composed of two piles, ≤ 0.5d in the direction of the connecting line between the two piles and ≤ 1/6d in the direction perpendicular to this line; for the wall foundation supported by a single pile, ≤ 0.5d in the direction parallel to the wall and ≤ 1/6d in the direction perpendicular to the wall.

当桩的中心距为桩管外径的 5 倍以内或小于 2 m 时，均应跳打，中间空出的桩须待邻桩混凝土达到设计强度的 50% 以后方可施打。

When the center distance of the pile is within 5 times the pile tube's outer diameter or smaller than 2 m, piles shall be alternately driven. The skipped piles in the middle are not driven until the concrete of adjacent piles reaches 50% of the design strength.

②振动沉管灌注桩。

② Vibro-driven tube-sinking cast-in-situ pile.

振动沉管灌注桩是采用激振器或振动冲击锤将钢套管沉入土中成孔而成的灌注桩，其沉管原理与振动沉桩完全相同。

The vibro-driven tube-sinking cast-in-situ pile is formed by sinking the steel casing into the soil with a vibration exciter or a vibration impact hammer. The tube-sinking mechanism is the same as pile vibrosinking.

a. 施工工艺。

a. Construction process.

振动沉管采用振动锤或振动冲击锤沉管，利用桩机强迫振动频率与土的自振频率相同时产生的共振而沉管。沉桩前，将桩管下端活瓣合拢或套入桩靴，对准桩位，徐徐放下桩管压入土中，勿使其偏斜，即可开动激振器沉管。桩管受振后与土体之间摩阻力减小，同时利用振动锤自重在桩管上加压，桩管即能沉入土中。桩管下沉到设计要求深度后，停止振动，立即用吊斗向套管内灌满混凝土，并再次开动激振器，边振动边拔管，同时在拔管过程中继续向管内灌注混凝土。如此反复，直至桩管全部拔出地面后即形成混凝土桩身。

The vibration hammer or vibration impact hammer is used for vibro-driving. The tube is sunk by the resonance generated when the forced vibration frequency of the pile driver coincides with the natural vibration frequency of the soil. Before the pile sinking, close the flap at the lower end of the pile tube or sleeve it into the pile shoe, and aim it at the pile position. Slowly lower the pile tube and press it into the soil without deflection. Then, start the vibration exciter to sink the tube. Upon vibration, the frictional resistance between the pile tube and the soil mass decreases, and with the self-weight of the vibration hammer, the pile tube can sink into the soil. After the pile tube sinks to the requirement of design depth, stop the vibration and fill the tube with concrete using a bucket. Start the vibration exciter again, pull out the tube while vibrating, and continue concreting inside the tube. Repeat these steps until the pile tube is totally pulled out of the ground to form a concrete tube.

振动灌注桩可采用单振法、反插法、复振法施工。

The vibro-driven cast-in-situ piles can be built by the single vibration method, reverse insertion method, and re-vibration method.

单振法：在沉入土中的桩管内灌满混凝土，开激振器 5~10 s，开始拔管，边振边拔。每拔 0.5~1.0 m，停拔振动 5~10 s，如此反复，直到桩管全部拔出。在一般土层内的拔管速度宜为 1.2~1.5 m/min，在软弱土层中，不得大于 0.8~1.0 m/min。单振法施工速度快，混凝土用量少，但桩的承载力低，适用于含水量较少的土层。

Single vibration method: Fill the pile tube in the soil full of concrete, run the vibration exciter for 5-10 s, and pull out the tube while vibrating. For every 0.5-1.0 m, stop pulling and vibrate for 5-10 s, and repeat until the entire pile tube is pulled out. The tube drawing speed should be 1.2-1.5 m/min for general soil layers and shall not exceed 0.8-1.0 m/min in weak soil layers. The single vibration method has a high construction speed and consumes a small amount of concrete, but the piles have a low bearing capacity, so it is suitable for soil layers with less water content.

反插法：在桩管内灌满混凝土后，先振动再开始拔管。每次拔管高度 0.5~1.0 m，再向下反插深度 0.3~0.5 m，如此反复进行并始终保持振动，直到桩管全部拔出地面。反插法能扩大桩的截面，从而提高桩的承载力，但混凝土耗用量较大，一般适用于饱和软土层。

Reverse insertion method: After the pile tube is full of concrete, vibrate before tube drawing. Each time pulled out for a height of 0.5-1.0 m, the tube is inserted backward for a depth of 0.3-0.5 m. Repeat and keep vibrating until the entire tube is pulled out of the ground. The reverse insertion method can expand the cross-section of the pile, thus improving its bearing capacity. However, it consumes more concrete and is usually suitable for saturated soft soil layers.

复振法：施工方法及要求与锤击沉管灌注桩的复打法相同。

Re-vibration method: The construction method and requirements are the same as the re-driving method of hammer-driven tube-sinking cast-in-situ piles.

b. 质量要求。

b. Quality requirements.

振动沉管灌注桩混凝土强度等级应不低于 C25；混凝土坍落度，在有筋时宜为 80~100 mm，无筋时宜为 60~80 mm；碎石粒径不大于 30 mm。

For vibro-driven tube-sinking cast-in-situ piles, the concrete shall have a strength of at least C25, a concrete slump of 80-100 mm (with reinforcement) or 60-80 mm(without reinforcement), and a particle gravel size of 30 mm or smaller.

桩的中心距不宜小于桩管外径的 4 倍，否则应跳打，相邻的桩施工时其间隔时间不得超过混凝土的初凝时间。

The center distance of piles should be at least 4 times the outer diameter of the pile tube, otherwise, piles shall be driven alternately. During the construction of adjacent piles, the interval time must not exceed the initial setting time of concrete.

在拔管过程中，桩管内应随时保持有不少于 2 m 高度的混凝土，以便形成足够的压力，防止混凝土在管内阻塞。

During the tube drawing, 2 m or higher concrete shall be kept in the pile tube, to maintain sufficient pressure and prevent concrete blocking in the tube.

为保证沉管灌注桩的承载力要求，必须严格控制最后两个管 2 min 内的沉管贯入度，其值按设计要求或根据试桩和当地长期的施工经验确定。

In order to meet the bearing capacity requirements of the tube-sinking cast-in-situ piles, the penetration of the last two tubes within 2 min must be strictly controlled, with the value determined as per the design requirements or according to the test pile and local long-term construction experience.

桩位允许偏差同锤击沉管灌注桩。

The allowable deviation of pile position is the same as that of hammer-driven tube-sinking cast-in-situ piles.

③施工中常见的问题及处理。

③ Common problems in construction and treatment.

a. 断桩。

a. Pile breaking.

断桩一般都发生在地面以下软硬土的交接处，并多数发生在黏土中，砂土及松土中则很少出现。主要原因是桩距过小，受邻桩施打时挤压的影响，桩身混凝土终凝不久就受到振动和外力；以及软硬土层间传递水平力大小不同，对桩产生剪应力等。处理方法是经检查有断桩后，应将断桩拔出，略增大桩的截面面积或加箍筋后，再浇筑混凝土；或者在施工过程中采取预防措施，如施工中控制桩中心距不小于 3.5 倍桩径，采用跳打法或者控制时间间隔的方法，使邻桩的混凝土达到设计强度等级的 50% 后，再施打中间桩等。

Pile breaking is likely to happen at the junction of soft and hard soil below the ground, mostly in clay, and rarely in sandy soil and loose soil. The main reasons include the small pile interval(pile squeezed as adjacent piles are driven), vibration and external force acting on the pile body just after final setting, and the different horizontal forces transmitted between soft and hard soil layers(generating shear stress on the pile). To solve this, pull out the broken pile, slightly increase the cross-sectional area of the pile or add stirrups before pouring concrete; or, take preventive measures during construction. For example, control the center distance of the pile to at least 3.5 times the pile diameter; apply the alternate driving method, or control the time interval for driving. Do not drive the middle pile until the concrete strength of adjacent piles reaches 50% of the design strength.

b. 瓶颈桩。

b. Bottleneck pile.

瓶颈桩是指桩的某处直径缩小，形似"瓶颈"（图2-10），其截面面积不符合设计要求。该问题多数发生在黏性土，土质软弱、含水率高的土以及饱和的淤泥或淤泥质软土层中。产生瓶颈桩的主要原因是在含水率较大的软弱土层中沉管时，土受挤压便会产生很高的孔隙水压，拔管后土便挤向新灌的混凝土，造成缩颈。拔管速度过快，混凝土量少、和易性差，混凝土出管扩散性差也会造成缩颈现象。处理方法：施工中保持管内混凝土略高于地面，使之有足够的扩散压力，拔管时采用复打或反插法，并严格控制拔管速度。

A bottleneck pile refers to a pile whose diameter is partially lessened to form a "bottleneck"(Figure 2-10), whose cross-sectional area does not meet the design requirements. This problem tends to occur in cohesive soil, soft soil, and wet soil, especially saturated silt or soft mucky soil layers. The main reason for the bottleneck pile is that when the tube sinks in the weak soil layer with high moisture content, the soil will be squeezed to generate a high pore water pressure. After tube drawing, the soil pushes tightly against the newly poured concrete, resulting in necking. Some other defects may also cause necking, such as quick tube drawing, insufficient concrete with poor workability, and poor diffusivity of concrete out of the tube. To solve this problem, keep the intra-tube concrete slightly higher than the ground during construction to maintain a sufficient diffusion pressure. During tube drawing, apply re-driving or reverse insertion method and strictly control the drawing speed.

c. 吊脚桩。

c. End-suspended pile.

吊脚桩是指桩的底部混凝土隔空或混进泥砂而形成松散层部分的桩。其产生的主要原因是预制钢筋混凝土桩尖承载力或钢活瓣桩尖刚度不够，沉管时被破坏或变形，从而导致水或泥砂进入桩管；或拔管时桩靴未脱落、活瓣未张开，混凝土未及时从管内流出等。处理方法是拔出桩管，填砂后重打；或"密振慢拔"，开始沉管时先反插几次再正常拔管。

End-suspended pile refers to a pile whose bottom concrete is hollow or mixed with mud and sand to form a loose layer. Main reasons: (1) The precast reinforced concrete pile tip has an insufficient bearing capacity, or the steel-flap pile tip has an insufficient rigidity, causing damage or deformation during tube sinking so that water or silt enters the pile tube. (2) During tube drawing, the pile shoe does not detach or the flap is not opened, and the concrete does not flow out of the tube in time. To solve this problem, pull out the pile tube, fill the sand, and re-drive it, or, apply dense vibration and slow drawing, and perform reverse insertion several times before normal drawing at the beginning of tube sinking.

d. 桩尖进水进泥。

d. Water and mud entering the pile tip.

桩尖进水进泥常发生在地下水位高或含水量大的淤泥和粉泥土土层中。产生的主要原因是钢筋混凝土桩尖与桩管结合处或钢活瓣桩尖闭合不紧密，或钢筋混凝土桩尖被打破或钢活瓣桩尖变形。处理方法：将桩管拔出，清除管内泥砂，修整桩尖活瓣变形缝隙，用黄砂回填桩孔后再重打；若地下水位较高，待沉管至地下水位时，先在桩管内灌入0.5 m厚度的水泥砂浆做封底，再灌1 m高度混凝土增压，然后再继续下沉管桩。

Water and mud entering the pile tip often occurs in sludge and silty soil layers with a high groundwater level or high water content. Main reasons: (1) The joint between the reinforced concrete pile tip and the pile tube, or the steel-flap pile tip is not tightly closed. (2) The reinforced

图2-10 瓶颈桩

Figure 2-10 Bottleneck Pile

concrete pile tip is broken, or the steel-flap pile tip is deformed. Solutions: Pull out the pile tube, and remove silt and sand from the tube, if trim the deformation joint of the flap at the pile tip, backfill the pile hole with yellow sand, and re-drive it; if the groundwater level is high, when the tube reaches the groundwater level, first pour 0.5 m of cement mortar into the pile tube for bottoming, then pour 1 m of concrete for pressurization. After that, continue tube sinking.

2.2.4 桩基础的检测与验收

2.2.4 Inspection and Acceptance of Pile Foundation

（1）桩基的检测

(1) Inspection of pile foundation

成桩的质量检验有两种基本方法：一种是静载试验法，又称破损试验（图 2-11）；另一种是动测试验法，又称无损试验（图 2-12）。

Two basic methods are used for pile quality inspection: One is the static load test method, also known as the damage test(Figure 2-11); the other is the dynamic test method, also known as the non-destructive test(Figure 2-12).

①静载试验法。

① Static load test method.

a. 试验目的。

a. Test purpose.

静载试验的目的是采用接近于桩的实际工作条件，通过静载加压，确定单桩的极限承载力。该试验作为桩的设计依据，对工程桩的承载力进行抽样检验和评价。

图2-11 桩基静载试验

Figure 2-11 Static Load Test of Pile Foundation

2-12 桩基动测试验

Figure 2-12 Dynamic Test of Pile Foundation

The static load test aims to determine the ultimate bearing capacity(use it as the design basis) of a single pile through static load compression under a condition close to the actual working conditions of the pile or to carry out sampling inspection and evaluation on the bearing capacity of engineering piles.

b. 试验方法。

b. Test method.

静载试验是根据模拟实际荷载情况，通过静载加压，得出一系列关系曲线，综合评定其容许承载力的一种试验方法。它能较好地反映单桩的实际承载力。荷载试验有多种，通常采用的是单桩竖向抗压静载试验、单桩竖向抗拔静载试验和单桩水平静载试验。

The static load test is a method to comprehensively evaluate the allowable bearing capacity of piles from a series of relation curves obtained by simulating the actual load conditions through static load pressurization. This method can properly reflect the actual bearing capacity of a single pile. Many kinds of load tests are designed, among which, the single-pile vertical compression static load test, single-pile vertical uplift static load test, and single-pile horizontal static load test are usually used.

c. 试验要求。

c. Test requirements.

预制桩在桩身强度达到设计要求的前提下，对于砂类土，不应少于 $10d$；对于粉土和黏性土，不应少于 $15d$；对于淤泥或淤泥质土，不少于 $25d$，待桩身与土体的结合基本趋于稳定，才能进行试验。现场灌注桩应在桩身混凝土强度达到设计等级的前提下，对于砂类土成桩不少于 $10d$；对于一般黏性土不少于 $20d$；对于淤泥或淤泥质土，不少于 $30d$，才能进行试验。对于地基基础设计等级为甲级或地质条件复杂，成桩质量可靠性低的灌注桩，应采用静载试验的

方法进行检验，检验桩数不应少于总数的 1% 且不应少于 3 根；当总桩数少于 50 根时，不应少于 2 根，其桩身质量检验时，抽检数量不应少于总数的 30%，且不应少于 20 根；其他桩基工程的抽检数量不应少于总数的 20%，且不应少于 10 根；对混凝土预制桩及地下水位以上且终孔后经过核验的灌注桩，抽检数量不应少于总数的 10%，且不应少于 10 根，每根柱子的承台不得少于 1 根。

Given the strength of the precast pile meets the design requirements, it shall be $\geq 10d$ for sandy soil, $\geq 15d$ for silt and cohesive soil, and $\geq 25d$ for sludge or mucky soil. The test must not be carried out until the combination of the pile body with the soil mass becomes stable. On-site cast-in-situ piles must not be tested until the concrete strength of the pile body reaches the design value, i.e., $\geq 10d$ for sandy soil, $\geq 20d$ for ordinary cohesive soil, and $\geq 30d$ for sludge or mucky soil. For cast-in-situ piles with Grade-A foundation design grade or facing complex geological conditions and low reliability of pile quality, the static load test shall be carried out for inspection, and at least 1% of the total piles (≥ 3 piles) shall be tested; when there are less than 50 piles, at least 2 piles shall be tested; for the quality inspection of the pile body, at least 30% (≥ 20 piles) of total piles shall be tested; for other pile foundation works, at least 20% (≥ 10 piles) of total piles shall be tested; for precast concrete piles and cast-in-situ piles above the groundwater level and verified after hole completion, at least 10% (≥ 10 piles) of total piles shall be tested, each column shall have at least 1 pile cap.

②动测试验法。

② Dynamic test method.

a. 特点。

a. Characteristics.

动测试验法，又称动力无损检测法，是检

测桩基承载力及桩身质量的一项新技术，可作为静载试验的补充。

The dynamic test method, also known as the dynamic non-destructive test, is a new technology to test the bearing capacity of the pile foundation and the quality of the pile body, as a supplement to the static load test.

一般静载试验装置比较笨重，装、卸操作费工费时，成本高，检测数量有限，并且容易破坏桩基。而动测法的试验仪器轻便灵活，检测速度快，单桩检测时间仅为静载试验的 1/50 左右，可大大缩短试验时间；检测不破坏桩基，试验结果也相对较准确，可进行桩基普查；费用低，单桩测试费为静载试验的 1/30 左右，可节省大量的人力、物力。

Generally, the device for static load test is heavy, labor- and time-consuming for handling, with high cost, limited test objects, and easy to destroy the pile foundation. On the contrary, the test instrument used for the dynamic test is light and flexible, allowing fast tests. The single-pile test time is only about 1/50 of the static load test, largely shortening the test time. The test does not destroy the pile foundation, and the test results are more accurate, suitable for general surveying of pile foundations. The cost is as low as about 1/30 of the static load test, saving a lot of manpower and materials.

　　b. 试验方法。

　　b. Test method.

动测法是相对于静载试验法而言，它是对桩—土体系进行适当的简化处理，建立起数学—力学模型，借助现代电子技术与量测设备采集桩—土体系在给定的动荷载作用下所产生的振动参数，结合实际桩土条件进行计算，所得的结构与相应的静载试验结果进行对比，在积累一定数量的动静载试验对比结果的基础上，找出两者之间的某种关系，并以此作为标准来确定桩基承载力的方法。单桩承载力的动测方法种类较多，国内常用的方法有动力参数法、锤击贯入法、水电效应法、共振法、机械阻抗法、波动方程法等。

Compared with the static load test, the dynamic test appropriately simplifies the pile-soil system and establishes a mathematical-mechanical model. With the help of modern electronic technology and measuring equipment, the vibration parameters generated by the pile-soil system under given dynamic loads are collected and calculated. The obtained structure is compared with the results of the static load test. After accumulating a proper number of comparison results from the dynamic and static load tests, operators figure out the relationship between the two and use it as a standard to determine the bearing capacity of the pile foundation. Many dynamic test methods are used to measure the bearing capacity of a single pile. The commonly used methods in China include the dynamic parameter method, hammer penetration method, hydroelectric effect method, resonance method, mechanical impedance method, and wave equation method.

　　c. 桩身质量检验。

　　c. Quality inspection of pile body.

在桩基动态无损检测中，国内外广泛使用的方法是应力波反射法，又称低（小）应变法。其原理是根据一维杆件弹性反射理论（波动理论）采用锤击振动力法检测桩体的完整性，即以波在不同抗阻和不同约束条件下的传播特性来判别桩身质量。

In the dynamic non-destructive test of pile foundations, the stress wave reflection method, also known as the low(small) strain method, is widely used at home and abroad. It tests the integrity of the pile body by the vibration force from hammering according to the elastic reflection theory(wave theory) of one-dimensional members, that is, to judge the quality of the pile body by the propagation characteristics of waves under different resistance and constraints.

（2）桩基验收

(2) Acceptance of pile foundation

①桩基验收资料。

① Acceptance data of pile foundation.

a. 工程地质勘查报告、桩基施工图、图纸会审纪要、设计交底记录、设计变更及材料代用通知单等。

a. Engineering geological survey reports, pile foundation construction drawings, minutes from the joint review of drawings, design disclosure records, design changes, and material substitution notices.

b. 经审定的施工组织设计、施工方案及执行变更情况记录。

b. Approved construction organization design, construction schemes, and implementation changes.

c.桩位测量放线图，包括桩位复核签证单。

c. Surveying and setting-out diagram of pile position, including pile position verification form.

d. 制作桩的材料试验检测记录，成桩质量检查报告。

d. Material test records and quality inspection

reports of piles.

e. 单桩承载力检测报告。

e. Single-pile bearing capacity test report.

f. 基坑挖至设计标高的桩基竣工平面图及桩顶标高图。

f. As-built plan of pile foundation(when the foundation pit is excavated to the design elevation) and elevation drawing of pile top.

②桩基允许偏差。

② Allowable deviation of pile foundation.

a. 预制桩允许偏差。

a. Allowable deviation of precast pile.

打（压）入桩（预制混凝土方桩、预应力管桩、钢桩）的桩位偏差必须符合表2-2的规定。斜桩倾斜角度的偏差不得大于倾斜角正切值的15%（倾斜角系桩的纵轴线与铅垂线间夹角）。

The pile position deviation of driven(pressed) piles(precast concrete square piles, prestressed tubular piles, and steel piles) must comply with the provisions of Table 2-2. The deviation of the raking pile's tilt angle must not exceed 15% of the tangent value of the tilt angle(i.e. the included angle between the pile's longitudinal axis and plumb line).

表2-2　预制桩（钢桩）桩位允许偏差

Table 2-2 Allowable Deviation of Precast Pile (Steel Pile) Position

序号 S/N	项目 Item	规范允许范围/mm Allowable range by specification / mm
1	盖有基础梁的桩：①垂直于基础梁的中心线；②沿基础梁的中心线 Pile covered with foundation beam: ① vertical to the centerline of foundation beam; ② along the centerline of foundation beam	100+0.01H 150+0.01H
2	桩数为1~3根桩基中的桩 Number of piles: 1~3 piles in the pile foundation	100
3	桩数为4~16根桩基中的桩 Number of piles: 4~16 piles in the pile foundation	1/2桩径或边长 1/2 pile diameter or side length

序号 S/N	项目 Item	规范允许范围/mm Allowable range by specification / mm
4	桩数大于16根桩基的桩：①最外边的桩；②中间桩 More than 16 piles in the pile foundation: ① outermost pile; ② middle pile	1/3桩径或边长 1/3 pile diameter or side length 1/2桩径或边长 1/2 pile diameter or side length

注：H为施工现场地面标高与桩顶设计标高的距离。

Note: H represents the distance from the ground elevation to the designed pile top elevation on the construction site.

b. 灌注桩允许偏差。

b. Allowable deviation of cast-in-situ piles.

灌注桩的桩位偏差必须符合表 2-3 的规定，桩顶标高至少要比设计标高高出 0.5 m，桩底清孔质量按不同的成桩工艺有不同的要求，应按规范要求执行。每浇筑 50 m³ 混凝土必须有一组试件，小于 50 m³ 的桩，每根桩必须有一组试件。

The pile position deviation of cast-in-situ piles must comply with the provisions of Table 2-3. The pile top elevation shall be at least 0.5 m higher than the design elevation. The hole cleaning quality at the pile bottom has different requirements according to different pile forming processes and shall be implemented according to the specification requirements. A group of test pieces must be taken for every 50 m³ of concrete or for every pile smaller than 50 m³.

表 2-3　灌注桩平面位置和垂直度允许偏差对比表

Table 2-4 Comparison of Allowable Deviations of Plane Position and Verticality of Cast-in-situ Piles

序号 S/N	成孔方法 Hole-forming method		桩径允许偏差/mm Allowable deviation of pile diameter /mm	垂直度允许偏差/% Allowable deviation of verticality /%	桩位允许偏差/mm Allowable deviation of pile position /mm	
					1~3根、单排桩基垂直于中心线方向和群桩基础的边桩 1-3 border piles that single-row pile foundation is perpendicular to the centerline direction and the pile group foundation	条形桩基沿中心线方向和群桩基础的中间桩 Strip pile foundation along the centerline direction and middle piles of the pile group foundation
1	泥浆护壁灌注桩 Slurry-supported cast-in-situ pile	$D \leqslant 1000$ mm $D > 1000$ mm	±50 ±50	<1 <1	D/6，且不大于100 100+0.01H D/6, and not greater than 100 100+0.01H	D/4，且不大于150；150+0.01H D/4, and not greater than 150; 150+0.01H

序号 S/N	成孔方法 Hole-forming method		桩径允许偏差/mm Allowable deviation of pile diameter /mm	垂直度允许偏差/% Allowable deviation of verticality /%	桩位允许偏差/mm Allowable deviation of pile position /mm	
					1~3根、单排桩基垂直于中心线方向和群桩基础的边桩 1-3 border piles that single-row pile foundation is perpendicular to the centerline direction and the pile group foundation	条形桩基沿中心线方向和群桩基础的中间桩 Strip pile foundation along the centerline direction and middle piles of the pile group foundation
2	套管成孔灌注桩 Cast-in-situ casing pile	$D \leqslant 500$ mm	−20	<1	70	150
		$D > 500$ mm	−20	<1	100	150
3	干成孔灌注桩 Dry-forming cast-in-situ pile		−20	<1	70	150
4	人工挖空桩 Manual digging pile	混凝土护壁 Concrete-supported wall	+50	<0.5	50	150
		钢套管护壁 Steel sleeve-supported wall	+50	<1	100	200

注：桩径允许偏差的负值是指个别断面。

Notes: For the allowable deviation of pile diameter, the negative value refers to individual sections.

采用复打、反插法施工的桩，其桩径允许偏差不受上表限制。

For piles built by re-driving or reverse insertion method, the allowable deviation of the pile diameter is not subject to the above table.

H 为施工现场地面标高与桩顶设计标高的距离，D 为设计桩径。

H represents the distance from the ground elevation to the designed pile top elevation on the construction site, D represents the designed pile diameter.

2.2.5 桩基工程的安全技术措施

2.2.5 Safety Technical Measures for Pile Foundation Works

①机具进场要注意危桥、陡坡、陷地和防止碰撞电杆、房屋等，以避免造成事故。

① When machines and tools are mobilized, pay attention to dangerous bridges, steep slopes, and subsidence, and keep them away from poles and houses to avoid accidents.

②施工前应全面检查机械，发现问题要及时解决，严禁"带病"作业。

② Before construction, all machinery shall be inspected, with problems solved in time. It is strictly prohibited to work with defects.

③在打桩工程中遇到地坪隆起或下陷时，应随时对机架及路轨进行调整垫平。

③ In case of floor uplift or subsidence during piling works, the rack and rail shall be adjusted and leveled in time.

④机械司机应持证上岗，施工操作时要思想集中，服从指挥信号，不得随意离开岗位，并经常注意机械运转情况，发现异常情况要及时纠正。

④ Mechanical drivers shall hold certificates, concentrate and obey command signals during construction, and must not leave their posts unless authorized. They shall keep watching the operation of machinery and correct abnormalities in time.

⑤悬挂振动桩锤的起重机，其吊钩上必须有防松脱的保护装置。振动桩锤悬挂钢架的耳环上应加装保险钢丝绳。

⑤ For the crane suspended with the vibrating pile hammer, its hook must have a protective device against loosening. Safety wire rope shall be installed on the earrings of the vibrating pile hammer, hanging the steel frame.

⑥钻孔灌注桩在已钻成的孔尚未浇筑混凝土前，必须用临时盖板封严；钢管桩打桩后必须及时加盖临时桩帽；预制混凝土桩送桩进入土层后的桩孔必须及时用砂子或者其他材料填满，以免发生人身安全事故。

⑥ The cast-in-situ bored pile must be sealed with a temporary cover before the drilled hole is poured with concrete. After the steel pipe pile is driven, a temporary pile cap must be added in time. After the precast concrete pile is driven into the soil layer, the pile hole must be filled with sand or other materials in time to avoid personal safety accidents.

⑦在进行冲抓锥或冲孔锤操作时不允许任何人进入落锤区施工范围内，以防砸伤。

⑦ During the operation of the grab-type drill bit or punching hammer, no one is allowed to enter the construction area of the drop hammer to avoid injury.

⑧成孔钻机操作时要注意钻机安定平稳，防止钻架突然倾倒或钻具下落发生事故。

⑧ The drilling rig must work stably during hole forming to prevent accidents caused by sudden toppling of the drilling frame or falling of the drilling tool.

⑨压桩时，非工作人员应离机 10 m 以上。起重机的起重臂下严禁站人。

⑨ During pile pressing, irrelevant persons shall keep 10 m away from the machine. No one is allowed to stand under the crane boom.

⑩夯锤下落时，在吊钩尚未降至夯锤吊环附近前，操作人员不得提前下坑挂钩。从坑中提夯锤时，严禁挂钩人员站在锤上随锤提升。

⑩ When the hammer ram falls, the operator must not go down the pit to hook until the hook is lowered to the vicinity of the rammer ring. When the hammer ram is lifted from the pit, the hooking operator is not allowed to stand on and ascend with the hammer.

砌筑工程
Masonry Works

学习目标:

Learning objectives:

掌握砌筑工程中所用脚手架和垂直运输设施的构造及使用要求;明确砌筑工程施工的准备工作内容和要求;掌握砖砌体、中小型砌块砌体的施工方法和施工工艺;掌握砌筑工程的质量要求及安全防护措施;能编制砌筑工程施工方案。

Be able to master the structure and operating requirements of scaffold and vertical transportation facilities used in masonry works; know the preparation contents and requirements of masonry construction; master the construction methods and process of brick masonry and small- and medium-sized block masonry; master the quality requirements and safety protection measures of masonry works; have the ability to prepare the construction schemes of masonry works.

技能抽查要求:

Skill checking requirements:

能按规范要求并正确使用常用检测工具对扣件式钢管脚手架工程的施工质量进行检查验

收；能正确填写脚手架工程施工质量检查验收记录表；掌握砖墙及砖基础的施工工艺，会使用砌筑工具，顺利完成砌筑工作；能正确使用常用检测工具对砌体工程质量进行检查验收。

Be able to inspect and accept the construction quality of steel tubular scaffold with couplers by correctly using common testing tools according to the specification requirements; correctly fill in the record forms for construction quality inspection and acceptance of scaffold works; master the construction technology of brick walls and brick foundations, and successfully complete the masonry work with masonry tools; inspect and accept the quality of masonry works by correctly using common testing tools.

建筑八大员岗位资格考试要求：

Qualification examination requirements for eight major posts of construction engineering:

掌握砖砌体、中小型砌块砌体的砌筑方法和施工工艺；掌握砌筑工程施工的质量与安全要求；掌握扣件式钢管、门式和碗扣式脚手架的构造、搭设和拆除要求；掌握物料提升机、塔吊和施工电梯的安装、使用和拆除要求。

Master the building method and construction technology of brick masonry and small- and medium-sized block masonry; master the quality and safety requirements for masonry works; master the construction, erection, and removal requirements of steel tubular scaffold with couplers, gate scaffold, and bowl-buckle scaffold; master the installation, use, and removal requirements of hoists, tower cranes, and construction elevators.

3.1 脚手架

3.1 Scaffold

脚手架是为了便于施工活动和安全操作而搭建的一种临时设施。砌筑用脚手架是砌筑过程中工人进行操作和堆放材料的一种临时设施。砌筑施工时，不利用脚手架所能砌到的高度为1.2~1.4 m，它被称为可砌高度。

Scaffold is a temporary facility to facilitate construction activities and safe operation. Scaffold for masonry is a temporary facility for workers to operate and stack materials during masonry. During masonry construction, the height that can be reached without scaffolds is 1.2-1.4 m, which is called the height of bricklaying.

脚手架的种类很多，按其搭设的位置可分为外脚手架和里脚手架两大类；按其所用材料分为木、竹和金属脚手架；按其构造形式可分为多立杆式、框式、吊挂式、悬挑式、升降式，以及用于楼层间操作的工具式脚手架等。

There are many types of scaffolds, including external scaffolds and internal scaffolds according to their erection positions; wooden, bamboo, and metal scaffolds according to their materials; multi-upright tube scaffolds, frame scaffolds, hanging scaffolds, cantilever scaffolds, lifting scaffolds, and implementary scaffolds for inter-floor operation according to their structural forms.

脚手架是砌体工程的辅助工具，在建筑物施工中，都需要搭设脚手架，当建筑物竣工后，脚手架应全部拆除，不留任何痕迹。脚手架与施工安全有着密切的联系，必须符合如下具体要求：

Scaffolds are auxiliary tools for masonry works and are mandatory in building construction. Upon completion of buildings, scaffolds shall be completely removed without any traces. Scaffolds are closely related to construction safety and must meet the following specific requirements:

①脚手架的各部分材料要有足够的强度，应能安全地承受上部的施工荷载和自重。

① The materials of each part of the scaffold shall have sufficient strength to safely bear its dead weight and the construction load of the upper part.

②脚手架要有足够的稳定性，不发生过大的变形。

② Scaffolds shall be sufficiently stable, without excessive deformation.

③脚手架板道上要留有足够的面积，以满足工人操作、堆放材料和运输的要求。

③ The scaffold floor shall provide sufficient area to allow workers' operation, stacking, and transportation.

④脚手架必须安全可靠，符合高空作业的要求。脚手架的连接、护身栏杆、挡脚板、安全网等的布设应按有关规定执行。

④ Scaffolds must be safe and meet the requirements for work at heights. The connection, protective railings, toe boards, and safety nets of scaffolds shall follow relevant regulations.

⑤脚手架属于周转性重复使用的临时设施，要力求构造简单，装拆方便，损耗小。

⑤ As temporary reusable facilities, scaffolds must feature a simple structure, convenient assembly/disassembly, and small loss.

⑥要因地制宜，就地取材，尽量节约架子用料。

⑥ It is recommended to consider local

conditions, use local materials, and save scaffold materials as much as possible.

3.1.1 扣件式钢管脚手架

3.1.1 Steel Tubular Scaffold with Couplers

扣件式钢管脚手架通过立杆、水平杆、剪刀撑、抛撑、扫地杆、连墙件以及脚手板等组成，扣件将其相连，如图3-1所示。具有承载能力大、拆装方便、搭设高度大、周转次数多、摊销费用低等优点。

The steel tubular scaffold with couplers consists of the upright tube, horizontal tube, diagonal bracing, throwing support, bottom horizontal tube, connecting tube, and scaffold floor, which are connected by couplers, as shown in Figure 3-1. It has the advantages of large bearing capacity, easy disassembly and assembly, large erection height, easy reuse, and low cost.

（1）组成部件

(1) Components

①钢管。一般均采用外径 48.3 mm，壁厚 3.6 mm 的钢管，立杆、纵向水平杆、斜杆的钢管长度 4~6.5 m，横向水平杆的钢管长度 2.1~2.3 m。

① Steel pipe. Generally, steel pipes with an outer diameter of 48.3 mm and a wall thickness of 3.6 mm are used. The steel pipe is 4-6.5 m long for upright tubes, longitudinal horizontal tubes, and diagonal tubes, and is 2.1-2.3 m long for transverse horizontal tubes.

立杆：平行于建筑物并垂直于地面，将脚手架荷载传递给底座。

Upright tube: Parallel to the building and

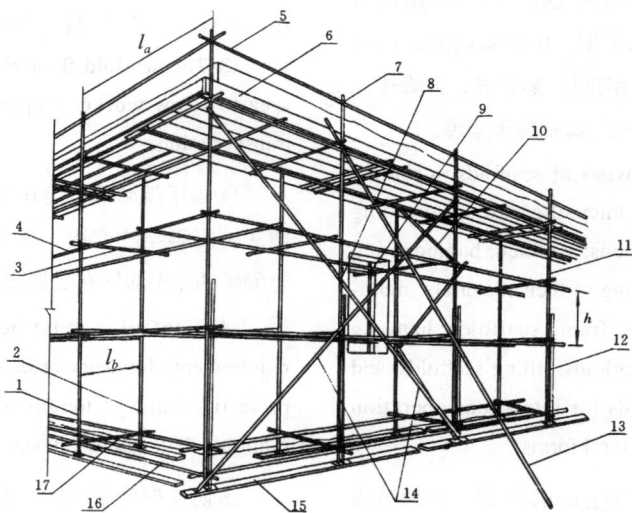

1.外立杆；2.内立杆；3.横向水平杆；4.纵向水平杆；5.栏杆；6.挡脚板；7.直角扣件；8.旋转扣件；9.连墙件；10.横向斜撑；11.主立杆；12.副立杆；13.抛撑；14.剪刀撑；15.垫板；16.纵向水平扫地杆；17.横向水平扫地杆

1. outer upright tube; 2. inner upright tube; 3. transverse horizontal tube; 4. longitudinal horizontal tube; 5. railing; 6. toe board; 7. right-angle coupler; 8. rotary coupler; 9. connecting tube; 10. transverse diagonal bracing; 11. main upright tube; 12. auxiliary upright tube;13. throwing support; 14. diagonal bracing; 15. subplate; 16. longitudinal bottom horizontal tube; 17. transverse bottom horizontal tube

图3-1　扣件式钢管脚手架

Figure 3-1 Steel Tubular Scaffold with Couplers

perpendicular to the ground, to transfer scaffold loads to the base.

纵向水平杆（大横杆）：平行于建筑物并在纵向水平连接各立杆，承受、传递荷载给立杆。

Longitudinal horizontal tube(large crossbar): Parallel to the building and horizontally connected to each upright tube in the longitudinal direction, to bear and transfer loads to the upright tube.

横向水平杆（小横杆）：垂直于建筑物并在横向连接内、外大横杆，承受、传递荷载给大横杆。

Transverse horizontal tube(small crossbar): Perpendicular to the building and connected to the inner and outer large crossbars transversely, to bear and transfer loads to the large crossbar.

剪刀撑：设在脚手架外侧面并与墙面平行的十字交叉斜杆，可增强脚手架的纵向刚度。

Diagonal bracing: A cross diagonal bar arranged on the outer side of the scaffold and parallel to the wall, to enhance the longitudinal rigidity of the scaffold.

连墙件：连接脚手架与建筑物，承受并传递荷载，且可防止脚手架横向失稳。

Connecting tube: Connecting the scaffold with the building, to bear and transfer loads and eliminate lateral instability.

水平斜拉杆：设在有连墙件的脚手架内、外立柱间的步架平面内的"之"字形斜杆，可增强脚手架的横向刚度。

Horizontal diagonal tube: Zigzag diagonal rod set in the step frame plane between the inner and outer columns of the scaffold with connecting tubes, to enhance the transverse rigidity of the scaffold.

纵向水平扫地杆：采用直角扣件固定在距底座上皮不大于 200 mm 处的立杆上，起约束立杆底端在纵向发生位移的作用。

Longitudinal bottom horizontal tube: Fixed

with right-angle couplers to the upright tube at 200 mm or less from the upper surface of the base, to restrain the bottom end of the upright tube from longitudinal displacement.

横向水平扫地杆：采用直角扣件固定在靠紧纵向扫地杆下方的立杆上的横向水平杆，起约束立杆底端在横向发生位移的作用。

Transverse bottom horizontal tube: A transverse horizontal tube fixed with right-angle couplers to the upright tube close to the bottom of the longitudinal bottom horizontal tube, to restrain the bottom end of the upright tube from transverse displacement.

②扣件。扣件用于钢管之间的连接，基本形式有三种，如图 3-2 所示。对接扣件用于两根钢管的对接连接；旋转扣件用于两根钢管呈任意角度交叉的连接；直角扣件用于两根钢管呈垂直角度交叉的连接。

② Coupler. Couplers are used to connect steel pipes in three basic forms, as shown in Figure 3-2. Butt couplers are used for the butt connection of two steel pipes; rotary couplers are used for the connection of two steel pipes crossing at any angle; right-angle couplers are used for the connection of two steel pipes in square crossing.

（a）对接扣件　（b）旋转扣件　（c）直角扣件
(a) Butt coupler (b) Rotary coupler (c) Right-angle coupler
图3-2　扣件形式
Figure 3-2 Coupler Type

③脚手板。脚手板是提供施工作业条件并承受和传递荷载给水平杆的板件，可采用钢、木、竹材料制成。

③ Scaffold floor. Scaffold floor refers to

boards that provide construction conditions and bear and transfer loads to horizontal tubes, which can be made of steel, wood, and bamboo.

④安全网。用于保证施工安全，减少灰尘、噪声、光污染，包括平网和立网两类。

④ Safety net. It is used to keep construction safe and reduce dust, noise, and light pollution, including horizontal nets and vertical nets.

⑤底座。设在立杆下端，承受并传递立杆荷载给地基，底座一般采用厚 8 mm、边长 150~200 mm 的钢板作底板，上焊 150~200 mm 高的钢管。底座有内插式和外套式，如图 3-3 所示。

⑤ Base. It is set at the lower end of the upright tube to bear and transfer loads from the upright tube to the foundation. The base is normally made of steel plates with a thickness of 8 mm and a side length of 150-200 mm as baseplate, with 150 mm high steel pipes welded on it. The base can be an inserted type or a jacketed type, as shown in Figure 3-3.

（2）构造形式

(2) Structure form

扣件式钢管脚手架分为双排和单排两种形式。

Steel tubular scaffold with couplers has two types: Double-row scaffold and single-row scaffold.

双排式沿墙外侧设两排立杆，横向水平杆两端支撑在纵向水平杆上。多高层建筑均可采用双排式，当建筑高度超过 50 m 时，其需专门设计。单排式沿墙外侧仅设一排立杆，其横向水平杆一端与纵向水平杆连接，另一端支撑在墙上，仅适应荷载较小、高度较低（＜25 m）、墙体有一定强度的多层房屋。

The double-row scaffold has two rows of upright tubes set along the outer side of the wall, with both ends of the transverse horizontal tube supported on the longitudinal horizontal tube. This kind of scaffold is applicable to high-rise buildings and needs to be customized when the building height exceeds 50 m. The single-row scaffold has only one row of upright tubes set along the outer side of the wall, with one end of the transverse horizontal tube connected with the longitudinal horizontal tube, and the other end resisting the wall. It is only suitable for multi-floor houses with small loads, low height (< 25 m), and adequate strength for the wall.

①立杆。立杆的横距为 1.0~1.5 m，纵距为 1.2~2.0 m，每根立杆底部应设置底座或垫板，与基底相连。基底面层土质应夯实、整平，其

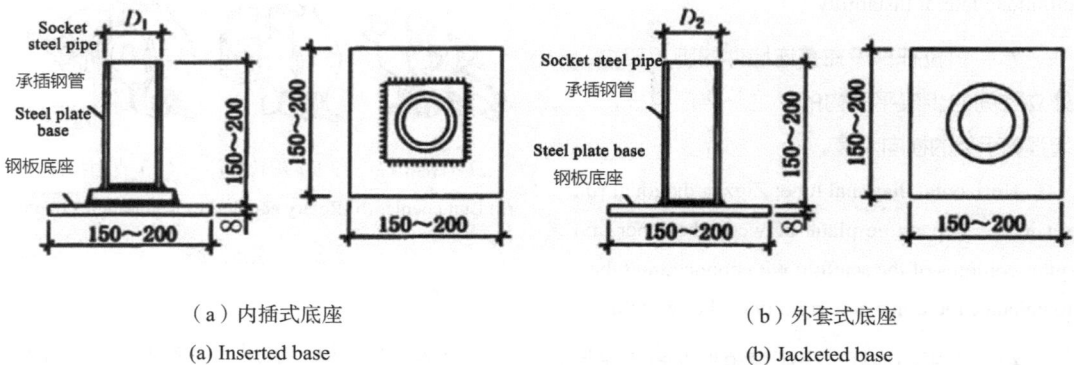

（a）内插式底座

(a) Inserted base

（b）外套式底座

(b) Jacketed base

图3-3　脚手架底座

Figure 3-3 Scaffold Base

上浇筑厚度 ≥ 100 mm 的 C15 素混凝土垫层，做好地面排水。脚手架立杆对接、搭接应符合下列规定：当立杆采用对接接长时，立杆的对接扣件应交错布置，两根相邻立杆的接头不应设置在同步内，同步内隔一根立杆的两个相隔接头在高度方向错开的距离不宜小于 500 mm；各接头中心至主节点的距离不宜大于步距的 1/3。

① Upright tube. The upright tube has a horizontal distance of 1.0-1.5 m and a longitudinal distance of 1.2-2.0 m. The bottom of each upright tube shall be provided with a base or subplate to connect with the foundation. The soil on the base surface shall be tamped and leveled, with ≥ 100 mm thick C15 plain concrete cushion poured on it, and ground drainage shall be properly done. The butt joint and overlapping of the scaffold's upright tubes shall meet the following requirements: When an upright tube is extended via butt joint, butt couplers of the upright tube shall be staggered. The joints of two adjacent upright tubes shall not be set in the same step. For two joints spaced by one upright tube in the same step, the staggering distance in the height direction should not be less than 500 mm. The distance from the center of each joint to the main node should not exceed 1/3 of the step.

②纵向水平杆（大横杆）。纵向水平杆设置在立杆内侧，其长度不应小于 2 跨，大横杆与立杆要用直角扣件扣紧，大横杆接长应采用对接扣件连接或搭接，并应符合下列规定：两根相邻纵向水平杆的接头不应设置在同步或同跨内；不同步或不同跨两个相邻接头在水平方向错开的距离不应小于 500 mm；各接头中心至最近主节点的距离不应大于纵距的 1/3。

② Longitudinal horizontal tube(large crossbar). The longitudinal horizontal tube is set at the inner side of the upright tube, with a length of not less than 2 spans. The large crossbar and upright tube shall be fastened with right-angle couplers, and the large crossbar shall be extended with butt couplers for connection or overlapping in line with the following requirements: The joints of two adjacent longitudinal horizontal tubes shall not be set in the same step or span; Two adjacent joints not in the same step or span shall be staggered at least 500 mm horizontally; The distance from the center of each joint to the nearest main node shall not exceed 1/3 of the longitudinal distance.

③横向水平杆（小横杆）。小横杆设置在立杆与大横杆的相交处，用直角扣件与大横杆扣紧，且应贴近立杆布置，小横杆距离立杆轴心线的距离不应大于 150 mm。

③ Transverse horizontal tube(small crossbar). The small crossbar is set at the intersection of the upright tube and the large crossbar, fastened with the large crossbar with right-angle couplers, and placed close to the upright tube. The distance from the small crossbar to the axis of the upright tube shall not exceed 150 mm.

④水平扫地杆。

④ Bottom horizontal tube.

a. 纵向水平扫地杆。它是连接立杆下端的纵向水平杆，作用是约束立杆底端，防止纵向发生位移。通常位于距底座下皮 200 mm 处。

a. Longitudinal bottom horizontal tube. It is a longitudinal horizontal tube connecting the lower end of the upright tube, used to constrain the bottom end of the upright tube from longitudinal displacement. It is typically located 200 mm below the bottom surface of the base.

b. 横向水平扫地杆。它是连接立杆下端的横向水平杆，作用是约束立杆底端，防止横向发生位移。通常位于纵向水平扫地杆上方，如图 3-4 所示横向扫地杆。

b. Transverse bottom horizontal tube. It is a transverse horizontal tube connecting the lower end of the upright tube, used to constrain the bottom end of the upright tube from transverse displacement.

It is usually located above the longitudinal bottom horizontal tube, as shown in Figure 3-4.

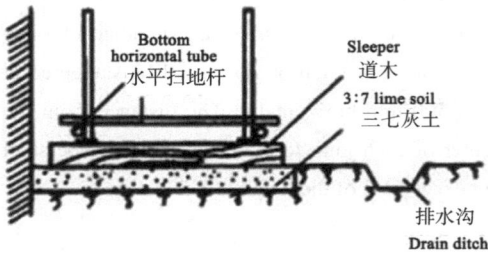

图3-4　横向水平扫地杆

Figure 3-4 Transverse Bottom Horizontal Tube

⑤脚手板。

⑤ Scaffold floor.

脚手板铺设在脚手架杆件上，用于直接承受施工荷载。

The scaffold floor is paved on the scaffold members to directly bear the construction loads.

作业层脚手板应铺满、铺稳、铺实。冲压钢脚手板、木脚手板、竹串片脚手板等，应设置在三根横向水平杆上。当脚手板长度小于 2 m 时，可采用两根横向水平杆支承，但应将脚手板两端与其可靠固定，严防倾翻。脚手板的铺设应采用对接平铺或搭接铺设。脚手板对接平铺时，接头处必须设两根横向水平杆，脚手板外伸长应取 130~150 mm，两块脚手板外伸长度的和不应大于 300 mm。脚手板搭接铺设时，接头必须支在横向水平杆上，搭接长度不应小于 200 mm，其伸出横向水平杆的长度不应小于 100 mm。

The scaffold floor on the operation layer should be fully paved, steady, and firm. The drawing steel scaffold floor, wooden scaffold floor, and bamboo-string scaffold floor shall be set on three transverse horizontal tubes. The scaffold floor shorter than 2 m may be supported by two transverse horizontal tubes, with both ends reliably fixed with the tubes to prevent rollover. The scaffold

floor shall be paved by butt jointing or overlapping. When scaffold floors are butt jointed for flat paving, two transverse horizontal tubes must be set at the joints. The scaffold floor shall be extended for a length of 130-150 mm, and the sum of the extension lengths of two scaffold floors shall not exceed 300 mm. When scaffold floors are overlapped, the joint must be supported by the transverse horizontal tube, with an overlapping length of 200 mm or longer, and the length extending out of the transverse horizontal tube shall be at least 100 mm.

⑥支撑体系。脚手架的支撑体系包括剪刀撑和横向支撑。设置支撑体系的目的是使脚手架成为一个几何稳定的构架，加强脚手架整体刚度，增大其抵抗侧向力的能力，避免出现节点的可变状态和过大的位移。

⑥ Support system. The support system of the scaffold includes diagonal bracing and transverse support. The support system makes the scaffold a geometrically stable frame, strengthens the overall rigidity, increases its ability against lateral forces, and avoids variable states and excessive displacement of nodes.

a. 剪刀撑。它设置在脚手架外侧面，用旋转扣件与立杆连接，形成与墙面平行的十字交叉斜杆。每道剪刀撑的宽度不应小于 4 跨，且不应小于 6 m，斜杆与地面呈 45°~60° 夹角。高度在 24 m 以下的单、双排脚手架，必须均在外侧立面两端、转角及中间间隔不超过 15 m 的立面上，各设置一道剪刀撑，并应由底至顶连续设置，且每片架子不少于三道。高度 ≥ 24 m 的双排脚手架应在外侧立面连续设置剪刀撑。剪刀撑的连接除顶层可采用搭接外，其余各接头必须采用对接扣件连接。搭接长度不低于 1 m，用不少于两个旋转扣件连接。

a. Diagonal bracing. It is set on the outer side of the scaffold and connected with the upright tube by rotary couplers to form a crossing diagonal bar parallel to the wall surface. Each diagonal bracing shall be at least 4-span(or 6 m) wide, and the angle between the diagonal bar and the ground shall be 45°-60°. For single-row and double-row scaffolds lower than 24 m, a diagonal brace must be provided at both ends of the outer facade, at corners, and on the middle facade with an interval of not over 15 m. They shall be continuously set from bottom to top, and each face of the scaffold shall have at least three diagonal braces. Diagonal bracing shall be continuously set on the outer facade of double-row scaffolds with a height ≥ 24 m. For the connection of diagonal bracing, all joints must be connected by butt couplers except for the top layer, where the joints can be overlapped. The overlapping length shall not be less than 1 m, and at least 2 rotary couplers shall be used for the connection.

b. 横向支撑。指在同节间由底至顶层呈"之"字形连续布置。高度 ≥ 24 m 的封闭型脚手架，拐角应设置横向支撑，中间应每隔6跨设置一道高度 < 24 m 的封闭型双排脚手架，可不设横向支撑。开口型双排脚手架的两端均必须设置横向支撑，并在中间每隔6跨加设一道横向支撑。

b. Transverse support. They are laid continuously in a zigzag shape from the bottom to the top within the same segment. For enclosed scaffolds with a height ≥ 24 m, transverse supports shall be set at corners, with one support set every 6 spans in the middle. For enclosed double-row scaffolds with a height < 24 m, transverse supports may be saved. Both ends of the open-type double-row scaffold must be provided with transverse supports, with another one added every 6 spans in the middle.

⑦连墙件。连墙构造对外脚手架的安全至关重要，因连墙件数量不足或构造不符合要求造成的事故屡有发生。连墙构造主要是刚性连接。常见的刚性连墙构造有八种：

⑦ Connecting tube. The connecting member is critical for the safety of external scaffolds.

Insufficient connecting tubes or unqualified structure has triggered accidents frequently. The connecting member is mainly a rigid connection. There are eight types of rigid connecting members:

a. 单杆穿墙夹固式：单根小横杆穿过墙体，在墙体两侧用 ≥ 0.6 m 的短钢管塞以垫木固定。

a. Single-tube through-wall clamping: A small single crossbar passes through the wall and is fixed with short steel pipes of ≥ 0.6 m on both sides of the wall, plugged with crossers.

b. 双杆穿墙夹固式：方法同上，穿墙杆为上下或相邻的一对小横杆。

b. Double-tube through-wall clamping: The same as above, but the through-wall tube is a pair of small crossbars in up/down or adjacent positions.

c. 单杆窗口夹固式：单根小横杆穿过门窗洞口，在洞口两侧用适长的（立放或平放均大于洞口尺寸0.5 m）钢管塞以垫木固定。

c. Single-tube window clamping: A small single crossbar passes through the door/window opening and is fixed with a proper length of steel pipe(vertically or horizontally placed, both 0.5 m greater than the opening size) on both sides of the opening, plugged with crossers.

d. 双杆窗口夹固式：方法同上，穿过洞口的小横杆为上下或相邻的一对。

d. Double-tube window clamping: The same as above, but the through-wall tube is a pair of small crossbars in up/down or adjacent positions.

e. 单杆箍柱式：单根长度适中的小横杆紧贴结构柱，用两根短钢管将其固定于柱侧。

e. Single-tube banded column: A small single crossbar with appropriate length is closely attached to the structural column and fixed to the side of the column with two short steel pipes.

f. 双杆箍柱式：用两根长度适中的水平横

杆和两根短钢管贴紧结构柱固定。

f. Double-tube banded column: Hoop the structural column tightly with two horizontal crossbars with appropriate length and two short steel pipes.

g. 埋件连固式：在砼墙体中埋设连墙件，用扣件与脚手架立杆或大横杆固定。

g. Embedded connection: Connecting tubes are embedded in the concrete wall and fixed to the scaffold's upright tube or large crossbar with couplers.

h. 绑捆连固式：采用绑或挂的方式固定螺栓套管连接件。

h. Binding connection: Bushing studs for connection are fixed by binding or hanging.

（3）承力结构

(3) Load-bearing structure

脚手架的承力结构可分为作业层、横向构架、纵向构架三部分。

The load-bearing structure of a scaffold includes three parts: Operation layer, transverse frame, and longitudinal frame.

①作业层：直接承受施工荷载。荷载由脚手板传给小横杆，再传给大横杆和立杆。

Operation layer: Directly bearing construction load. The load is transferred from the scaffold floor to the small crossbar, then to the large crossbar and upright tube.

②横向构架：由立杆和横向水平杆组成，是脚手架直接承受和传递垂直荷载的部分。

Transverse frame: Composed of upright tubes and transverse horizontal tubes, it is the part that directly bears and transfers vertical loads.

③纵向构架：由各横向构架通过纵向水平杆相互之间连成的一个整体，它一般沿房屋的四周形成一个连续封闭的结构。

Longitudinal frame: Composed of transverse frames interconnected through longitudinal horizontal tubes. It generally forms a continuous closed structure along the periphery of the house.

脚手架传力路径：荷载→脚手板→横向水平杆→纵向水平杆→立杆→基础。

Force transmission path of scaffold: Load → scaffold floor→ transverse horizontal tube → longitudinal horizontal tube → upright tube → foundation.

（4）搭设工艺流程

(4) Process flow of scaffold erection

扣件式钢管脚手架搭设工艺流程：夯实平整场地→材料准备→设置垫板与底座→纵向水平扫地杆→搭设立杆→横向水平扫地杆→搭设纵向水平杆→搭设横向水平杆→搭设剪刀撑→固定连墙件→搭设防护栏杆→铺设脚手板→绑扎安全网。

Erection process of steel tubular scaffold with couplers: Site tamping and leveling → material preparation → setting of subplate and base → longitudinal bottom horizontal tube → erect upright tube → transverse bottom horizontal tube → erect longitudinal horizontal tube → erect transverse horizontal tube → erect diagonal bracing → fix connecting tubes → install protective railings → pave scaffold floor→ bind safety nets.

3.1.2 碗扣式钢管脚手架

3.1.2. Steel Tubular Bowl-buckle Scaffold

碗扣式钢管脚手架是一种多功能的工具式脚手架，由主部件、辅助构件、专用构件三大类组成，全系列分为23类、53种规格。它除能作为一般单双排脚手架、支撑架外，还可用作支撑柱、物料提升架、悬挑脚手架、爬升脚手架等。

This is a multi-functional implementary scaffold. It is composed of main components, auxiliary components, and special components, covering 23 categories and 53 specifications. Besides the role of general single-row and double-row scaffolds and brackets, it can also be used as a support pillar, material hoisting frame, cantilever scaffold, and climbing scaffold.

碗扣分上碗扣和下碗扣。下碗扣焊在钢管上；上碗扣对应地套在钢管上，其销槽对准焊在钢管上的限位销即能上下滑动。连接时，只需将横杆接头插入下碗扣内，将上碗扣沿限位销扣下，并顺时针旋转，靠上碗扣螺旋面使之与限位销顶紧，从而将横杆和立杆牢固地连在一起，形成框架结构。碗扣式接头可同时连接 4 根横杆，横杆可组成各种角度，因而可以搭设各种形式的脚手架，特别适合扇形表面及高层建筑施工和装修两用外脚手架，还可作为模板的支撑。其支撑形式、构造要求等参照扣件式钢管脚手架（图 3-5）。

The bowl-buckle includes the upper bowl buckle and lower bowl buckle. The lower bowl buckle is welded on the steel pipe, and the upper bowl buckle is sleeved on the steel pipe, with its keyway aligned with the limit pin welded on the steel pipe for vertical sliding. During connection, insert the crossbar joint into the lower bowl buckle, pull down the upper bowl buckle along the limit pin, rotate it clockwise, to resist the spiral surface of the upper bowl buckle tightly with the limit pin. This secures the crossbar with the upright tube to form a frame structure. The bowl-buckle joint can connect 4 crossbars at the same time, and the crossbars can form different angles, allowing the erection of scaffolds in various forms. It is especially suitable for scaffolds with fan-shaped surfaces, as well as for external scaffolds for the construction and decoration of high-rise buildings, and can also be used as formwork support. For the support form and structural requirements, refer to the steel tubular scaffold with couplers(Figure 3-5).

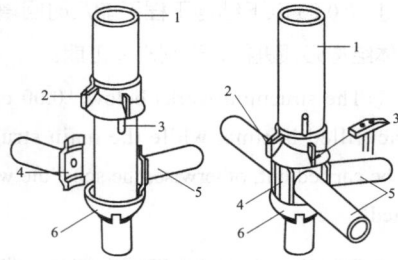

1.立杆；2.上碗扣；3.限位销；4.横杆接头；5.横杆；6.下碗扣

1.upright tube; 2.upper bowl buckle; 3. limit pin; 4. crossbar joint; 5. crossbar; 6. lower bowl buckle

图3-5　碗扣式脚手架细节

Figure 3-5 Detail of Bowl-buckle Scaffold

3.1.3 钢梁悬挑脚手架

3.1.3 Steel Beam Cantilever Scaffold

（1）钢梁悬挑脚手架的构造

(1) Structure of steel beam cantilever scaffold

悬挑脚手架是指通过水平构件将架体所受竖向荷载传递到主体结构上的外脚手架，如图 3-6 所示。悬挑脚手架适用于下列三种情况：

Cantilever scaffold refers to the external scaffold that transfers the vertical load from the scaffold to the main structure through horizontal members, as shown in Figure 3-6. Cantilever scaffolds are suitable for the following three conditions:

图3-6　钢梁悬挑脚手架

Figure 3-6 Steel Beam Cantilever Scaffold

① ±0.00 以下结构工程不能及时回填土，而主体结构必须进行，否则影响工期。

① The structural works below ±0.00 cannot be backfilled in time, while the main structure must be carried out, otherwise, the schedule will be affected.

②高层建筑主体结构四周有裙房，脚手架不能支承在地面上。

② There are podiums around the main structure of a high-rise building, where scaffolds cannot be erected on the ground.

③超高建筑施工中，脚手架搭设高度超过了容许搭设高度，可将整个脚手架按允许搭设高度分成若干段，每段脚手架支承在建筑结构向外悬挑的结构上。

③ During the construction of super-high buildings, if the scaffold height exceeds the allowable erection height, the whole scaffold will be divided into several sections within the allowable erection height, with each section supported on the outside cantilever structure of the building structure.

钢梁悬挑脚手架主要构配件有悬挑梁、钢管、扣件、脚手板、安全网等，如图 3-7 所示。钢梁悬挑脚手架主要由悬挑梁（工字钢、槽钢）和钢管扣件式脚手架组成，一次性悬挑脚手架高度不宜超过 20 m。

The steel beam cantilever scaffold mainly consists of the cantilever beam, steel pipes, couplers, scaffold floor, and safety net, as shown in Figure 3-7. The steel beam cantilever scaffold is mainly composed of the cantilever beam(I-steel and channel steel) and steel tubular scaffold with couplers. The height of the disposable cantilever scaffold should not exceed 20 m.

钢悬挑梁宜优先选用工字钢，因为工字钢具有截面对称性、受力稳定性好等优点，钢梁截面高度不应小于 160 mm。锚固型钢悬挑梁

图3-7　钢梁悬挑脚手架的主要配构件
Figure 3-7 Main Components of Steel Beam Cantilever Scaffold

的 U 形钢筋拉环或锚固螺栓直径不宜小于 20 mm。钢悬挑梁悬挑长度应按设计确定，固定端长度不应小于悬挑长度的 1.25 倍。型钢悬挑梁固定端应采用 2 个（对）及以上 U 形钢筋拉环或螺栓与建筑结构楼板固定。

Due to its sectional symmetry and good stress stability, I-steel is the preferred material for the steel cantilever beam, and the section height of the steel beam should not be less than 160 mm. The U-shaped steel pull ring or anchor bolt of the anchored steel cantilever beam should have a diameter of not less than 20 mm. The cantilever length of the steel cantilever beam shall be determined according to the design, and the length of the fixed end shall not be smaller than 1.25 times the cantilever length. The fixed end of the section-steel cantilever beam shall be fixed to the floor slab of the building structure with 2(pairs) or more U-shaped steel pull rings or bolts.

一次性悬挑脚手架高度不宜超过 20 m。脚手架构造措施参照扣件式钢管脚手架（图 3-8）。

The height of the disposable cantilever scaffold should not exceed 20 m. For scaffold construction measures, refer to the steel tubular scaffold with couplers(Figure 3-8).

图3-8　悬挑钢梁的楼面锚固

Figure 3-8 Floor Anchorage of Steel Cantilever Beam

（2）钢梁悬挑脚手架搭设的工艺流程

(2) Erection Process of Steel Beam Cantilever Scaffold

钢梁悬挑脚手架搭设工艺流程：预埋U形螺栓→安装水平悬挑梁→安装纵向水平扫地杆、立杆→安装横向水平扫地杆→安装纵向水平杆→安装横向水平杆→安装剪刀撑→安装连墙件→安装防护栏杆→铺脚手板→扎安全网。

Erection process of steel beam cantilever scaffold: Embed U-bolt in advance → install horizontal cantilever beam → install longitudinal bottom horizontal tube and upright tube → install transverse bottom horizontal tube → install longitudinal horizontal tube → install transverse horizontal tube → install diagonal bracing → install connecting tube → install protective railings → pave scaffold floor → bind safety nets.

3.1.4 脚手架的拆除与安全技术

3.1.4 Scaffold Removal and Safety Technology

（1）脚手架的拆除

(1) Scaffold removal

①脚手架拆除应按专项方案施工，拆除前应做好下列准备工作：全面检查脚手架的扣件连接、连墙件、支撑体系等是否符合构造要求；根据检查结果补充完善脚手架拆除专项方案中的拆除顺序和措施，经审批后方可实施。

① Scaffold removal shall follow a special scheme, with the following preparations done before removal: Check in all aspects to find out whether the coupler connection, connecting tubes, and support system of the scaffold meet the structural requirements; supplement and improve the removal sequence and measures in the special scheme for scaffold removal according to the inspection results, and implement them after approval.

②单、双排脚手架拆除作业必须由上而下逐层进行，严禁上下同时作业。连墙件必须随脚手架逐层拆除，严禁先将连墙件整层或数层拆除后再拆脚手架。

② Single/double-row scaffold removal must proceed floor by floor from the top down instead of removing the upper and lower parts simultaneously. The connecting tubes must be removed floor by floor along with the scaffold instead of removing the whole or several floors of connecting tubes before scaffold removal.

③当单、双排脚手架拆至下部最后一根长立杆的高度（约6.5 m）时，应先在适当位置搭设临时抛撑加固后，再拆除连墙杆。当单、双

排脚手架采取分段、分立面拆除时，对不拆除的脚手架两端，应先按相关规范设置连墙件和横向支撑加固。

③ When the single/double-row scaffold is dismantled to a height equal to that (about 6.5 m) of the last upright tube at the bottom, temporary throwing supports shall be set at appropriate positions to fix the scaffold before removing the connecting tubes. When single/double-row scaffolds are dismantled in sections and facades, the two ends of the scaffold reserved shall be reinforced with connecting tubes and transverse supports according to relevant specifications.

④不允许将拆除的构配件从高空抛至地面。

④ It is not allowed to throw dismantled components and fittings to the ground from a high altitude.

（2）脚手架的安全技术

(2) Safety technology of scaffold

①扣件钢管脚手架安装与拆除人员必须是经考核合格的专业架子工。架子工应持证上岗。

① Workers responsible for the installation and removal of steel tubular scaffold with couplers must be qualified professional scaffolders. Scaffolders shall hold certificates for their posts.

②搭、拆脚手架的人员必须戴安全帽、系安全带、穿防滑鞋。

② Workers erecting and removing the scaffold must wear safety helmets, safety belts, and anti-skid shoes.

③作业层上的施工荷载应符合设计要求，不得超载。不得将模板支架、缆风绳、泵送混凝土和砂浆的输送管等固定在架体上。严禁在架体上悬挂起重设备，严禁拆除或移动架体上的安全防护设施。

③ The construction load on the operation floor shall conform to the design requirements, and no overloading is allowed. Formwork supports, hawser cables, and delivery pipes for pump concrete and mortar must not be fastened to the scaffold. It is strictly prohibited to hang lifting equipment or remove or move safety protection facilities on the scaffold.

④操作层脚手板应铺设牢靠、严实，并应用安全平网双层兜底，施工层以下每隔10 m应设安全平网封闭。

④ The scaffold floor on the operation floor shall be paved firmly and tightly, double bottoming with horizontal safety nets, which shall be set every 10 m below the construction floor for sealing.

⑤单、双排脚手架、悬挑式脚手架沿墙体外应用密目式安全网全封闭，密目式安全网宜设置在脚手架外立杆的内侧，并应与架体结扎牢固。

⑤ Single/double-row scaffolds and cantilever scaffolds shall be fully enclosed with fine-mesh safety nets along the wall periphery. Such nets shall be set on the inner side of the upright tubes outside the scaffold and shall be firmly tied to the scaffold body.

⑥在脚手架使用期间，严禁拆除下列杆件：主节点处的纵、横向水平杆，纵、横向水平扫地杆、连墙件。

⑥ During the use of scaffolds, it is strictly prohibited to remove the following members: Longitudinal and transverse horizontal tubes, longitudinal and transverse bottom horizontal tubes, and connecting tubes at the main nodes.

⑦临街搭设脚手架时，外侧应有防止坠物伤人的防护措施。

⑦ For scaffolds facing the street, the outer side shall have protective measures to prevent falling objects.

⑧在脚手架上进行电、气焊作业时，应有防火措施和专人看守。

⑧ For electrical and gas welding work on the scaffold, fireproof measures shall be taken and special personnel shall be assigned for supervision.

⑨工地临时用电线路的架设及脚手架接地、避雷措施等，应按现行行业标准的有关规定执行。脚手架与支模架要分开搭设，不能将两者混搭在一起，脚手架不能当作支模架使用。

⑨ The erection of temporary power lines on the construction site, as well as scaffold grounding and lightning protection measures, shall follow relevant provisions of the prevailing industry standards. Scaffolds and formwork supports shall be erected separately and cannot be mixed together. Scaffolds must not be used as formwork supports.

3.2 砖砌体施工
3.2 Construction of Brick Masonry

3.2.1 砖砌体施工的基本要求
3.2.1 Basic Requirements for Brick Masonry Construction

砌体结构工程施工前，应编制砌体结构工程施工方案，砌筑顺序应符合规定。

Before construction, the construction scheme of masonry structure works shall be prepared, and the construction sequence shall comply with the regulations.

砌体工程所用的材料应有产品的合格证书及产品性能检测报告。块材、水泥、钢筋、外加剂等还应有材料主要性能的进场复验报告。

All materials used in masonry works shall bear qualification certificates and performance inspection reports of the products. Block materials, cement, steel bars, and admixtures shall also be provided with mobilization re-inspection reports to prove the main properties of the materials.

（1）砖墙的组砌形式
(1) Brick wall laying method

一块砖有三组两两相等的面，面积最大的面叫大面；较细长的一面叫条面；最短小的一面叫丁面。

A brick has six sides, with each two sides equal. The largest side is called the big side; the slender side is called the strip side; the shortest side is called the header.

砖砌入墙体后，条面与墙长平行的叫顺砖；丁面与墙长平行的叫丁砖。

After bricks are laid into the wall, the one with its strip side parallel to the wall length is called a stretcher brick; the one with its header surface parallel to the wall length is called a header brick.

常用的组砌形式有：一顺一丁、三顺一丁、梅花丁、两平一侧、全顺、全丁。如图3-9所示。

Ordinary bricklaying forms include one stretcher and one header, three stretchers and one header, quincunx header, two flat and one vertical, full stretchers, and full headers, as shown in Figure 3-9.

烧结多孔砖宜采用一顺一丁或梅花丁的砌筑形式，上下皮垂直灰缝相互错开1/4砖长。

Fired perforated bricks should be laid in the form of one stretcher and one header or quincunx header, with the vertical mortar joints of the upper

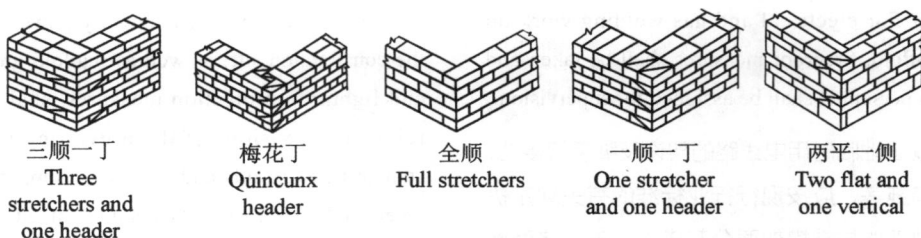

| 三顺一丁
Three
stretchers and
one header | 梅花丁
Quincunx
header | 全顺
Full stretchers | 一顺一丁
One stretcher
and one header | 两平一侧
Two flat and
one vertical |

图3-9　砖砌体的组砌形式

Figure 3-9 Bricklaying Forms

and lower bricks staggered by 1/4 of the brick length.

一顺一丁也称满丁满条组砌法，由一皮顺砖、一皮丁砖组砌而成，上下皮之间竖向灰缝相互错开1/4砖长。这种组砌方法整体性较好，砌筑效率较高。

The walls laid by "one stretcher and one header" method appear as one layer of strecher bricks and one layer of header bricks. The vertical mortar joints between the upper and lower bricks are staggered by 1/4 of the brick length. This bricklaying method features good integrity and high masonry efficiency.

三顺一丁组砌法是采用三皮顺砖间隔一皮丁砖组砌而成，上下皮顺砖搭接半砖长，丁砖与顺砖搭接1/4砖长。山墙与檐墙的丁砖层不能在同一皮砖上，以利于错峰搭接。这种组砌方法砌筑效率高，墙面易于平整。

The "three stretchers and one header" method is laying three layers of strecher bricks and adding one layer of header bricks. The upper and lower stretchers are overlapped by half of the brick length, and the header and stretcher are overlapped by 1/4 of the brick length. The header layers of the gable wall and the eave wall must not lie on the same brick to facilitate staggered overlapping. This bricklaying method has high masonry efficiency, allowing easy leveling of the wall surface.

梅花丁砌法是在同一皮砖上采用两块顺砖间隔一块丁砖，上下皮砖的竖向灰缝错开1/4砖长。这种砌法整体性较好，灰缝整齐美观，但砌筑效率较低。

For the quincunx header method, two stretchers and one header are laid on the same brick layer, with the vertical mortar joints of the upper and lower bricks staggered by 1/4 brick length. This bricklaying method features good integrity and neat mortar joints but has low masonry efficiency.

全顺砌法是全部采用顺砖砌筑，每皮砖上下搭接1/2砖长，适用于半砖墙的砌筑。

For the full stretcher method, all bricks laid are stretchers, with 1/2 brick length overlapped above and below each brick layer, suitable for the masonry of half-brick walls.

全丁砌法是全部采用丁砖砌筑，每皮砖上下搭接1/4砖长，适用于圆形烟囱和窨井的砌筑。

For the full header method, all bricks laid are headers, with 1/4 brick length overlapped above and below each brick layer, suitable for the masonry of circular chimneys and inspection wells.

（2）砌砖墙的一般规定

(2) General provisions of brick wall laying

砖的品种、强度等级必须符合设计要求，砖应提前1天浇水湿润，避免砖过多吸收砂浆中的水分而影响黏结力，烧结普通砖、空心砖的含水率宜为10%~15%，灰砂砖、粉煤灰砖的含水率宜为5%~8%（现场用"断砖法"检查，砖截面四周浸水深度15~20 mm时为符合要求的含水率）。

The type and strength of bricks must meet the design requirements. Bricks shall be watered 1 day earlier to avoid excessive absorption of moisture from mortar to affect the cohesion. The fired common bricks and hollow bricks should have a moisture of 10%-15%, and lime-sand bricks and fly ash bricks should have a moisture of 5%-8%(by the "brick breaking method" on site to identify the required moisture with a water immersion depth of 15-20 mm around the section of brick).

在有冻胀环境和条件的地区，地面或防潮层以下不宜采用多孔砖。

Where frost heaving could occur, perforated bricks should not be used on the ground or below the damp-proof course.

3.2.2 砖砌体施工前的准备

3.2.2 Construction Preparation of Brick Mansonry Construction

（1）砖的准备

(1) Preparation of bricks

砖的品种、强度等级必须符合设计要求，并应规格一致。用于清水墙、柱表面的砖，外观要求应尺寸准确、边角整齐、色泽均匀，无裂纹、掉角、缺棱和翘曲等现象。在砌砖前应提前1~2天浇水湿润，以使砂浆和砖能很好地黏结。严禁砌筑前临时浇水，以免因砖表面存有水膜而影响砌体质量。烧结类砌块的相对含水率为60%~70%，吸水率较大的轻骨料混凝土小型空心砌块、蒸压加气混凝土砌块的相对含水率为40%~50%。

Bricks must meet the design requirements for their types and strength and shall have consistent specifications. Bricks used for the surface of drywalls and columns shall have accurate sizes, regular edges and corners, uniform color, and free of cracks, missing corners/edges, or warps. Before bricklaying, bricks shall be watered 1-2 day(s) earlier to allow tight bonding with mortar. Temporary watering before masonry is strictly prohibited, otherwise, the water film on the brick surface will affect the masonry quality. The relative moisture content is 60%-70% for fired blocks and 40%-50% for small hollow blocks made of lightweight aggregate concrete and autoclaved aerated concrete blocks with large water absorption.

（2）施工机具的准备

(2) Preparation of construction machines and tools

砌筑前，一般应按照施工设计的要求组织垂直和水平运输机械、砂浆搅拌机械进场，进行安装、调试等工作。垂直运输可采用塔式起重机、井架、龙门架和人货两用的施工电梯，水平运输多采用手推车或机动翻斗车。对多、高层建筑，还可以用灰浆泵输送砂浆。同时，还要准备脚手架、砌筑工具（如皮数杆、托线板）等。

Before masonry, vertical and horizontal transportation machinery, mortar mixers shall be mobilized, installed, and commissioned according to the requirements of the construction design. Tower cranes, derricks, gantries, and construction elevators(for both people and goods) can be used for vertical transportation, while trolleys or diesel dumpers are often used for horizontal transportation. For multi-storey and high-rise buildings, mortar pumps can also be used to transport mortar. Also, scaffolds and masonry tools(e.g. height poles and cord gauges) shall be prepared.

3.2.3 砖砌体的施工工艺

3.2.3 Construction Process of Brick Masonry

砖砌体的砌筑工艺流程为：抄平放线→摆砖→立皮数杆→盘角、挂线→砌筑、勾缝→清理。

Construction process of brick masonry:

Leveling and setting-out → brick placing → erection of height poles → corner making and plumb-line hanging → bricklaying and pointing → cleaning.

（1）抄平放线

(1) Leveling and setting-out

砌筑前应在基础防潮层或楼面上定出各层标高，并用水泥砂浆或 C15 细石混凝土抄平，使各段砖墙底部标高符合设计要求。然后根据龙门板上给定的轴线及图纸上标注的墙体尺寸，在基础顶面上用墨线弹出墙的轴线和墙的宽度线，并定出门洞口位置线。二楼以上墙的轴线可以用经纬仪或垂球将轴线引测上去。

Before masonry, the elevation of each layer shall be determined on the damp-proof course of the foundation or on the floor, with cement mortar or C15 fine aggregate concrete used for leveling, so that the bottom elevation for each section of the brick wall meets the design requirements. Then, according to the given axis on the batter board and the wall size marked on the drawing, the axis and the width line of the wall are snapped with ink lines on the top surface of the foundation, and the position line of the door opening is determined. The axis of the wall above the second floor can be measured with a theodolite or plumb bob.

（2）摆砖

(2) Brick placing

摆砖，又称撂底，是指在弹好线的基础顶面上按选定的组砌方式用干砖试摆。摆砖的目的是核对所放的墨线在门窗洞口、墙垛等处是否符合砖的模数，以尽可能减少砍砖，使砌体灰缝均匀，组砌得当。

Brick placing, also known as bottoming, refers to the trial laying of dry bricks on the foundation top with snapped lines according to the selected bricklaying method. Brick placing aims to check whether the ink lines placed at the door/window openings and wall piers meet the modulus of bricks, so as to minimize brick cutting and form uniform mortar joints for the masonry.

（3）立皮数杆

(3) Erection of height poles

皮数杆是指在其上画有每皮砖和砖缝厚度，以及门窗洞口、过梁、楼板、梁底、预埋件等标高位置的一种木制标杆。砌筑时用来控制墙体竖向尺寸及各部位构件的竖向标高，并保证灰缝厚度的均匀性。皮数杆一般设置在房屋的四大角以及纵横墙交接处，如墙面过长时，应每隔 10~15 m 立一根，如图 3-10 所示。

Height pole refers to a wooden benchmark post that bears the thickness of each brick and brick joint, as well as the elevations of door/window openings, lintels, floor slabs, beam bottoms, and embedded parts. During masonry, it is used to control the vertical size of the wall and the vertical elevations of components in each part and keep the thickness of mortar joints uniform. Height poles are usually set at four corners of the room and the junction of vertical and horizontal walls. If the wall is too long, a height pole shall be erected every 10-15 m, as shown in Figure 3-10.

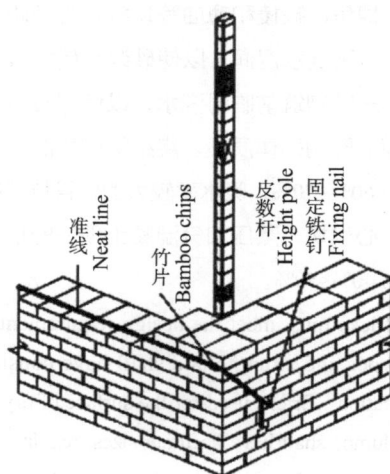

图3-10　皮数杆

Figure 3-10 Height Pole

（4）盘角、挂线

(4) Corner Making and plumb-line hanging

墙角是控制墙面横平竖直的主要依据，所以，一般砌筑时应先砌墙角，墙角砖层高度必须与皮数杆相符合，做到"三皮一吊，五皮一靠"，墙角必须双向垂直。墙角砌好后，即可挂线，作为砌筑中间墙体的依据，以保证墙面平整。一般一砖墙、一砖半墙可单面挂线，一砖半以上的墙则应双面挂线。

Wall corners, as the main reference for controlling the horizontal and vertical straightness of the wall, shall be built first during masonry. The brick height at the wall corner must be consistent with the height pole following the rule of "plumb tested every three layers and ruler measured every five layers". The wall corner must be vertical in both directions. After the wall corner is completed, plumb lines can be hung as the reference for building the middle wall to keep the flatness of the wall surface. Generally, the one-brick wall and one-brick-and-a-half wall can have plumb lines hung on one side, and walls longer than one-brick-and-a-half wall shall have plumb lines hung on both sides.

（5）砌筑、勾缝

(5) Bricklaying and pointing

虽然砌砖的操作方法各地不一，但均应保证砌筑质量要求，常用"铺浆法"和"三一砌砖法"。"铺浆法"即在墙顶上铺一段砂浆，然后用砖挤入砂浆中一定厚度之后把砖放平，达到"下齐边、上齐线、横平竖直"的要求。这种砌法可以连续挤砌几块砖减少烦琐的动作，使灰缝饱满，效率高，保证砌筑质量。采用铺浆法砌筑砌体，铺浆长度不得超过 750 mm，当施工期间气温超过 30℃ 时，铺浆长度不得超过 500 mm。"三一砌砖法"即"一铲灰，一块砖，一挤揉"，并随手将挤出的砂浆刮去的砌筑方法。

这两种砌法的优点是灰缝容易饱满，黏结力好，墙面整洁。

Although bricklaying methods vary from place to place, the masonry quality requirements shall be met. The commonly used methods include the mortar spreading method and the "three ones" method. The mortar spreading method works in such a way as laying a section of mortar on the wall top, then pressing bricks into the mortar for a certain depth and leveling them up to "align the lower and upper edges for being horizontal and vertical". This masonry method allows continuous squeeze of several bricks to reduce tedious labor, capable of achieving full mortar joints and high efficiency to ensure masonry quality. For masonry buildings by the mortar spreading method, the spread-mortar length must not exceed 750 mm, or 500 mm when the temperature exceeds 30°C. The "three ones" method refers to the bricklaying operation of "one shovel of mortar, one brick, one squeeze", and scraping off the extruded mortar. These two masonry methods can achieve full mortar joints, good adhesive properties, and clean wall surfaces.

勾缝是砌清水墙的重要工序，可以用砂浆随砌随勾缝，叫作原浆勾缝；也可以砌完墙后再用 1∶1.5 水泥砂浆或加色砂浆勾缝，称为加浆勾缝。勾缝具有保护墙面和增加墙面美观度的作用。为了确保勾缝质量，勾缝前应清除墙面黏结的砂浆和杂物，并洒水湿润。在砌完墙以后，应画出 1 cm 的灰槽，灰缝可勾成凹、平、斜或凸形状。勾缝完毕后，应进行墙面、柱面和落地灰的清理，如图 3-11 所示。

Pointing is an important process of drywall building, in which mortar can be used for pointing along with bricklaying, i.e., pointing with original mortar. Also, 1∶1.5 cement mortar or colored mortar can be used for pointing after the wall is completed, i.e., pointing with additional mortar.

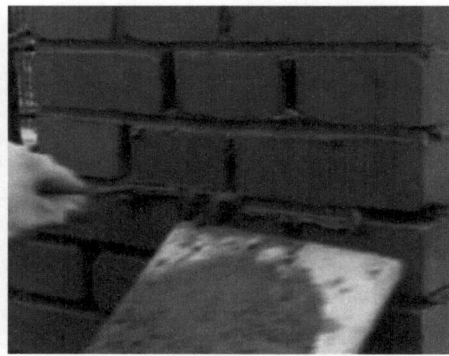

图3-11 勾缝

Figure 3-11 Pointing

Pointing can protect the wall surface, making it more aesthetic. For the quality of pointing, the wall surface shall be cleaned to remove attached mortar and sundries and be watered before pointing. After the wall is completed, a 1 cm mortar trough shall be made, in concave, flat, inclined, or convex shapes. After pointing, the wall and column surface shall be cleaned, with falling mortar removed, as shown in Figure 3-11.

3.2.4 砖墙砌筑的质量要求

3.2.4 Quality Requirements for Brick Wall Masonry

砖墙砌筑的质量要求：灰缝横平竖直，砂浆饱满均匀，砌砖错缝搭接，交接接槎可靠。

Quality requirements for brick wall masonry: The mortar joints shall be horizontal and vertical, with full and uniform mortar; bricks shall be staggered and overlapped, with reliable joints at junctions.

（1）灰缝横平竖直

(1) Horizontal and vertical mortar joints

砖砌体抗压性能好，而抗剪和抗拉性能差，砌筑时应使砌体均匀受压，不产生剪切水平推力，保证砌体灰缝应横平竖直。

The brick masonry sees a good compressive performance, but with poor shear and tensile performance. During masonry, bricks shall be uniformly compressed to eliminate shear horizontal thrust, and the mortar joints of bricks shall be horizontal and vertical.

砌筑时应做到每块砖要放平，每皮砖要在一个平面上，每块砖的位置要放准，上下对齐，保证墙面平整垂直，避免游丁走缝，影响墙体外观质量。砌筑过程中应"三皮一吊，五皮一靠"，把砌筑误差消灭在操作过程中，以保证墙面的垂直平整。

During masonry, each brick shall be placed horizontally, and a layer of bricks shall be aligned on the same plane. Bricks shall be put in accurate positions and aligned up and down to ensure a flat and vertical wall surface and avoid misalignment that affects the appearance of the wall. In the process of masonry, the rule of "plumb tested every three layers and ruler measured every five layers" shall be followed to eliminate masonry errors during operation and keep the verticality and flatness of the wall surface.

（2）砂浆饱满均匀

(2) Full and uniform mortar

水平灰缝要求砂浆饱满，厚薄均匀，保证砖块均匀受力并使砖块紧密结合，以避免砖块因受力不均而产生弯曲和剪切破坏。

Horizontal mortar joints shall be full of mortar and uniform in thickness, so that bricks can be uniformly stressed and closely bonded, protecting bricks against bending and shear failure resulting from uneven stress.

（3）砌砖错缝搭接

(3) Staggered and overlapped bricklaying

砌筑时砖块排列应遵循上下错缝、内外搭接的原则，避免出现连续的垂直通缝，以提高砌体的整体性、稳定性和承载能力。

During masonry, bricks shall be staggered in the up/down direction for internal and external overlapping, to avoid continuous vertical straight joints and improve the integrity, stability, and bearing capacity of the masonry.

错缝或搭接长度一般不小于 60 mm，若小于 25 mm 则可视为通缝。

The staggered joint or overlapping length is usually not smaller than 60 mm. If less than 25 mm, it can be considered a straight joint.

（4）交接接槎可靠

(4) Reliable joints at junctions

砖墙的转角处和交接处应同时砌筑，当无可靠措施时，严禁内外墙分砌施工。对不能同时砌筑但又必须留置的临时间断处应砌成斜槎，斜槎的高度不宜超过一步高，斜槎的水平投影长度不应小于高度的 2/3，如图 3-12 所示。

The corners and junctions of brick walls shall be built at the same time. When no reliable measures are available, the inner and outer walls must not be built separately. Temporary discontinuities that cannot be built simultaneously but must be reserved shall be built into inclined joints with a height limited to one step and a horizontal projection length equal to or longer than 2/3 of the height, as shown in Figure 3-12.

砖砌体接槎施工时，必须将接槎处的表面清理干净，浇水湿润，并填实砂浆，保持灰缝平直。

During the construction of brick joints, the joint surface must be cleaned up, watered, and filled with mortar to keep the mortar joint straight.

3.2.5 砖墙砌筑的施工要点

3.2.5 Key Points of Brick Wall Masonry

全部砖墙应平行砌筑，砖层必须水平，砖层正确位置用皮数杆控制，基础和每层楼砌完后必须校对一次水平、轴线和标高，在允许偏差范围内，其偏差应在基础或楼板顶面调整。

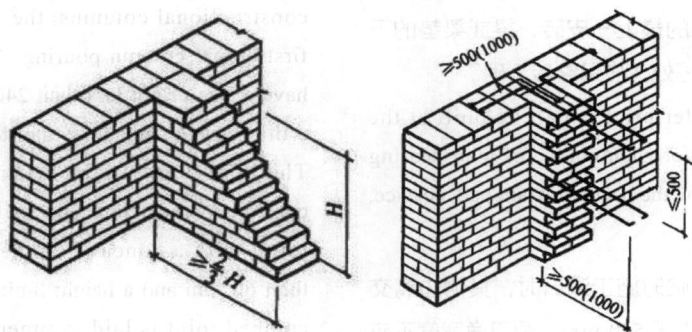

图3-12　普通砖砌体斜槎与普通砖砌体直槎（单位：mm）

Figure 3-12 Inclined Joint of Ordinary Brick Masonry and Straight Joint of Ordinary Brick Masonry

All brick walls shall be built in parallel. Brick courses must be horizontal and laid at correct positions under the control of the height pole. After the foundation and each floor are completed, the level, axis, and elevation must be calibrated once and must fall into the allowable deviation range, with the deviation adjusted on the top surface of the foundation or floor slab.

砖墙的水平灰缝和竖向灰缝宽度一般为10 mm，但不小于 8 mm，也不应大于 12 mm。水平灰缝的砂浆饱满度不得低于 80%，竖向灰缝宜采用挤浆或加浆方法，使其砂浆饱满。砖柱水平灰缝和竖向灰缝饱满度不得低于 90%，每检验批抽查应不少于 5 处。检查时，用百格网检查砖底面与砂浆的黏结痕迹面积，每处检测 3 块，取其平均值。

Horizontal and vertical mortar joints of the brick wall are usually 10 mm wide, within the range of 8~12 mm. The mortar fullness of horizontal mortar joints shall be at least 80%, and vertical mortar joints should be full of mortar by shoving or adding method. The fullness of horizontal and vertical mortar joints of brick columns shall be at least 90%, and at least 5 positions shall be checked for each inspection lot. During the inspection, check the area of bonding trace between the brick bottom and mortar with a grille screen, test 3 pieces at each position, and take the average value.

每层承重墙的最上一皮砖、梁或梁垫的下面及挑檐、腰线等处，应是整砖丁砌。

Whole header bricks shall be built at the uppermost brick for each course of load-bearing wall, the part below the beam or beam pad, cornice, and waistline.

砖墙中留置临时施工洞口时，其侧边离交接处的墙面不应小于 500 mm，洞口净宽度不应超过 1 m。

When a temporary construction opening is reserved in the brick wall, its side shall be kept at least 500 mm from the wall surface at the junction, and the net width of the opening shall not exceed 1 m.

砖墙相邻工作段的高度差，不得超过一个楼层的高度，也不宜大于 4 m。工作段的分段位置应设在变形缝或门窗洞口处。砖墙临时间断处的高度差，不得超过一步脚手架的高度。砖墙每天砌筑高度以不超过 1.5 m 为宜。

The height difference between adjacent working sections of the brick wall shall not exceed the height of one floor or 4 m. The partition of working sections shall be set at the deformation joint or door/window openings. The height difference between temporary discontinuities of the brick wall shall not exceed the height of a step of the scaffold. The masonry height of the brick wall should not exceed 1.5 m per day.

钢筋混凝土构造柱施工：构造柱需先砌筑墙，后浇筑柱，尺寸不宜小于 240 mm×240 mm，其厚度不宜小于墙厚度。砖砌体与构造柱的连接处应砌成马牙槎，马牙槎凹凸尺寸不宜小于 60 mm，高度不应超过 300 mm，马牙槎先退后进，对称砌筑。墙与柱连接处应沿高度方向每隔 500 mm 设 2φ6 水平拉结筋，伸入墙内不宜少于 1 000 mm，如图 3-13 所示。

For the construction of reinforced concrete constructional columns, the wall shall be built first before column pouring. The column should have a size of not less than 240 mm×240 mm and a thickness of not less than the wall thickness. The joint between the brick masonry and the constructional column shall be built into a combed joint, with a concave-convex size of not less than 60 mm and a height limited to 300 mm. The combed joint is laid symmetrically, retreating before advancing. 2φ6 horizontal tie bars are set every 500 mm along the height direction at the connection between the wall and column, and extend into the wall for a length of not less than 1 000 mm, as shown in Figure 3-13.

柱内竖向受力钢筋一般采用 HPB300 级钢筋，中柱不宜小于 4φ12，边柱不宜小于 4φ14，其箍筋一般采用 φ6@200，楼层上下 500 mm 范围内宜采用 φ6@100，构造柱竖向受力钢筋应在基础梁和楼层圈梁中锚固（图 3-14）。

HPB300 rebars are usually used as the vertical stressed reinforcement in the column, with a size of 4φ12 or larger for the middle column and 4φ14 or larger for the side column. Its stirrup is generally φ6@200, and φ6@100 should be used within 500 mm above and below the floor. The vertical stressed reinforcement of the constructional column should be anchored in the foundation beam and ring beam of the floor(Figure 3-14).

框架梁的填充墙砌至梁底应预留 18~20 cm，间隔 14 天左右时间后再用实心砖斜砌挤紧，砂浆要饱满。间隔一周是为了让新砌砌体完成墙体自身沉缩，斜砌可减少灰缝收缩，以防止梁底由于墙体沉缩造成开裂，如图 3-15 所示。

When the filler wall of the frame beam reaches the beam bottom, a space of 18-20 cm shall be reserved, and after 14 days, compacted with full mortar via inclined masonry of solid bricks. The one-week interval allows the new masonry to complete wall settlement, and the inclined masonry can reduce the shrinkage of mortar joints to prevent the beam bottom from cracking due to wall settlement, as shown in Figure 3-15.

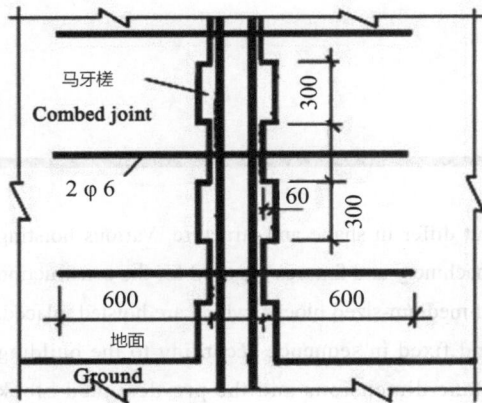

图3-13　砖墙与构造柱连接

Figure 3-13 Connection Between Brick Wall and Structural Column

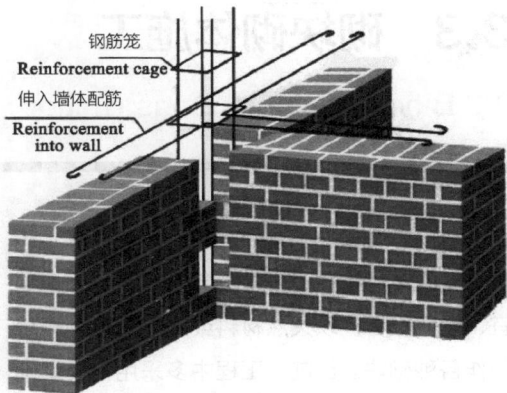

图3-14　构造柱的构造

Figure 3-14 Structure of Structural Column

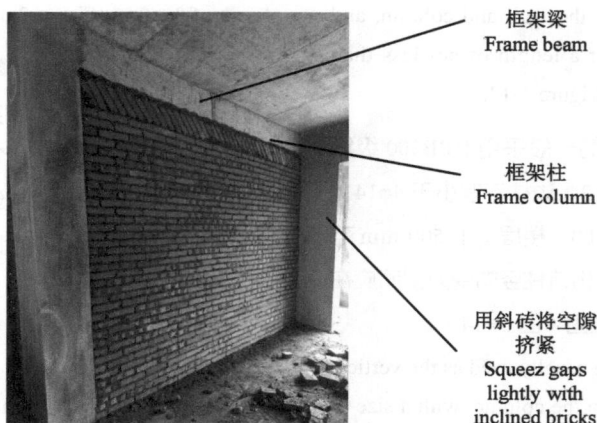

图3-15 梁底的整砖斜砌

Figure 3-15 Inclined Masonry of Whole Bricks at Beam Bottom

3.3 砌块砌体施工

3.3 Block Masonry Construction

砌块作为一种墙体材料，用其代替黏土砖是墙体改革的重要途径。近年来，各地因地制宜、就地取材，以天然材料或工业废料为原料制作各种砌块。目前，工程中多采用中小型砌块。小型砌块施工与传统的砖砌体施工工艺相似，手工砌筑，但在形状、构造上有一定差异。中型砌块施工采用各种吊装机械及夹具，按照建筑平面尺寸及预先设计的砌块排列次序吊装、就位、固定。

Masonry block as a wall material to replace clay brick, it is an important way of wall reform. In recent years, various masonry blocks have been made from natural materials or industrial wastes according to local conditions. Currently, small- and medium-sized blocks are widely used in projects. Small blocks, similar to the process of traditional brick masonry, are built manually but differ in shape and structure. Various hoisting machinery and fixtures are used for the construction of medium-sized blocks, which are hoisted, placed, and fixed in sequence according to the building plane dimensions and the pre-designed block arrangement.

3.3.1 施工准备

3.3.1 Construction preparation

（1）施工机械准备

(1) Preparation of construction machinery

吊装方案：

Lifting scheme:

一是采用塔式起重机进行砌块和砂浆的运输，以及楼板等其他构件的吊装，而用台灵架吊装砌块，台灵架在楼层上的转移靠塔吊来完成。

First, tower cranes are used for the transportation of blocks, mortar, the hoist of floor slabs, and other components, with blocks masonry by the tower frame; the transfer of the tower frame on the floor is done by the tower crane.

二是以井架进行材料的垂直运输，以杠杆车进行楼板吊装，所有的预制构件及材料则用砌块车和手推车进行水平运输，台灵架负责砌块的吊装。

Second, derricks are used for the vertical transportation of materials, and the lever cart is used for floor slab hoisting. All prefabricated components and materials are transported horizontally by block carts and trolleys, and the tower frame is used to hoist blocks.

（2）材料准备

(2) Material preparation

砌块的生产龄期不应小于 28 天，砌筑时应清除砌块表面的污物和芯柱所用砌块孔洞的底部毛边。

The age of masonry blocks shall be at least 28 days. During masonry, dirt on the block surface and the bottom burrs at the holes of blocks used for the core column shall be removed.

一般情况下，砌块不用浇水，但当天气炎热干燥时，可提前浇水润湿。对于粉煤灰砌块和轻骨料混凝土小型砌块可提前浇水湿润，对加气混凝土砌块应向砌筑面适量浇水。

Generally, the building blocks are not watered, but in hot and dry days, they can be watered in advance. Fly ash blocks and small blocks made of lightweight aggregate concrete can be watered in advance, and aerated concrete blocks shall be appropriately watered on the masonry surface.

（3）绘制砌块排列图

(3) Draw the block layout diagram

在砌块吊装前应在基础平面按每片纵、横

墙分别绘制砌块排列图，放出第一皮砌块的轴线、边线和洞口线，对于空心砌块还应放出分块线，以指导砌块的吊装施工。

Before the hoisting of blocks, a block layout diagram shall be drawn on the foundation plane to illustrate each longitudinal and transverse wall, by setting out the axis, sideline, and opening line of the first block. For hollow blocks, the partitioning line shall also be set out to guide the hoisting of blocks.

3.3.2 施工工艺

3.3.2 Construction process

砌块施工的工艺流程主要有：铺灰→砌块就位→校正→灌缝→浇灌芯柱。

Process of block construction: Mortar spreading → block in place → correction → pointing → core column concreting.

3.3.3 施工要点

3.3.3 Key points of construction

①施工时所用的砌块的龄期不应小于 28 天，砌筑时不得浇水。

① The age of blocks used in construction shall not be less than 28 days, and watering is not allowed during masonry.

②砌块的砌筑应立皮数杆、拉准线，从转角处或定位处开始，内外墙同时砌筑，纵横墙交错搭接，如图 3-16 所示。

② Blocks shall be built by erecting a height pole and pulling neat lines, starting from the corner or positioning point. The inner and outer walls shall be built at the same time, with the longitudinal and transverse walls overlapped in a staggered manner, as shown in Figure 3-16.

③砌块的砌筑应遵循"对孔、错缝、反砌"的规则进行，即上皮砌块的孔洞对准下皮砌块的孔洞，则上下皮砌块的壁、肋可较好传递竖

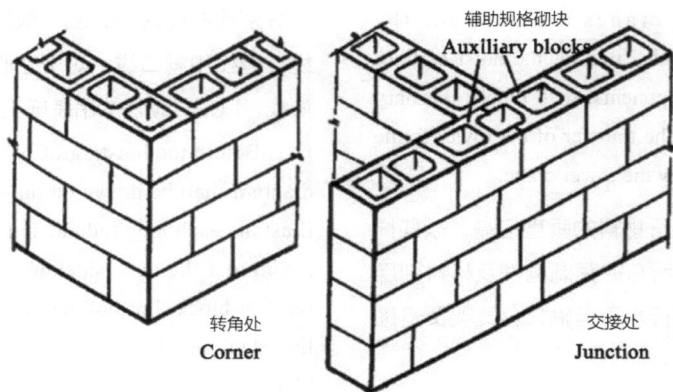

图3-16 小砌块墙转角处及T字交接处砌法

Figure 3-16 Masonry Method at the Corner of Small Block Wall and T-junction

向荷载，保证砌体的整体性和强度。错缝（搭砌）可增强砌体的整体性。将砌块生产时的底面朝上，便于铺放砂浆和保证水平灰缝的饱满度。

③ The masonry of blocks shall follow the rule of "hole alignment, staggered joint, and reverse construction". That is, the holes of the upper block shall be aligned with the holes of the lower block, allowing better transmission of vertical loads via the walls and ribs of the upper and lower blocks, to keep their integrity and strength. Staggered joints(lap joints) can enhance the integrity of the masonry. Turning the block bottom upward during production can facilitate mortar laying and ensure the fullness of horizontal mortar joints.

上下皮小砌块竖向灰缝错开 190 mm，特殊情况无法对孔砌筑时，普通砼小砌块错缝长度不小于 90 mm，轻骨料砼砌块错缝长度不小于 120 mm。无法满足此规定时，应在水平灰缝中设置 4φ4 钢筋网片，网片每端均应超过该竖向灰缝长度 400 mm。

The vertical mortar joints of the upper and lower small blocks are staggered by 190 mm. When the holes cannot be aligned under special circumstances, the staggered joint of small ordinary concrete blocks shall not be shorter than 90 mm, and the staggered joint of lightweight aggregate concrete blocks shall not be shorter than 120 mm. When this requirement fails to be met, 4φ4 reinforcing mesh shall be set in the horizontal mortar joints, and each end of the mesh shall exceed the length of the vertical mortar joint by 400 mm.

④小砌块砌体的临时间断处应砌成斜槎，斜槎长度不小于高度的 2/3。转角处及抗震设防区严禁留置直槎。非抗震设防区的内外墙临时间断处留斜槎有困难时，可从砌体面伸出 200 mm 砌成阴阳槎，并每三皮砌块设拉结筋或钢筋网片，接槎部位延至门窗洞口，如图 3-17 所示。

④ The temporary discontinuities of small block masonry shall be built into inclined joints, with a length of not less than 2/3 of the height. Straight joints are not allowed at corners and seismic fortification areas. When it is difficult to reserve inclined joints at temporary discontinuities of the inner and outer walls in non-seismic fortification areas, it can extend 200 mm from the masonry surface to form internal and external joints, with tie bars or reinforcement meshes set every three blocks, and the joint part shall extend to the door/window openings, as shown in Figure 3-17.

⑤承重墙体严禁使用断裂砌块。

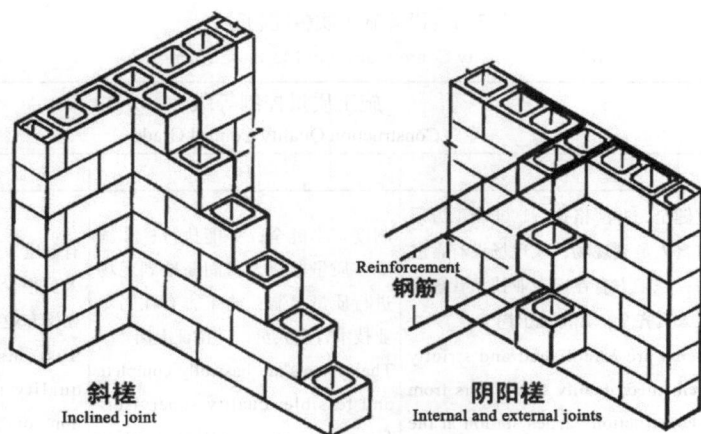

图3-17　小砌块砌体斜槎和直槎

Figure 3-17 Inclined and Straight Joints of Small Block Masonry

⑤ Broken blocks are not allowed for load-bearing walls.

⑥需移动砌体中的砌块或砌块被撞动时，应重新铺砌。

⑥ When blocks in the masonry are impacted or need to be moved, they shall be re-paved.

⑦砌块的日砌筑高度一般控制在 1.4 m 或一步架内。

⑦ The daily masonry height of blocks is usually limited to 1.4 m or controlled within the one-step frame.

3.4　砌筑工程的质量及安全技术

3.4 Quality and Safety Technology of Masonry Works

3.4.1 砌筑工程的质量要求

3.4.1 Quality Requirements for Masonry Works

（1）砌体施工质量控制等级

(1) Quality control grade of masonry construction

砌体施工质量控制等级分为三级，其标准应符合表 3-1 的要求。

The quality control grade of masonry construction is divided into three grades, and its standards shall meet the requirements of Table 3-1.

表 3-1 砌体施工质量控制等级

Table 3-1 Quality Control Grade of Masonry Construction

项目 Item	施工质量控制等级 Construction Quality Control Grade		
	A	B	C
现场质量管理 Site quality management	制度健全，并严格执行；非施工方质量监督人员到现场，或现场设有常驻代表；施工方有在岗专业技术管理人员，人员齐全，并持证上岗 Systems are established and strictly implemented; quality supervisors from non-construction parties station at the site, or resident representatives are available at the site; the construction party provides on-post professional technical managers in sufficient numbers and with certificates	制度基本健全，并能执行；非施工方质量监督人员间断地到现场进行质量控制；施工方有在岗专业技术管理人员，并持证上岗 The systems are basically complete and feasible; quality supervisors from non-construction parties visit the site from time to time for quality control; the construction party provides on-post professional technical managers with certificates	有制度；非施工方质量监督人员很少到现场；施工方有在岗专业技术管理人员 Systems are available; quality supervisors from non-construction parties are rarely present at the site; the construction party provides on-post professional technical managers
砂浆、混凝土强度 Strength of mortar and concrete	施工质量高，试块强度满足验收规定，离散性小 Construction quality is high, test blocks are made according to the specifications, with small discreteness and strength meeting the acceptance regulations	施工质量较高，试块强度满足验收规定，离散性小 Construction quality is relatively high, test blocks are made according to the specifications, with small discreteness and strength meeting the acceptance regulations	施工质量一般，试块强度满足验收规定，离散性大 Construction quality is mediorce, test blocks are made according to the specifications, with large directeness and strength meeting the acceptance regulations
砂浆拌合方式 Mortar mixing method	机械搅拌；配合比计量控制严格 Mechanical mixing; mix proportion measurement under strict control	机械搅拌；配合比计量控制一般 Mechanical mixing; mix proportion measurement under normal control	机械或人工拌合；配合比计量控制较差 Mechanical or manual mixing; mix proportion measurement under poor control
砌筑工人 Masonry worker	中级工以上，其中高级工不少于30% Intermediate workers or above, including at least 30% of senior workers	高、中级工不少于70% At least 70% of intermediate and senior workers	初级工以上 Above Junior workers

注：1.砂浆、混凝土强度项目中离散性大小根据强度标准差确定；2.配筋砌体不得为C级。

Notes: 1. The discreteness in the strength item of mortar and concrete is determined according to the standard deviation of strength; 2. Reinforced masonry shall not be Grade C.

（2）砌体结构工程检验批验收

(2) Inspection lot acceptance of masonry structure works

砌体结构工程检验批验收时，其主控项目应全部符合规范规定；一般项目应有 80% 及以上的抽检处符合规范规定；有允许偏差的项目，最大超差值为允许偏差的 1.5 倍。

During the inspection lot acceptance of masonry structure works, the main control items shall all meet the specifications; 80% or more of the sampled general items shall meet the specifications; for items with allowable deviations, the out-of-tolerance value must not exceed 1.5 times the allowable deviation.

（3）砌体工程所用的材料

(3) Materials for masonry works

砌体工程所用的材料应有产品的合格证书、产品性能检测报告。水泥进场时应对其品种、等级、包装或散装仓号、出厂日期等进行检查，并应对其强度、安定性进行复验，其质量必须符合现行国家标准的有关规定。

All materials used in masonry works shall be accompanied by product qualification certificates and product performance inspection reports. When cement is mobilized, its variety, grade, packaging or bulk warehouse number, and delivery date shall be inspected, with its strength and stability re-inspected, and its quality must comply with prevailing national standards.

（4）验收标准

(4) Acceptance criteria

同一验收批砂浆试块强度平均值应大于或等于设计强度等级值的 1.1 倍；同一验收批砂浆试块抗压强度的最小一组平均值应大于或等于设计强度等级值的 85%。

For the same acceptance batch, the average strength of mortar test blocks shall be ≥ 1.1 times the design strength, and the minimum average value of compressive strength for mortar test blocks group shall be ≥ 85% of the design strength.

3.4.2 砌筑工程的安全与防护措施

3.4.2 Safety and Protection Measures for Masonry Works

①在砌筑操作前，必须检查施工现场各项准备工作是否符合安全要求，如道路是否畅通，机具是否完好牢固，安全设施和防护用品是否齐全，经检查符合要求后才可施工。

① Before masonry operation, operators must check whether all preparations on the construction site meet the safety requirements, by finding out whether the roads are unblocked, whether machines and tools are in good condition and secured, and whether the safety facilities and protective articles are complete. Construction must not start until the inspection meets the requirements.

②砌筑基础时，应检查和注意基坑土质的变化情况。堆放砖石材料应离开坑边 1 m 以上。

② When laying the foundation, check and notice the change in the soil quality of the foundation pit. Stacked masonry materials shall be kept at least 1 m away from the pit edge.

③当砖墙高度超过地坪 1.2 m 以上时，应搭设脚手架。架上堆放材料不得超过规定荷载，堆砖高度不得超过三皮侧砖，同一块脚手板上的操作人员不应超过两人，且应按规定搭设安全网。

③ When the height of the brick wall exceeds 1.2 m above the ground, scaffolds shall be erected. Materials stacked on the scaffold must not exceed the specified load, with the brick-stacking height not exceeding three bricks laid by the side. One scaffold floor can hold not more than two operators. Safety nets shall be erected as required.

④不允许站在墙顶上做画线、刮缝及清扫墙面或检查大角垂直等工作。不允许用不稳固的工具或物体在脚手板上垫高操作。

④ It is not allowed to stand on the top of the wall to draw lines, scrape joints, clean the wall surface, or check the verticality of large corners. It is not allowed to raise the scaffold floor with unstable tools or objects.

⑤砍砖时应面向墙面，工作完毕后应将脚手板和砖墙上的碎砖、灰浆清扫干净，防止掉落伤人。正在砌筑的墙上不允许有人行走。

⑤ When cutting bricks, operators shall face the wall. After the work is done, broken bricks and mortar on the scaffold floor and brick wall shall be cleaned up to prevent falling and hurting people. No one is allowed to walk on the wall under construction.

⑥山墙砌完后，应立即安装临时支撑，防止倒塌。

⑥ After the gable wall is built, temporary supports shall be installed immediately to prevent collapse.

⑦雨天或每日下班时，应做好防雨准备，以防雨水冲走砂浆，致使砌体倒塌。

⑦ On rainy days or after work every day, rainproof preparations shall be made to prevent rainwater from washing away mortar and causing masonry collapse.

⑧砌石墙时不准在墙项或架上整修石材，以免振动墙体，影响石墙质量或导致石片掉下伤人。不允许徒手移动墙上的石块，以免压破或擦伤手指。不允许勉强在超过胸部的墙上进

行砌筑，以免将墙体碰撞倒塌或上石时失手掉下石块造成安全事故。石块不得往下抛掷。运石上下时，脚手板要钉装牢固，并钉防滑条及扶手栏杆。

⑧ When building a stone wall, operators must not repair stones on the wall top or frame in case the wall is vibrated to affect the quality or hurt people by falling stone chips. Operators are not allowed to move stones on the wall with bare hands to avoid crushing or scratching their fingers. Operators are not allowed to force to build the wall with a height above the chest, to avoid safety accidents caused by wall collapse due to collision or by falling stones not being held tightly. Stones must not be thrown down. When stones are moved up and down, the scaffold floor must be nailed firmly, and anti-slip strips and handrails shall be installed.

⑨凡脚手架、井架、门架搭设好后，须经专人验收合格后方准使用。

⑨ After the scaffold, derrick, and portal frame are erected, they must be accepted by special personnel before use.

3.5 砌体工程冬期施工

3.5 Masonry Works in Winter

当室外日平均气温连续 5 天稳定低于 5°C 时，砌体工程施工应按照《砌体结构工程施工质量验收规范》及《建筑工程冬期施工规程》中的有关规定进行。

When the outdoor daily average temperature drops below 5°C for 5 consecutive days, the masonry works shall be carried out in accordance with relevant provisions as specified in the Code for Acceptance of Construction Quality of Masonry Engineering and the Code for Winter Construction of Building Engineering.

3.5.1 砌体冬期施工规定

3.5.1 Regulations on Winter Construction of Masonry

①冬期施工对材料的要求。

① Material requirements for winter construction.

a. 砌筑前，应清除块材表面的污物、冰霜等。遭受冻结的砖和砌块不得使用。

a. Before construction, dirt and frost on the block surface shall be removed. Frozen bricks and blocks must not be used.

b. 砌筑砂浆应采用普通硅酸盐水泥拌制，不得用无水泥拌制的砂浆。

b. Masonry mortar shall be mixed with ordinary portland cement, and mortar without cement must not be used.

c. 石灰膏、电石膏等应防止受冻，如遭冻结，经融化后方可使用。

c. Lime putty and calcium carbide shall be subject to anti-freezing measures. If frozen, they shall be melted before use.

d. 拌制砂浆用的砂，不得含有冰块和直径大于 10 mm 的冻结块。

d. The sand for mixing mortar must not contain ice or frozen blocks with a diameter greater than 10 mm.

e. 拌制砂浆时，水的温度不得超过 80℃，砂的温度不得超过 40℃，一般应采用两步投料法。

e. For mortar mixing, the water temperature must not exceed 80°C, and the sand temperature must not exceed 40°C. Generally, the two-step feeding method is adopted.

f. 冬期砌筑砂浆的稠度可适当增加，一般比常温施工时增大 10~30 mm，但不宜超过 130 mm。砖砌体所用砂浆的稠度为 80~130 mm，加气混凝土砌块砌筑砂浆稠度为 130 mm。

f. In winter, the consistency of masonry mortar can be appropriately increased, usually, 10-30 mm higher than that in normal temperature, but should not exceed 130 mm. The consistency of mortar is 80-130 mm for brick masonry and 130 mm for aerated concrete block masonry.

②冬期搅拌砂浆的时间应适当延长，一般比常温时增加 0.5~1 倍。

② The time for mixing mortar in winter shall be appropriately extended, normally 0.5-1 times longer than the mixing time in normal temperature.

③冬期施工的砌筑砂浆应随拌随用，普通砂浆的储存时间是 15 min，掺盐砂浆 20 min，砂浆使用温度不应低于 5℃。

③ Masonry mortar for winter construction shall be used immediately after mixing. The storage time shall be 15 min or 20 min for ordinary mortar and salt mortar respectively, and the use temperature of mortar shall not be lower than 5°C.

④砖砌体应按"三一砌筑法"施工，并采用一顺一丁或梅花丁的砌筑形式。

④ The brick masonry shall be built with the "three-one" masonry method and in the form of "one stretcher and one header" or "quincunx header".

⑤每日砌筑后应及时在砌筑表面覆盖保温材料，砌体表面不得残留砂浆。在继续砌筑时，应扫净砌体表面，然后再续砌。

⑤ After daily construction, the masonry surface shall be covered with thermal insulation materials in time, with no mortar remaining on the masonry surface. When construction continues, the masonry surface shall be cleaned up first.

⑥冬期施工的砌筑工程要加强质量控制，在现场制作的砂浆试块除按常温规定的要求外，还应增留不少于一组与砌体同条件养护的试块，28 天后检验其强度。

⑥ In winter construction, quality control shall be enhanced for masonry works. For mortar test blocks made at the site, besides the requirements specified for normal temperature, at least one group of blocks cured under the same conditions as the masonry shall be reserved to test its strength after 28 days.

⑦普通砖、多孔砖和空心砖在气温高于 0℃ 条件下砌筑时，应浇水湿润；在气温低于、等于 0℃ 条件下砌筑时，可不浇水，但必须增大砂浆稠度。

⑦ When common bricks, perforated bricks, and hollow bricks are laid at a temperature above 0°C, they shall be watered. When the temperature reaches or drops below 0°C, watering may be saved, but the mortar consistency must be increased.

3.5.2 砌体冬期施工方法

3.5.2 Construction Method of Masonry in Winter

砌体工程的冬期施工方法有外加剂法、冻结法和暖棚法等，应优先选用外加剂法。

The winter construction methods of masonry works include the admixture method, freezing method, and warm-shed method. The admixture method is preferred.

4

混凝土结构工程
Concrete Structure Works

学习目标:

Learning objectives:

了解模板的类型及特点,掌握模板的安装与拆除的方法及要求;掌握钢筋的配料、代换的计算方法;掌握混凝土施工配合比换算;掌握混凝土施工工艺、质量控制方法及安全生产技术要求;能进行混凝土工程施工质量的检查、评定。

Learn about the types and characteristics of formwork, and master the methods and requirements for formwork erection and removal; master the calculation methods for reinforcement detailing and replacement; master the conversion of concrete working mix proportion; master the construction technology, quality control methods and work safety technical requirements for concrete works; be able to inspect and evaluate the construction quality of concrete works.

技能抽查要求:

Skill checking requirements:

掌握模板工程施工工艺;能顺利完成简单构件模板支设技能操作;掌握钢筋工程施工

工艺，正确计算钢筋下料长度，规范填写钢筋下料单，能顺利完成钢筋加工和钢筋绑扎技能操作；掌握混凝土工程施工工艺，能进行混凝土施工配合比换算，熟悉混凝土试块的留置和混凝土养护工作，能顺利完成混凝土浇筑等操作；熟悉《混凝土结构工程施工质量验收规范》（GB 50204—2015），能按国家规范要求列出检测项目和项目允许偏差，并正确使用常用检测工具对混凝土梁、柱等的施工质量进行检查验收。

Master the construction technology of formwork works; be able to carry out the skilled operation in formwork erection for simple members; master the construction technology of reinforcement works, correctly calculate blanking lengths for reinforcement, fill in the reinforcement detailing sheet in a standardized manner, and be able to carry out the skilled operation in reinforcement processing and binding; master the construction technology of concrete works, be able to carry out conversion to the concrete working mix proportion, be familiar with the retention of concrete test tubes and concrete curing, and be able to carry out concrete pouring and other skilled operations; be familiar with the Code for Quality Acceptance of Concrete Structure Construction (GB 50204—2015), be able to list the inspection items and their allowable deviations according to the requirements of national specifications, and correctly use common inspection tools to inspect and accept the construction quality of concrete beams and columns.

建筑八大员岗位资格考试要求：

Qualification examination requirements for eight major posts of construction engineering:

掌握梁、板、柱、墙模板的构造、安装工艺与拆除要求；掌握钢筋配料、代换、加工、连接和安装要求；掌握混凝土施工配合比计算和施工工艺；掌握混凝土工程施工质量与安全要求；掌握混凝土表面缺陷的修补方法。

Master the structures, installation processes and removal requirements of formwork for beams, slabs, columns and walls; master the requirements for reinforcement detailing, replacement, processing, connection and placement; master the calculation of concrete working mix proportion and construction process; master the construction quality and safety requirements of concrete works; master the repair methods of concrete surface defects.

4.1 模板工程

4.1 Form Works

模板是按设计要求塑造混凝土结构形状和尺寸的模具。

Formwork is used to build concrete structures in shape and size up to the design requirements.

模板工程是混凝土工程重要的组成部分，占混凝土结构工程总价的 20% ~ 30%，劳动量的 30% ~ 40%，工期的 50% 左右，决定着混凝土结构工程的施工方法和施工机械的选择，直接影响工期和造价。

Form works consist a very important part of concrete works, accounting for 20%-30% of the total cost of concrete structural works, 30%-40% of the total workload, and about 50% of the construction period, which plays a decisive role in the selection of construction method and construction machinery for concrete structural works and directly affects the construction period and cost.

模板工程的施工包括模板的选材、选型、设计、制作、安装、拆除和周转等过程。

The construction of form works includes multiple processes of material selection, type selection, design, fabrication, erection, removal and turnover of forms.

4.1.1 模板的安装工艺

4.1.1 Installation Process of Formwork

模板工程系统包括模板系统和支撑系统两大部分，此外还有适量的紧固连接件。

The formwork system includes two main parts, i.e., forms and supports, and an appropriate number of fasteners and connectors.

模板必须有一定的承载能力和稳定性，此外模板的尺寸还必须准确，确保成型的混凝土结构有准确的形状、位置和尺寸。

Formwork must have a certain load-carrying capacity and stability. In addition, the formwork dimensions must be accurate to ensure that the formed concrete structures are accurate in terms of shape, position and size.

（1）基础模板

(1) Foundation formwork

基础模板的特点是高度不大而体积较大，一般上下分级且各级尺寸不同。

Foundation formworks are characterized by small height and large volume, generally leveled vertically and varied in size at different levels.

阶梯形基础模板（图 4-1）由四块侧板拼钉而成，其中两块侧板的尺寸与相应的台阶侧面尺寸相等。另两块侧板长度应比相应的台阶侧面长度大 150 ~ 200 mm，高度与其相等。安装时要保证上、下模板不发生相对位移。

The stepped foundation formwork(Figure 4-1) is made of four side boards, of which two are in the same dimensions as the side of the corresponding step. The length of the other two shall be 150-200 mm longer than the side of the corresponding step, and the height shall be equal to it. It is necessary to ensure that the upper and lower forms are erected to prevent relative displacement.

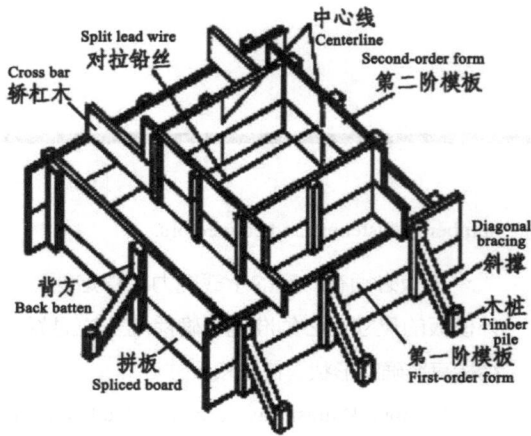

图4-1 阶梯形基础模板

Figure 4-1 Stepped Foundation Formwork

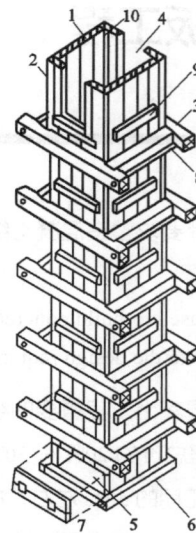

1.内拼板；2.外拼板；3.柱箍；4.梁缺口；5.清理孔；6.木框；7.盖板；8.拉紧螺栓；9.拼条；10.三角木条

1. inner spliced board; 2. outer spliced board; 3. column clamp; 4. beam opening; 5. cleaning hole;6. wooden frame; 7. cover board; 8. tension bolt; 9. splicing strip; 10. triangular wooden strip

图4-2 柱模板

Figure 4-2 Column Formwork

（2）柱模板

(2) Column formwork

柱子是竖向构件，其特点是断面尺寸不大但比较高，安装施工时必须注意保证柱模板的竖向稳定。

Columns are a kind of vertical member, characterized by small sectional dimensions and a large height, and shall be erected to ensure their vertical stability.

柱模板由两块相对的内拼板、两块相对的外拼板和柱箍组成。拼板上端应根据实际情况开有与梁模板连接的缺口，底部开有清理孔，沿高度每隔2 m开有浇筑孔。柱模板如图 4-2 所示。柱模板的安装过程及要求：

Column formwork consists of two opposite inner spliced boards, two opposite outer spliced boards and column clamps. The spliced board shall be provided with an opening at the upper end for connection to the beam forms according to the actual conditions, and be perforated with a cleaning hole at the bottom and a pouring hole every 2 m vertically. Column formwork is shown in Figure 4-2. The erection process and requirements of column formwork are as follows:

①弹线及定位：先在基础顶面（楼面）弹出柱轴线及边线，同一柱列则先弹两端柱，再拉通线弹中间柱的轴线及边线。按照边线先把底部木框固定好，然后再对准边线安装柱模板。

① Line snapping and positioning: Snap the column axis and sideline on the top surface of the foundation(floor) in the sequence of the columns at both ends first and then the middle columns against a reference line pulled through for a column row. Fix the bottom wooden frame aligned to the sideline, and then erect the column forms aligned to the sideline.

②柱箍的设置：为防止混凝土浇筑时模板发生鼓胀变形，柱箍应根据柱模断面大小经计算确定，下部的间距应小些，往上可逐渐增大间距。当柱截面尺寸较大时，应考虑在柱模内设置对拉螺栓。

② Setting of column clamps: In order to prevent the formwork from bulging deformation during concrete pouring, the column clamps shall be detailed by calculation according to the sectional size of the column formwork, be placed at a small spacing at the lower part and at a gradually increasing spacing upward. In the case of a large column sectional size, split bolts shall be used in the column formwork.

③柱模板须在底部留设清理孔，沿高度每2 m 开有混凝土浇筑孔和振捣孔。

③ Column formwork shall be provided with a cleaning hole at the bottom, a concrete pouring hole and a vibrating hole every 2 m vertically.

④对于通排柱模板，应先装两端柱模板，校正固定后，再在柱模板上口拉通线校正中间各柱模板。

④ For the column formwork in a row, the column forms at both ends shall be erected first, and then those in the middle shall be aligned by pulling through a reference line across the upper opening of the column form.

（3）梁模板

(3) Beam formwork

梁的特点是跨度大而宽度不大，梁底一般是架空的。

Beams are characterized by a large span and small width, and are generally overhead at the bottom.

梁模板（图4-3）主要由底板、侧板、夹木及支架系统组成。底板用长条模板加拼条拼成，或用整块板条。

Beam formwork(Figure 4-3) is mainly composed of base board, side board, wooden clamp and support system. The base board is assembled with long strip forms and splicing strips, or is made of an integral batten.

梁模板安装过程及要求：

The erection process and requirements of beam formwork are as follows:

①梁模板的安装应在复核梁底标高、校正轴线位置无误后进行。

① The installation of beam formwork shall

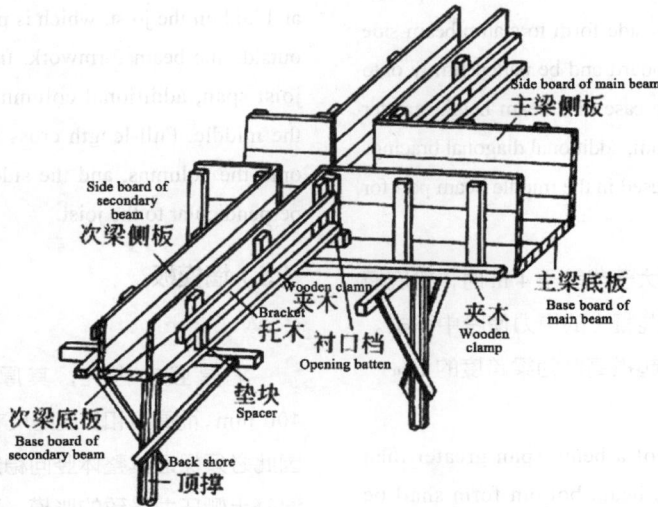

主梁侧板
Side board of main beam

次梁侧板
Side board of secondary beam

夹木
Wooden clamp

主梁底板
Base board of main beam

托木
Bracket

衬口档
Opening batten

夹木
Wooden clamp

次梁底板
Base board of secondary beam

垫块
Spacer

顶撑
Back shore

图4-3　梁模板

Figure 4-3 Beam Formwork

not be erected until the beam bottom elevation and axis location are checked and corrected.

②梁底板下用顶撑（琵琶撑）支设，顶撑间距视梁的断面大小而定，一般为 0.8 ～ 1.2 m，顶撑之间应设水平拉杆和剪刀撑，使之互相拉撑成为一整体，水平拉杆沿竖向间距一般不大于 2 m，为确保顶撑支设的坚实，应在用混凝土硬化的地面上设置垫板和楔子。

② The back shores(jack shores) are used to support under the beam base board, at a spacing of generally 0.8-1.2 m depending on the beam section. Horizontal ties and diagonal bracings shall be set between the back shores to make them into an integral structure with mutual bracing. The horizontal ties are generally spaced not more than 2 m vertically. In order to ensure firm bracing of the back shores, the backing plate and wedge shall be set on the ground of hardened concrete.

③把侧板放上，两头钉于衬口档上。梁侧模下方应设置夹木，将梁侧板与底板夹紧，并钉牢在顶撑上。当梁高度大于或等于 700 mm 时，应在梁中部另加斜撑或对拉螺栓固定。

③ The side board is placed and nailed at both ends onto the opening batten. A wooden clamp shall be set under the beam side form to clamp beam side board to the bottom board and be nailed firmly onto the back shore. In the case of a beam height greater than or equal to 700 mm, additional diagonal bracings or split bolts shall be used in the middle beam part for fixing.

④当梁的跨度大于或等于 4 m 时，梁底模要起拱，防止由于混凝土的重力使跨中下垂。如设计无规定时，起拱高度为梁跨度的 1‰ ～ 3‰。

④ In the case of a beam span greater than or equal to 4 m, the beam bottom form shall be deflected upwards to prevent the midspan from sagging due to the gravity action of concrete. The upward deflection height shall be 1‰-3‰ of the beam span, unless otherwise specified in the design.

（4）楼板模板

(4) Floor slab formwork

楼板的特点是面积大而厚度比较薄，侧向压力小，板底架空。

Floor slabs are characterized by a large area, small thickness and small lateral pressure, and are overhead at the slab bottom.

楼板模板及其支架系统，主要承受混凝土的自重垂直荷载和其他施工荷载，要保证模板不变形。如图 4-4 所示，楼板模板的底模用木板（胶合板或定型模板）拼成，铺设在楞木上，楞木搁置在梁模板外的托木上。当楞木的跨度较大时，中间应加设立柱。立柱上钉通长的杠木，侧模板应垂直于楞木方向铺设。

Floor slab formwork and its support system mainly bear the vertical load from the dead weight of concrete and other construction loads, so it is necessary to guarantee the formwork free from deformation. As shown in Figure 4-4, the bottom form of floor slab formwork is assembled with wooden boards(plywood or typified formwork) and laid on the joist, which is placed on the bracket outside the beam formwork. In the case of a large joist span, additional columns shall be set up in the middle. Full-length cross bars shall be nailed onto the columns, and the side form shall be laid perpendicular to the joist.

（5）墙模板

(5) Wall formwork

墙是竖向构件，其厚度一般为 200 ～ 400 mm，而高度和长度的尺寸远远大于其厚度，因此必须保证其整体竖向稳定性并防止新浇注混凝土侧压力导致的胀模。采用胶合板作为现浇混凝土墙体的模板可以减少模板组拼工作量，减少混凝土外露表面的接缝，浇筑的混凝土表

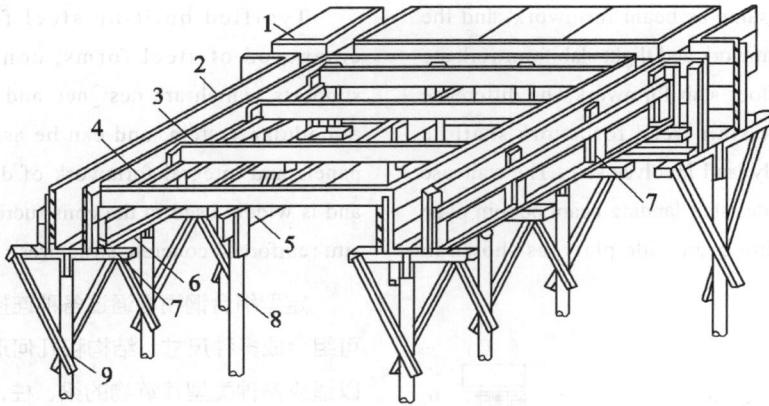

1.楼板模板；2.梁侧模板；3.楞木；4.托木；5.杠木；6.夹木；7.短撑；8.杠木撑；9.琵琶撑

1. floor formwork; 2. beam side form; 3. joist; 4. bracket; 5. cross bar; 6. wooden clamp; 7. short bracing;

8. cross bar bracing; 9. jack shore

图4-4　梁及楼板模板

Figure 4-4 Beam and Floor Slab Formwork

面质量好。

Walls are a kind of vertical member, 200 mm-400 mm in thickness generally, and in a height and length far greater than its thickness, so it is necessary to ensure its overall vertical stability and prevent formwork expanding deformation caused by the lateral pressure of newly poured concrete. Plywood formwork used in the cast-in-place concrete wall can reduce the workload of formwork assembly, and reduce the joints on the exposed surface of concrete, so as to ensure good quality of the poured concrete surface.

墙模板由两侧模板及其支撑体系构成，如图 4-5 所示。

Wall formwork consists of the forms on both sides and their supporting systems, as shown in Figure 4-5.

（6）楼梯模板

(6) Staircase formwork

楼梯梁模板构造与梁模板基本相同，休息平台模板和梯段板模板的构造与楼板相似，不同点是楼梯板的模板要倾斜支设，且在楼梯板上还

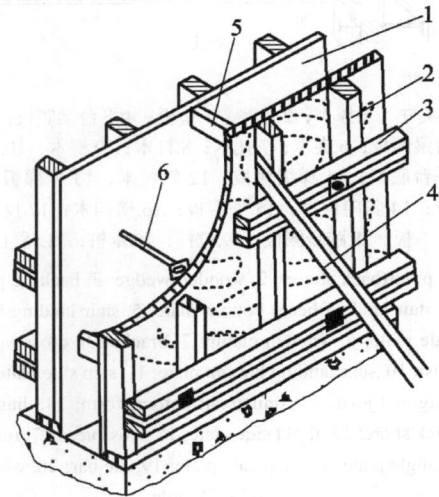

1.胶合板；2.小楞（立档）；3.大楞（横档）；
4.斜撑；5.撑头；6.穿墙螺栓

1. plywood; 2. small joist(jamb); 3. large joist(cross bar);
4. diagonal bracing; 5. top bracing; 6. split bolt

图4-5　胶合板墙体模板

Figure 4-5 Plywood Wall Formwork

有踏步。楼梯模板包括平台梁底板、平台梁侧板等，如图 4-6 所示。

The staircase beam formwork is basically

structured the same as beam formwork, and the stair landing formwork and flight slab formwork are similar to the floor slab formwork. The difference lies in that the stair tread formwork shall be erected obliquely, and involves steps. The staircase formwork includes stair landing beam bottom plate and stair landing beam side plate, as shown in Figure 4-6.

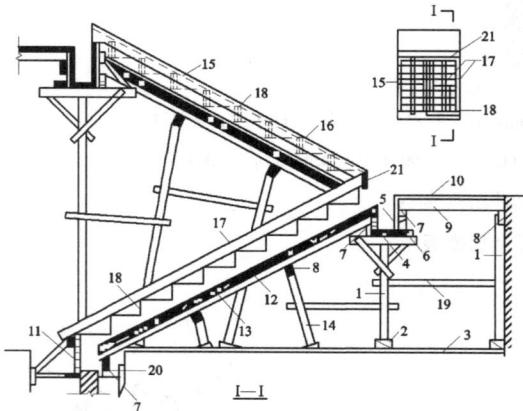

1.支柱（顶撑）；2.木楔；3.垫板；4.平台梁底板；5.平台梁侧板；6.夹木；7.托木；8.杠木；9.楞木；10.休息平台底板；11.梯级侧板；12.斜楞木；13.楼梯板底模板；14.斜向顶撑；15.外帮板；16.横档木；17.反三角板；18.踏步侧板；19.拉杆；20.木桩；21.轿杠

1. pillar(back shore); 2. wooden wedge; 3. backing plate; 4. stair landing beam bottom plate; 5. stair landing beam side plate; 6. wooden clamp; 7. bracket; 8. cross bar; 9. joist; 10. stair landing bottom plate; 11. step side plate; 12. diagonal joist; 13. stair tread bottom form; 14. diagonal back shore; 15. flight side plate; 16. crosspiece;17. reverse triangle plate; 18. step side plate; 19. .tie bar; 20. wooden pile; 21. pole

图4-6　楼梯模板

Figure 4-6 Staircase Formwork

（7）定型组合钢模板

(7) Typified built-up steel formwork

定型组合钢模板，由钢模板、连接件和支撑件组成，按照模数设计制作，能拼装成不同尺寸的面板和整体模架，在现浇钢筋混凝土结构施工中应用比较广泛。

Typified built-up steel formwork is composed of steel forms, connectors and supports, which are designed and fabricated in a modular system, and can be assembled into panels and integral formwork of different sizes, and is widely used in the construction of cast-in-situ reinforced concrete structures.

定型组合钢模板通过各种连接件和支承件可组合成多种尺寸、结构和几何形状的模板，以适应各种类型建筑物的梁、柱、板、墙、基础和设备等施工的需要，组装灵活，通用性强，拆装方便；周转次数多，每套钢模可重复使用50～100次；加工精度高，浇筑混凝土的质量好，成型后的混凝土尺寸准确，棱角整齐。

Typified built-up steel formwork is assembled by various connectors and supports into various sizes, structures and geometric shapes to meet the construction needs of beams, columns, slabs, walls, foundations and equipment for various types of buildings, featuring flexible assembly, strong adaptability and convenient disassembly and assembly; large turnover times, with each set of steel formwork recyclable for 50-100 times; high processing accuracy, good quality of poured concrete, accurate size and neat edges and corners of formed concrete.

除了采用人工散装散拆的施工方法，定型组合钢模板还可事先按设计要求预组拼成梁、柱、墙、楼板的大型面板，也可用其拼装成大模板、滑模、隧道模和台模等，然后用机械整体吊装就位。

In addition to the manual erection and removal of the typified built-up steel formwork, the methods of pre-assembly into large panels of beams, columns, walls and floors according to the design requirements, or assembly into large formwork, sliding formwork, tunnel formwork and bench formwork, and then integral hoisting in place mechanically can be adopted.

（8）大模板

(8) Large formwork

大模板是指单块模板的高度相当于楼层的层高、宽度约等于房间的宽度或进深的大块定型模板，适用于全现浇高层或多层剪力墙结构、框剪结构。一般一个墙面使用一两块大模板，由于重量重，一般需起重吊装机械配合进行安装、拆除施工。其优点是模板安装和拆除工序简单，墙面平整；缺点是一次投资大，通用性较差。

Large formwork refers to the large-block typified formwork with a single-piece height equivalent to the floor height and a width approximately equal to the width or depth of the room, which is suitable for full cast-in-situ high-rise or multi-storey shear wall structures and frame-shear structures. Generally, one or two sets of large formwork are used for a wall. Due to the large weight, they are generally erected and removed with the help of lifting machinery. Its advantages are simple erection and removal procedures and flat wall surface; its disadvantages are a large one-shot investment and poor adaptability.

①大模板的组成。

① Composition of large formwork.

大模板由面板、加劲肋、竖肋、支撑系统和操作平台等组成，如图 4-7 所示。

Large formwork consists of a panel, stiffening rib, vertical rib, support system and operation platform, as shown in Figure 4-7.

②大模板的施工要点。

② Key points of large formwork construction.

a. 在拟建工程附近，起重吊装工作半径范围内，留出一定面积的堆放区，以便直接吊运就位。

a. Near the proposed works, a stacking area shall be designated within the working radius of lift for directly lifting in place.

b. 大模板吊装前，针对大模板及其施工工

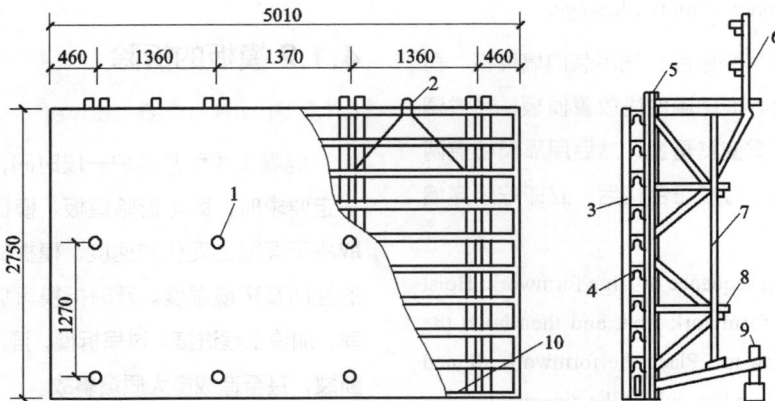

1.穿墙螺栓孔；2.吊环；3.面板；4.横肋；5.竖肋套管；
6.护身栏杆；7.支撑立杆；8.支撑横杆；9.地脚螺丝；10.封板

1. split bolt hole; 2. lifting ring; 3. panel; 4. transverse rib; 5. vertical rib sleeve; 6. protective railing;
7. supporting pole; 8. supporting cross bar; 9. anchor screw; 10. closure plate

图4-7　大模板

Figure 4-7 Large Formwork

程特点，应组织全体施工人员熟悉图纸、流水段划分及大模板拼装位置，做好施工技术和安全交底。

b. Before large formwork hoisting, according to the large formwork characteristics and the actual engineering features, all construction personnel shall be organized to get familiar with the drawings, division in the operation flow process, and the large formwork assembly position, and properly make construction technology and safety disclosure.

c. 内外墙体钢筋绑扎完毕后，立即进行门窗洞口模板、水电预留安装，办理隐蔽工程检查验收手续。并在大模板下部抹好找平砂浆，以便模板就位并防止漏浆。

c. After the reinforcement binding is finished for the inner and outer walls, the formwork of door and window openings and the embedded installation for water and electricity systems shall be carried out immediately before the inspection and acceptance procedures of concealed works. Leveling mortar shall be plastered at the lower part of the large formwork to facilitate the formwork placement and prevent mortar leakage.

d. 大模板吊装顺序：先吊装内墙模板，再吊装外墙模板。根据墙位线放置模板，通过调整大模板斜支撑使其垂直，然后用靠尺检查两侧模板垂直度，待校正合格后，立即拧紧穿墙螺栓。

d. Hoisting sequence of large formwork: Hoist the inner wall formwork first, and then hoist the outer wall formwork. Place the formwork aligned to the wall position line, adjust the diagonal bracing to keep the large formwork vertical, and then check the verticality of the formwork on both sides with a guiding rule, and make correction if any before tightening the split bolts immediately.

（9）滑升模板

(9) Sliding formwork

滑升模板（简称为滑模）是一种工具式模板，是在混凝土连续浇筑过程中，可使模板面紧贴混凝土面滑动的模板。采用滑模施工要比常规施工节约70%左右的模板和脚手板等，可以节约30%～50%的劳动力，比常规施工的工期短、速度快，可以缩短30%～50%的施工周期；结构整体性好，抗震效果明显，适用于高层或超高层抗震建筑物和高耸构筑物施工。

Sliding formwork is a kind of tool formwork, and is used to make the form face slide close against the concrete surface during continuous concrete pouring. Construction with sliding forms can save consumption of formwork and scaffold boards by about 70%, save manpower by 30%-50% and shorten the construction period by 30%-50% compared with conventional construction; can bring positive effect to structural integrity and create an obvious seismic effect, suitable for the construction of high-rise buildings and high-rise or super-high-rise structures to withstand earthquakes.

4.1.2 模板的拆除

4.1.2 Formwork Removal

混凝土成型并养护一段时间，当强度达到一定要求时，即可拆除模板。模板的拆除日期取决于混凝土硬化的速度、模板的用途、结构的性质及环境温度。及时拆模可提高模板周转率，加快工程进度。过早拆模，混凝土会变形、断裂，甚至造成重大质量事故。

Formwork shall not be removed until the concrete poured reaches the design strength after curing for a period of time. The time to remove formwork depends on the concrete hardening rate, formwork purpose, structure nature, and ambient temperature. Timely removal of formwork can improve the turnover rate of formwork and speed up construction progress. Removal of formwork too early may cause concrete deformation, fracture and even major quality accidents.

一般现浇混凝土结构模板的拆除日期取决于结构的性质、模板的用途和混凝土硬化速度，工程结构设计中对拆模时混凝土的强度也有具体规定。如果未做具体规定，应遵守下述规定。

Generally, the time to remove formwork for cast-in-situ concrete structures depends on the structure nature, formwork purpose and concrete hardening rate. The strength of concrete for formwork removal is generally specified in the engineering structure design. If not specified, the following shall be observed.

（1）侧模板拆除

(1) Removal of side form

侧模板在混凝土强度能保证其表面及棱角不因拆模而受损坏时，方可拆除。对后张法预应

力混凝土结构构件，侧模宜在预应力张拉前拆除。

Side form shall not be removed until the concrete reaches the strength to ensure its surface, edges and corners will not be damaged due to formwork removal. For the post-tensioned prestressed concrete structural members, the side form should be removed before prestress tensioning.

（2）底模板及支架拆除

(2) Removal of bottom form and support

底模板及支架应在与结构同条件养护的试块强度达到设计要求时方能拆模。若设计无具体要求时，混凝土强度应符合表4-1的规定。

The bottom form and support shall not be removed until the test tube cured under the same conditions as the structure reaches the design strength. If the design contains no specific requirements, the concrete strength shall comply with the provisions of Table 4-1.

表4-1 底模拆除时的混凝土强度要求

Table 4-1 Requirement on Concrete Strength for Removal of Bottom Form

构件类型 Type of Member	构件跨度/m Member Span/m	达到设计的强度的百分率/% Percentage of Design Strength/%
板 Slab	≤2	≥50
	>2，≤8	≥75
	>8	≥100
梁、拱、壳 Beam, arch and shell	≤8	≥75
	>8	≥100
悬臂构件 Cantilever member	—	≥100

（3）模板拆除施工的一般要求

(3) General requirements for formwork removal

①拆模顺序一般按先支后拆、后支先拆，先拆除非承重模板（如侧模），后拆除承重模板（底模）进行，并应从上而下进行拆除。

① The formwork removal sequence is generally as follows: removal of forms erected earlier or later;

the non-bearing forms(such as side forms) to be removed before the bearing forms(bottom forms); the upper forms to be removed before lower ones.

②拆模时不要用力过猛，不得抛扔，尽量避免混凝土表面或模板受到损坏。拆完后，应及时运走，按类别及尺寸分别堆放，以便下次使用。

② Do not remove the formwork forcefully

or throw it away, and avoid damage to the concrete surface or formwork as far as possible. The removed forms shall be moved away in time and stacked separately by category and size for reuse.

③多层楼板模板的支柱拆除中，上一层楼板正在浇筑混凝土时，下一层楼板的支柱不得拆除，再下一层楼板的支柱，仅可部分拆除。

③ For the formwork of multi-storey floors, when the current floor is undergoing concrete pouring, the pillars of the next lower floor shall not be removed, and only the pillars of the further next lower floor can be partially removed.

④在拆除模板的过程中，如发现混凝土有影响结构安全的质量问题时，应暂停拆除，先处理问题，然后再拆模。

④ In the process of formwork removal, if any concrete quality problem is found affecting structural safety, formwork removal shall be suspended for removing such problem.

⑤已拆除模板及其支架结构的混凝土，应在其强度达到设计强度标准值后，才允许承受全部使用荷载。

⑤ The concrete structure with the formwork and its support structure removed shall not bear all service loads until its strength reaches the standard value of design strength.

4.1.3 模板工程施工质量检查与验收

4.1.3 Construction Quality Inspection and Acceptance of Formwork

在浇筑混凝土之前，应对模板工程施工质量进行检查验收。模板及其支架应具有足够的承载能力、刚度和稳定性，能可靠地承受浇筑混凝土的重量、侧压力以及施工荷载。模板安装和浇筑混凝土时，应对模板及其支架进行观察和维护。发生异常情况时，应按施工技术方案及时进行处理。

Formwork construction quality shall be inspected and accepted before concrete pouring. Formwork and its support shall have sufficient bearing capacity, rigidity and stability to reliably bear the cast concrete weight, lateral pressure and construction load. Formwork and its support shall be observed and maintained in the process of formwork erection and concrete pouring. Any abnormality arising shall be handled in time according to the construction technical scheme.

模板工程的施工质量检验应分主控项目和一般项目。

The construction quality inspection of formwork shall be divided into critical items and general items.

检验批合格质量应符合下列规定：主控项目的质量经抽样检验合格；一般项目的质量经抽样检验合格；当采用计数检验时，除有专门要求外，一般项目的合格率应达到80%及以上，且不得有严重缺陷；具有完整的施工操作依据和质量验收记录。

The acceptable quality of the inspection lot shall meet the following requirements: The quality of the critical items and general items is accepted based on the sampling inspection; when an inspection by attributes is adopted, except for special requirements, the pass rate of general items shall reach 80% or above, and any serious defect is unacceptable; the inspection lot shall be accompanied with complete construction operation basis and quality acceptance records.

（1）主控项目

(1) Critical item

①安装现浇结构的上层模板及其支架时，下层楼板应具有承受上层荷载的承载能力，或加设支架；上、下层支架的立柱应对准，并铺设垫板。

① When erecting the upper formwork and

support of a floor in a cast-in-place structure, the lower floor shall be ensured in sufficient capacity to bear the load from the upper floor, or additional supports shall be erected; the pillars used for supports shall be aligned between the two floors and shall be laid with backing plates.

检查数量：全数检查。

Inspection quantity: All.

检验方法：对照模板设计文件和施工技术方案观察。

Inspection method: Observe against the formwork design documents and construction technical scheme.

②在涂刷模板隔离剂时，不得玷污钢筋和混凝土接槎处。

② The joint between reinforcement and concrete shall not be stained by the brushed formwork release agent.

检查数量：全数检查。

Inspection quantity: All.

检验方法：观察。

Inspection method: Visual inspection.

③底模及其支架拆除时的混凝土强度应符合规范要求。

③ The bottom form and its supports shall not be removed until the concrete strength reaches the specified value.

检查数量：全数检查。

Inspection quantity: All.

检验方法：检查同条件养护试件强度试验报告。后浇带模板的拆除和支顶应按施工技术方案执行。

Inspection method: Check the strength test report of the specimen cured under the same condition. The post-cast strip formwork shall be removed and jacked as per the construction

technical scheme.

检查数量：全数检查。

Inspection quantity: All.

检验方法：观察。

Inspection method: Visual inspection.

（2）一般项目

(2) General item

①模板安装应满足下列要求：

① Formwork shall be erected in compliance with the following requirements:

a. 模板的接缝处不应漏浆；在浇筑混凝土前，木模板应浇水湿润，但模板内不应有积水。

a. Formwork joints shall be free from mortar leakage; before pouring concrete, wooden formwork shall be wettened by watering and be free of ponding inside.

b. 模板与混凝土的接触面应清理干净并涂刷隔离剂，但不得采用影响结构性能或妨碍装饰工程施工的隔离剂。

b. The contact surface between formwork and concrete shall be cleared up and applied with the release agent without any impact on the structural performance or impedance to the finishing and fitment construction.

c. 浇筑混凝土前，模板内的杂物应清理干净。

c. Before concrete pouring, foreign matters within formwork shall be cleared.

d. 对清水混凝土工程及装饰混凝土工程，应使用能达到设计效果的模板。

d. Formwork used for fair-faced concrete and decorative concrete shall reach the design effect.

检查数量：全数检查。

Inspection quantity: All.

检验方法：观察。

Inspection method: Visual inspection.

②用作模板的地坪、胎模等应平整光洁，不得产生影响构件质量的下沉、裂缝、起砂或起鼓。

② The floor and moulding bed used as formwork shall be flat and smooth, and shall be free of sinking, cracking, dusting or bulging that may affect the quality of members.

检查数量：全数检查。

Inspection quantity: All.

检验方法：观察。

Inspection method: Visual inspection.

③对跨度大于 4 m 的现浇钢筋混凝土梁、板，其模板应按设计要求起拱；当设计无具体要求时，起拱高度宜为跨度的 1/1 000 ～ 3/1 000.

③ For the cast-in-situ reinforced concrete beams and slabs greater than 4 m in span, their formwork shall be deflected upwards in accordance with the design requirements; if there is no specific requirement in the design, the upward deflection height should be 1/1 000-3/1 000 of the span.

检查数量：在同一检验批内，梁，应抽查构件数量的 10%，且不少于 3 件；板，应按有代表性的自然间抽查 10%，且不少于 3 间；大

空间结构，板可按纵、横轴线划分检查面，抽查 10%，且不少于 3 面。

Inspection quantity: In the same inspection lot, 10% of the number of members and not less than 3 pieces shall be randomly inspected for beams; 10% of the representative natural rooms and not less than 3 rooms shall be randomly inspected for slabs; 10% of the inspection faces divided along the longitudinal and transverse axes and not less than 3 faces shall be randomly inspected for slabs in large space structures.

检验方法：水准仪或拉线、钢尺检查。

Inspection method: Inspection with a level, wire-pulling tool, or steel rule.

④固定在模板上的预埋件、预留孔和预留洞均不得遗漏，且应安装牢固，其偏差应符合表 4-2 的规定。现浇结构模板安装的偏差及检查方法应符合表 4-3 的规定。

④ The embedded parts and provided openings and holes shall be fixed firmly on the formwork with omission and the deviation meet the provisions of Table 4-2. The deviation and inspection method of formwork erection for the cast-in-situ structures shall comply with the provisions of Table 4-3.

表 4-2　预埋件和预留孔洞的允许偏差

Table 4-2 Allowable Deviations of Embedded and Provided Openings and Holes

项目 Item		允许偏差/mm Allowable Deviation/mm
预埋钢板中心线位置 Centerline position of embedded steel plates		3
预埋管、预留孔中心线位置 Centerline position of embedded pipes and provided openings		3
插筋 Steel dowel	中心线位置 Centerline position	5
	外露长度 Exposed length	+10, 0

项目 Item		允许偏差/mm Allowable Deviation/mm
预埋螺栓 Embedded bolt	中心线位置 Centerline position	2
	外露长度 Exposed length	+10, 0
预留孔 Provided opening	中心线位置 Centerline position	10
	尺寸 Dimension	+10, 0

注：检查中心线位置时，应沿纵、横两个方向量测，并取其中的较大值。

Note: When checking the centerline position, measurement shall be conducted longitudinally and transversely and the larger value shall be taken.

检查数量：在同一检验批内，对梁、柱和独立基础,应抽查构件数量的10%,且不少于3件;对墙和板，应按有代表性的自然间抽查10%，且不少于3间；对大空间结构，墙可按相邻轴线间高度5 m 左右划分检查面，板可按纵横轴线划分检查面，抽查10%，且均不少于3面。

Inspection quantity: In the same inspection lot, 10% of the number of members and not less than 3 pieces shall be randomly inspected for beams, columns and independent foundations; 10% of the representative natural rooms and not less than 3 rooms shall be randomly inspected for walls and slabs; 10% of the inspection faces divided at about 5 m in height between adjacent axes for walls and the inspection faces divided along the longitudinal and transverse axes for slabs and both not less than 3 faces shall be randomly inspected for large space structures.

检验方法：钢尺检查。

Inspection method: Inspection with a steel rule.

表 4-3　现浇结构模板安装的允许偏差及检验方法

Table 4-3 Allowable Deviations and Inspection Methods for Formwork Erection of Cast-in-situ Structures

项目 Item		允许偏差/mm Allowable Deviation/mm	检验方法 Inspection Method
轴线位置 Axis position		5	钢尺检查 Inspection with a steel rule
底模上表面标高 Elevation of upper surface of bottom form		±5	水准仪或拉线、钢尺检查 Inspection with a level wire-pulling tool, or steel rule
截面内部尺寸 Internal dimension of section	基础 Base	±10	钢尺检查 Inspection with a steel rule
	柱、墙、梁 Column, wall, beam	+4, −5	钢尺检查 Inspection with a steel rule
层高垂直度 Verticality in story height	≤5 m	6	经纬仪或吊线、钢尺检查 Inspection with a theodolite, plumbline, or steel rule
	>5 m	8	
相邻两板表面高低差 Surface level difference of adjacent slabs		2	钢尺检查 Inspection with a steel rule

项目 Item	允许偏差/mm Allowable Deviation/mm	检验方法 Inspection Method
表面平整度 Surface roughness	5	2 m靠尺和塞尺检查 Inspection with a 2 m guiding rule and feeler gauge

注：检查轴线位置时，应沿纵、横两个方向量测，并取其中的较大值。

Note: When checking the axis position, measurement shall be conducted longitudinally and transversely and the larger value shall be taken.

⑤侧模拆除时的混凝土强度应能保证其表面及棱角不受损伤。模板拆除时，不应对楼层形成冲击荷载。拆除的模板和支架宜分散堆放并及时清运。

⑤ Side forms shall not be removed until the concrete reaches the design strength so that the surface, edges and corners are not vulnerable to damage. Formwork shall be removed without any impact load on floors. The removed forms and supports shall be stacked in an uncentralized manner and shall be cleared and transported in time.

检查数量：全数检查。

Inspection quantity: All.

检验方法：观察。

Inspection method: Visual inspection.

4.2　钢筋工程

4.2 Steel Reinforcement(Bar) Works

4.2.1 钢筋加工

4.2.1 Processing of Steel Reinforcements

钢筋的加工包括冷拉、调直、除锈、切断、接长、弯曲等工作。

The processing of steel reinforcements includes cold stretching, straightening, derusting, cutting, lengthening and bending.

钢筋一般在钢筋车间或加工棚加工，然后运至现场安装或绑扎，冷拉和冷拔是钢筋常用的冷加工方法。

Steel reinforcements shall be generally processed in the steel reinforcement workshop, and then transported to the site for placement or binding. Cold stretching and cold drawing are the commonly used cold processing methods for steel reinforcements.

钢筋冷拉和冷拔，可以提高钢筋的强度设计值，节约钢材和满足预应力筋的需要。

Cold stretching and cold drawing can improve the design strength of steel reinforcements, save steel and meet the requirements of prestressed reinforcement.

4.2.2 钢筋连接

4.2.2 Connection of Steel Reinforcements

钢筋的连接方式可分为焊接连接、绑扎连接和机械连接。

Steel reinforcements are connected by three modes: binding, welding and mechanical connection.

焊接连接：成本较低，质量可靠，应优先采用。

Welding: It is preferred due to low cost and reliable quality.

绑扎连接：浪费钢筋且连接不可靠。

Binding: It's a waste of steel reinforcements and unreliable.

机械连接：设备要求简单，操作方便，但成本高。

Mechanical connection: This method has simple requirements for equipement, convenient operation, but high cost.

（1）焊接连接

(1) Welding

焊接是受力钢筋之间通过熔融金属直接传力。若焊接质量可靠，则不存在强度、刚度、恢复性能、破坏形态等方面的缺陷，是十分理想的连接形式，且其价格也远低于机械连接。

Welding works to directly transmit force between stressed bars by the action of metal melting. Reliable welding quality can eliminate defects in strength, rigidity, recovery performance, failure behavior, etc., so welding is a very ideal connection mode, at a cost far lower than that of mechanical connection.

①焊工必须持证操作，施焊前应进行现场条件下的焊接工艺试验，试验合格后，方可正式施焊。

① Welding operation shall only be conducted by welders with a qualification certificate. Formal welding shall not be carried out until the welding procedure qualification test is conducted under the site conditions and the procedure is proven.

②根据钢筋品种、规格、位置选定适宜的焊接工艺。

② An appropriate welding procedure shall be selected according to the reinforcement type,

specification and service position.

③钢筋焊接施工之前，应清除钢筋或钢板焊接部位和与电极接触的钢筋表面上的锈斑、油污、杂物等；钢筋端部若有弯折、扭曲时，应予以矫直或切除。

③ Before reinforcement welding, it is necessary to remove the rust spots, oil stains and foreign matters at the welding parts of the steel bars or the steel plates and on the reinforcement surface in contact with the electrode. The steel bars bent or twisted at the ends shall be straightened or processed by cutting.

④焊剂应存放在干燥的库房内，受潮时，使用前应经 250 ~ 300℃ 的高温烘焙 2 h。

④ Soldering flux shall be stored in a dry warehouse. Any damped flux shall be baked at 250-300°C for 2 h before use.

⑤焊机应经常维护保养和定期检修，确保正常使用。

⑤ The welding machine shall be regularly maintained and overhauled to ensure normal operation.

⑥带肋钢筋进行闪光对焊、电弧焊、电渣压力焊和气压焊时，宜将纵肋对纵肋安放和焊接。

⑥ Ribbed steel bars shall be connected by flash butt welding, electric arc welding, electroslag pressure welding and gas pressure welding, with the longitudinal ribs aligned.

（2）绑扎连接

(2) Binding

绑扎连接施工简便，但当钢筋较粗时，绑扎连接容易产生较宽的裂缝。

Binding is easy-to-operate. However, this connection mode is prone to resulting in wide cracks at the lap joints of steel bars in large diameters.

钢筋绑扎安装前，应先熟悉施工图纸，核

对钢筋配料单和料牌，研究钢筋安装与有关工种配合的顺序，准备绑扎用的铁丝、绑扎工具、绑扎架等。钢筋绑扎一般用 18~22 号铁丝，其中 22 号铁丝只用于绑扎直径 12 mm 以下的钢筋。

Before reinforcement binding and placement, it is necessary to get familiar with the working drawings, check the reinforcement detailing sheet and tags, study the sequence of reinforcement placement in cooperation with relevant types of work, and prepare iron wires, binding tools and binding frames. Generally, the grade 18-22 iron wires are used for reinforcement binding, of which the grade 22 iron wire is only used for binding steel bars with a diameter of less than 12 mm.

（3）机械连接

(3) Mechanical connection

钢筋机械连接是指通过连接件的机械咬合作用或钢筋端面的承压作用，将一根钢筋中的力传递至另一根钢筋的连接方法。钢筋机械连接常用套筒挤压连接、锥螺纹套筒连接和直螺纹套筒连接三种形式。

The mechanical connection of reinforcement is to transmit force from one steel bar to another through the mechanical engagement of the connector or under the compression effect at the end face of the steel bars. Steel bars are mechanically connected mainly by three means, i.e., sleeve squeezing connection, taper threaded sleeve connection and straight threaded connection.

4.2.3 钢筋配料

4.2.3 Batching of Steel Reinforcements

钢筋混凝土构件中的钢筋，由于设计及规范要求，有的需在中间弯折一定角度，有的则要求在两端做各种角度的弯钩。在加工有弯折或有弯钩的钢筋时，若仅是简单按照设计图纸中钢筋的标注尺寸（在施工中也称量度尺寸）逐段相加，并以此值作为下料长度，那么加工

成型后的钢筋长短是不合适的，造成这种情况的原因：一是弯曲后钢筋在弯曲处内皮收缩、外皮延伸，而轴线长度不变；二是在弯曲处形成圆弧。

Based on the design and specification requirements, some steel bars in reinforced concrete members need to be bent at a certain angle in the middle, and some need to be bent into hooks at various angles at both ends. For processing the steel bars with bents or hooks, if the marked dimensions of reinforcement in the design drawings(also called measured dimensions in construction) are simply summed and this sum is taken as the blanking length, the length of the processed and formed reinforcement is inappropriate, due to the following causes: the inner skin of the steel bar shrinks and the outer skin extends at the bending position, while the axis length remains unchanged; an arc is formed at the bend.

常用钢筋下料长度计算方法：

Calculation methods of common reinforcement blanking lengths:

直钢筋下料长度 = 构件长度 − 保护层厚度 + 弯钩增加长度

The blanking length of straight steel bar = member length − thickness of concrete cover + length increase of hook

弯起钢筋下料长度 = 直段长度 + 斜段长度 + 弯钩增加长度 − 弯曲调整值

Blanking length of bent steel bar = length of straight part + length of inclined part + length increase of hook − bending adjustment value

箍筋下料长度 = 箍筋周长 + 箍筋长度调整值

Blanking length of stirrup = stirrup perimeter + stirrup length adjustment value

注：上述钢筋需要搭接的话，还应增加钢筋搭接长度。

Note: An lap length shall be included for the

above steel bars to be lapped.

4.2.4 钢筋代换

4.2.4 Replacement of Steel Reinforcements

（1）代换原则

(1) Principles of replacement

当施工中遇有钢筋的品种或规格与设计要求不符时，可参照以下原则进行钢筋代换：

When the type or specification of reinforcement is found inconsistent with the design requirements during construction, the reinforcement may be replaced according to the following principles:

①等强度代换：当构件受强度控制时，钢筋可按强度相等原则进行代换。

① Equal strength replacement: When the members are subject to strength control, the reinforcement may be replaced according to the equal strength principle.

②等面积代换：当构件按最小配筋率配筋时，钢筋可按面积相等原则进行代换。

② Equal area replacement: When the members are reinforced according to the minimum reinforcement ratio, the reinforcement may be replaced according to the equal area principle.

③当构件受裂缝宽度或挠度控制时，代换后应进行裂缝宽度或挠度验算。

③ When the members are subject to crack width or deflection control, the crack width or deflection shall be checked after replacement.

（2）钢筋代换的注意事项

(2) Precautions for steel reinforcements replacement

①对抗裂性能要求高的构件，不宜用光面钢筋代换变形钢筋。

① For the members with high requirements for crack resistance, plain reinforcement should not be used to replace deformed reinforcement.

②钢筋代换时不宜改变构件的有效计算高度（h_0，单排筋改双排筋）。

② The effective calculated height of members should not be changed after reinforcement replacement (h_0, a single row of reinforcement changed to double rows of reinforcement).

③梁的纵向受力钢筋与弯起钢筋应分别代换，以保证正截面与斜截面强度。

③ The longitudinally stressed reinforcement and bent reinforcement of beams shall be replaced separately to ensure the strength of the normal section and oblique section.

④除满足强度外，还应满足规范规定的最小配筋率、钢筋间距、根数、最小钢筋直径及锚固长度等要求。

④ In addition to the strength requirement, reinforcement replacement shall also meet the requirements regarding minimum reinforcement ratio, reinforcement spacing, number, minimum reinforcement diameter and anchorage length specified in the specification.

⑤凡属重要构件或预应力构件，代换应征得设计院同意。

⑤ Reinforcement replacement is subject to approval by the design institute for important members or prestressed members.

4.2.5 钢筋配料单及料牌的填写

4.2.5 Filling of Steel Reinforcements Detailing List and Tag

钢筋配料单是根据施工设计图纸标定钢筋的品种、规格及外形尺寸、数量进行编号，并计算下料长度，用表格形式表达的技术文件。

A reinforcement detailing sheet is used as a technical document expressed in the form of a table, filled with numbers depending on the type, specification, overall dimension and quantity of reinforcement indicated on the construction design drawings, and the calculated blanking length.

钢筋配料单的作用：钢筋配料单是确定钢筋下料加工的依据，是提出材料计划、签发施工任务单和限领料单的依据。

The function of a steel reinforcement detailing list: The basis for reinforcement cutting and processing, and the basis for proposing a material plan, issuing a construction work order and a material requisition form.

配料单的形式：钢筋配料单一般用表格的形式反映，其内容由构件名称、钢筋编号、钢筋简图、尺寸、级别、数量、下料长度及重量等内容组成。

Form of a steel reinforcement detailing list: Generally in the form of a table, composed of items including member name, reinforcement number, reinforcement sketch, size, grade, quantity, blanking length and weight.

4.2.6 钢筋的绑扎与安装

4.2.6 Binding and Placement of Steel Reinforcement

（1）施工准备

(1) Preparation

核对成品钢筋与料单与图纸是否相符，确定绑扎先后顺序及方法，确定钢筋保护层厚度。

The finished reinforcement is checked against the detailing sheet and drawings for consistency, and the binding sequence and method, and the thickness of the concrete cover are defined.

（2）钢筋绑扎的相关规定

(2) Relevant provisions on steel reinforcement binding

①轴心受拉及小偏心受拉构件的纵向受力钢筋不得采用绑扎搭接接头；当受拉钢筋的直径 $d > 25$ mm 及受压钢筋的直径 $d > 28$ mm 时不宜采用绑扎搭接接头。

① Lap joints by binding shall not be used for the longitudinally stressed reinforcement used in the members subject to axial tension and small eccentric tension; when the tensile reinforcement in a diameter $d > 25$ mm and the compression reinforcement in a diameter $d > 28$ mm, lap joints by binding should not be used.

②钢筋接头宜设置在构件受力较小处，同一纵向受力钢筋宜少设接头，接头末端至钢筋弯起点的距离不应小于钢筋直径的 10 倍。

② Reinforcement joints should be set at the part of members subject to small stress, and one piece of longitudinally stressed reinforcement should be provided with joints as few as possible. The distance from the joint end to the bending starting point of the reinforcement should not be less than 10 times the reinforcement diameter.

③同一构件中相邻纵向受力钢筋的绑扎搭接接头宜相互错开，位于同一连接区段内（钢筋搭接长度的 1.3 倍）的受拉钢筋搭接接头面积百分率：对梁类、板类及墙类构件不宜大于 25%，对柱类构件不宜大于 50%。

③ Lap joints by binding of adjacent pieces of longitudinally stressed reinforcement should be staggered in one member; the area percentage of lap joints of tensile reinforcement located in the same connection section(1.3 times the lap length of reinforcement) should be not more than 25% for beam, slab and wall members, and not more than 50% for column members.

④在梁、柱类构件的纵向受力钢筋搭接长

度范围内，应按设计要求配置箍筋。当设计无具体要求时，应符合下列规定：

④ Stirrups shall be configured in accordance with design requirements within the lap joint length range of longitudinally stressed reinforcement in beam and column members. When there are no specific design requirements, the following provisions shall be met:

a. 箍筋直径不小于搭接钢筋较大直径的25%。

a. The diameter of the stirrup not less than 25% of the larger diameter of lapped steel bars.

b. 受拉搭接区段的箍筋间距不大于搭接钢筋较小直径的 5 倍，且不大于 100 mm。

b. The stirrup spacing in the tensile lap section not greater than 5 times the smaller diameter of lapped steel bars, and not greater than 100 mm.

c. 受压搭接区段的箍筋间距不大于搭接钢筋较小直径的 10 倍，且不大于 200 mm。

c. The stirrup spacing in the compression lap section not greater than 10 times the smaller diameter of lapped steel bars, and not greater than 200 mm.

⑤同一构件中受力钢筋的机械连接接头或焊接接头宜相互错开，位于同一连接区段内（机械连接为 35d，焊接为 35d 且 ≥ 500 mm）的纵向受力钢筋接头面积百分率应符合设计要求，当设计无具体要求时，应符合下列规定：

⑤ The mechanical connection joints or welded joints of the stressed reinforcement in the same member should be staggered, and the area percentage of joints in the longitudinally stressed reinforcement located in the same connection section(35d for mechanical connection, 35d and ≥ 500 mm for welding) shall meet the design requirements. When there are no specific requirements in the design, the following requirements shall be met:

a. 在受拉区不宜大于 50%。

a. The area percentage of joints should not be more than 50% at a tensile zone.

b. 接头不宜设在有抗震要求的框架梁端、柱端的箍筋加密区；无法避开时，如果必须在加密区搭接应采用机械连接或焊接；对等强度高质量的机械连接接头不大于 50%。

b. Joints should not be arranged in the stirrup dense area at the frame beam end and column end with seismic requirements; mechanical connection or welding shall be adopted if necessary to create lap joints in the stirrup dense area; the area percentage of the mechanical connection joints with equivalent strength and high quality shall not exceed 50%.

c. 直接承受动力荷载的结构构件中，不宜采用焊接接头；当采用机械连接接头时受拉区不宜大于 50%。

c. Welded joints should not be used in the structural elements directly bearing dynamic loads; the area percentage of mechanical joints used when necessary shall not be more than 50%.

⑥钢筋接头搭接处，应在中心和两端用铁丝扎牢。绑扎接头的搭接长度应符合设计要求且不得小于规范规定的最小搭接长度（受拉钢筋 300 mm，受压钢筋 200 mm）。

⑥ Lap joints of reinforcement shall be fastened with iron wires at the center and both ends. The length of lap joints by binding shall meet the design requirements and shall not be less than the minimum lap length specified in the specification (300 mm for tensile reinforcement and 200 mm for compression reinforcement).

⑦应特别注意板上部的负筋：保证其绑扎位置准确；防止施工人员踩踏，尤其是雨篷、挑檐、阳台等悬臂板，防止其拆模后断裂垮塌。

⑦ Special attention shall be paid to the negative reinforcement on the upper part of slabs:

to ensure that the binding position is accurate; to prevent the construction personnel from treading on it, especially cantilever slabs such as awnings, cornices and balconies, in case of breaking and collapse after formwork removal.

⑧钢筋在混凝土中的保护层厚度，可用水泥砂浆垫块（限制和淘汰）或塑料卡（推荐使用）垫在钢筋与模板之间进行控制，垫块应布置成梅花形，其相互间距不大于 1 m。上下双层钢筋之间的尺寸可用绑扎短钢筋来控制。

⑧ The thickness of the concrete cover of reinforcement may be controlled by placing cement mortar spacers(restricted and eliminated) or plastic clips(recommended) between reinforcement and formwork, in a quincunx shape and at a spacing of not more than 1 m. The double-layer reinforcement size between the upper and lower layers may be controlled by binding short steel bars.

⑨板、次梁与主梁交叉处，板的钢筋在上，次梁钢筋居中，主梁钢筋在下。当有圈梁、垫梁时，主梁钢筋在上。

⑨ At the intersection position of slabs, secondary beams and main beams, the slab steel bars shall be at the upper layer, the secondary beam steel bars shall be at the middle layer, and the main beam steel bars shall be at the lower layer. The main beam steel bars shall be at the upper layer at the intersection with ring beams or bearing beams.

4.2.7 钢筋隐蔽工程的验收
4.2.7 Acceptance of Steel Reinforcement Concealed Works

钢筋工程属隐蔽工程，浇筑混凝土前应组织对钢筋和预埋件进行验收，并做好隐蔽工程记录，相关各方签字确认，以备查证。

As a part of concealed works, steel reinforcements and embedded parts shall be inspected before concrete pouring, and records of concealed works shall be properly made and signed by relevant parties for verification.

4.3   混凝土工程
4.3 Concrete Works

混凝土工程施工是形成钢筋混凝土整体结构的最后一个施工环节，也是钢筋混凝土结构工程施工中一项重要的工种工程。混凝土的强度、耐久性、抗渗性等质量指标均与混凝土工程施工中的每一个工程过程直接相关。混凝土工程施工的过程包括：原材料的进场存放与检验，配合比的试配与最后确定，施工方案的编订与审批，混凝土的拌合与输送，新拌混凝土的浇筑与振捣密实，混凝土构件的养护与质量验收，混凝土结构的成品保护和缺陷修补等。

Concrete engineering construction is the last procedure and also an important type of work to build an integral reinforced concrete structure. Quality indicators such as concrete strength, durability and impermeability are directly related to each step in concrete engineering construction. The steps of concrete engineering construction include storage and inspection of incoming raw materials on the construction site, trial mixing and final selection of mix proportion, preparation and approval of a construction scheme, mixing and transport of concrete, pouring and vibrated dense of fresh concrete, curing and quality acceptance of concrete members, finished product protection and defect repair of concrete structures.

4.3.1 混凝土浇筑

4.3.1 Concreting

混凝土浇筑前，应对模板、钢筋、支架和预埋件进行检查并填写隐蔽工程质量验收记录。

Before concrete pouring, the formwork, reinforcement, support and embedded parts shall be inspected and the quality acceptance records of concealed works shall be completed.

（1）混凝土浇筑的一般规定

(1) General provisions for concrete pouring

①混凝土浇筑前不应发生离析或初凝现象，如已发生，须重新搅拌。混凝土运至现场后，其坍落度应满足表 4-4 的要求。

① Concrete shall be free of segregation or initial setting before pouring, and re-mixing must be conducted in case of any occurrence. The concrete transported to the site shall meet the slump requirements as shown in Table 4-4.

表 4-4　混凝土浇筑时的坍落度
Table 4-4 Slump During Concrete Pouring

序号 S/N	结构种类 Structure type	坍落度（mm） Slump (mm)
1	基础或地面等的垫层、无配筋的厚大结构（挡土墙、基础或厚大的块体）或配筋稀疏的结构 Bedding of foundation or ground, etc., thick and large structures without reinforcement(retaining wall, foundation or massive block) or structures with sparse reinforcements	10 ~ 30
2	板、梁及大、中型截面的柱子等 Slabs, beams and columns of large and medium-sized sections, etc.	30 ~ 60
3	配筋密列的结构（薄壁、斗仓、筒仓、细柱等） Structures with dense reinforcements(thin wall, hopper, silo, slender column, etc.)	50 ~ 70
4	配筋特密的结构 Structures with extreme dense reinforcements	70 ~ 90

②混凝土自高处倾落时，其自由倾落高度不宜超过 2 m，在竖向结构中浇筑混凝土的高度不得超过 3 m，否则应设串筒、溜槽、溜管或振动溜管等。

② Concrete shall be poured from a high place at a free-falling height not exceeding 2 m, and concrete shall be poured in a vertical structure of a height not exceeding 3 m, otherwise tumbling barrels, chutes, tubes or vibrating chutes shall be used for pouring.

③浇筑混凝土时应经常观察模板、支架、钢筋、预埋件和预留孔洞的情况。当发现有变形、移位时，应立即停止浇筑，并在已浇筑混凝土凝结前修整完好。

③ The formwork, supports, steel bars, embedded parts, and provided openings and holes shall be regularly inspected while pouring concrete. In case any deformation or displacement is found, pouring should be suspended immediately and the defect should be rectified before the poured concrete is set.

④混凝土的浇筑应分段、分层连续进行，

随浇随捣。上层混凝土应在下层混凝土初凝前浇筑完毕。

④ Concrete shall be poured continuously in sections and layers, and shall be tamped while pouring. The upper layer of concrete shall be poured before the initial setting of the lower layer of concrete.

⑤浇筑竖向结构混凝土前，底部应先填以50～100 mm厚，与混凝土成分相同的水泥砂浆。

⑤ Before concrete pouring of a vertical structure, the bottom shall be filled with cement mortar with a thickness of 50-100 mm and the same composition as the concrete.

⑥混凝土在浇筑及静置过程中，应采取措施防止产生裂缝。混凝土因沉降及干缩产生的非结构性的表面裂缝，应在混凝土终凝前予以修整。在浇筑与柱和墙连成整体的梁和板时，应在柱和墙浇筑完毕后停歇1～1.5 h，使混凝土获得初步沉实后，再继续浇筑，以防止接缝处出现裂缝。

⑥ Measures shall be taken to prevent cracking during concrete pouring and setting. The non-structural surface cracks of concrete due to settlement and dry shrinkage shall be rectified before the concrete is finally set. For concrete pouring of beams and slabs that are connected with columns and walls, a pause for 1-1.5 h shall be given after the concrete pouring of columns and walls is finished until the concrete is initially set, and then concrete pouring is continued so as to prevent cracking at the joints.

（2）施工缝的留设与处理

(2) Setting-up and treatment of construction joints

如果由于技术或施工组织上的原因，不能对混凝土结构一次连续浇筑完毕，并且停歇时间已超过混凝土的初凝时间，致使混凝土已初凝；当继续浇混凝土时，形成了接缝，即为施工缝。

A construction joint is formed when pouring is restored after the concrete initially set after a suspension longer than the initial setting time under the circumstance that the concrete structure cannot be poured continuously in one shot due to technical or construction organization reasons.

①施工缝的留设位置。

① Setting-up position for construction joint.

施工缝一般宜留在结构剪力较小且便于施工的部位。具体留设方法如下：

Generally, construction joints shall be set up at the parts subject to small structural shear and convenient for construction. The detailed setting-up methods are as follows:

a. 柱子上的施工缝留置在基础的顶面、梁或吊车梁牛腿的下面、吊车梁的上面、无梁楼板柱帽的下面，如图4-8所示。

a. For columns, construction joints are set up on the top surface of the foundation, under the bracket of beams or crane beams, over the crane beams and under the column cap of beamless floor slabs, as shown in Figure 4-8.

b. 和板连成整体的大截面梁，施工缝留置在板底面以下20～30 mm处。当板下有梁托时，留在梁托下部。

b. For the large-section beam connected with slabs, the construction joints are set up at 20-30 mm under the bottom surface of the slab. When there is a corbel under the slab, it is set up under the beam bracket.

c. 单向板，施工缝留置在平行于板的短边的任何位置。

c. For one-way slabs, the construction joints are set up at any position parallel to the short side of the slabs.

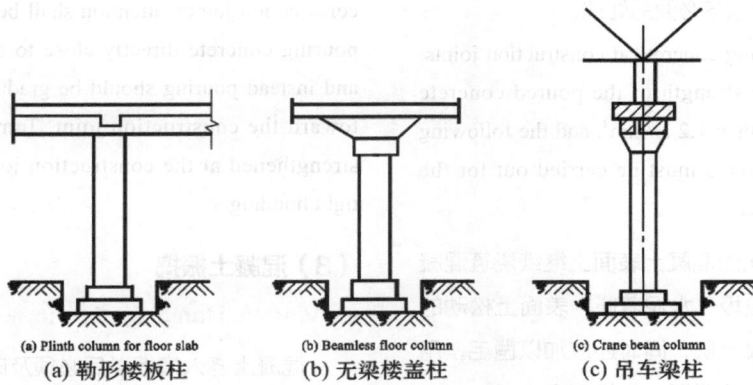

(a) Plinth column for floor slab　　(b) Beamless floor column　　(c) Crane beam column
(a) 勒形楼板柱　　　　　　　　(b) 无梁楼盖柱　　　　　　　　(c) 吊车梁柱

图4-8　柱施工缝位置

Figure 4-8 Positions of Column Construction Joints

d. 有主次梁的楼板，施工缝宜顺着次梁方向浇筑，施工缝应留置在次梁跨度的中间三分之一范围内，如图 4-9 所示。

d. For the floor slabs with main and secondary beams, pouring shall be conducted along the secondary beam direction, and the construction joints shall be set up within the range of one-third in the middle of the secondary beam span, as shown in Figure 4-9.

e. 双向受力楼板、大体积混凝土结构、拱、弯拱、薄壳、蓄水池、斗仓、多层钢架及其他结构复杂的工程，施工缝的位置应按设计要求留置。单向板的施工缝可留在平行于板的短边的任何位置。

e. For two-way stressed floor slabs, mass concrete structures, arches, camber arches, thin shells, reservoirs, hoppers, multi-storey steel frames and other works with complex structures, the construction joints shall be positioned according to the design requirements. The construction joints of one-way slabs shall be set up at any position parallel to the short side of the slabs.

f. 承受动力作用的设备基础，不应留置施工缝；当必须留置时，应征得设计单位同意。

f. Construction joints are generally not

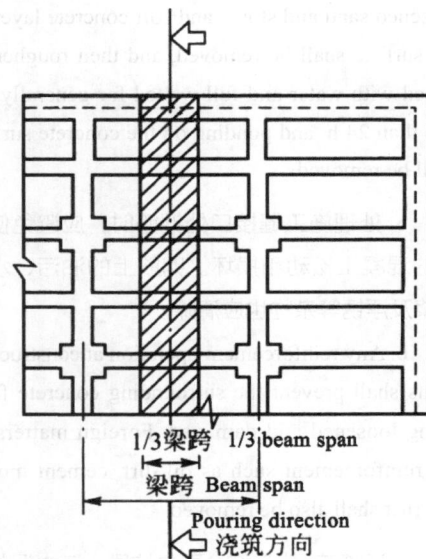

1/3梁跨　1/3 beam span
梁跨　Beam span
Pouring direction
浇筑方向

图4-9　有主次梁楼盖的施工缝位置

Figure 4-9 Positions of Construction Joints in Floors with Main and Secondary Beams

allowed for the equipment foundation under dynamic action; shall only be set up when necessary and subject to approval by the designer.

②施工缝的处理。

② Treatment of construction joint.

在施工缝处继续浇筑混凝土时，已浇筑的混凝土抗压强度不应小于 1.2 N/mm²。同时，必须对施工缝进行以下必要的处理。

When pouring concrete at construction joints, the compressive strength of the poured concrete shall not be less than 1.2 N/mm^2, and the following necessary treatments must be carried out for the construction joints.

a. 在已硬化的混凝土表面上继续浇筑混凝土前，应清除垃圾、水泥薄膜、表面上松动的砂石和软弱混凝土层，同时还应加以凿毛，用水冲洗干净并充分湿润，一般不宜少于 24 h，残留在混凝土表面的积水应予以清除。

a. Before continuing pouring concrete on the hardened concrete surface, waste, cement film, loosened sand and stone, and soft concrete layer on the surface shall be removed, and then roughened, rinsed with water and full wetted for generally not less than 24 h, and ponding on the concrete surface shall be removed.

b. 处理施工缝接口处钢筋时，应避免使周围的混凝土松动和损坏。钢筋上的油污、水泥砂浆及浮锈等杂物也应清除。

b. Any reinforcement operation at construction joints shall prevent the surrounding concrete from being loosened and damaged. Foreign matters on the reinforcement such as oil dirt, cement mortar and rust shall also be removed.

c. 浇筑前，应在结合面先抹刷一道水泥浆，或铺上一层 10 ~ 15 mm 厚的水泥砂浆，其配合比与混凝土内的砂浆成分相同。

c. Before pouring, a pass of cement slurry or cement mortar with a thickness of 10-15 mm shall be plastered on the joint face, with the same mix proportion as that of the mortar in the concrete.

d. 从施工缝处开始继续浇筑时，要注意避免直接靠近缝边浇筑混凝土，宜向施工缝处逐渐推进，应加强对施工缝接缝的捣实工作，使其与混凝土紧密结合。

d. When pouring is continued from construction joints, attention shall be paid to avoid pouring concrete directly close to the joint edge, and instead pouring should be gradually advanced toward the construction joint. Tamping shall be strengthened at the construction joint to achieve tight bonding.

（3）混凝土振捣

(3) Vibration tamping of concrete

混凝土浇入模板以后必须及时进行振捣密实，因为此时混凝土里面存在着许多空隙与气泡，同时混凝土也没有充满、填实整个模板。而混凝土结构的强度、抗冻性、抗渗性及耐久性等，都与混凝土的密实程度密切相关，所以对浇入模板的混凝土及时进行振捣密实是确保混凝土工程质量的一项关键工序。

Timely vibration tamping is necessary for the concrete poured into the formwork, since the newly poured concrete contains voids and bubbles, and the formwork is not filled with such concrete. In addition, the strength, frost resistance, impermeability and durability of concrete structures are closely related to the compactness of poured concrete, so timely vibration tamping of the concrete poured into the formwork is a key step to ensure the quality of concrete works.

混凝土的振捣方式分为人工振捣和机械振捣两种。人工振捣是利用捣锤或插钎等工具的冲击力来使混凝土密实成型，其效率低、效果差。机械振捣是将振动器的振动力传给混凝土，使之发生强迫振动破坏水泥浆的凝胶结构，降低了水泥浆的黏度和骨料之间的摩擦力，提高了拌和物的流动性，使混凝土密实成型，效率高、密度大、质量好。混凝土振捣机械按其工作方式分为内部振动器、表面振动器、外部振动器和振动台等，如图 4-10 所示。

Concrete is vibrated manually or mechanically. Manual vibration is to leverage

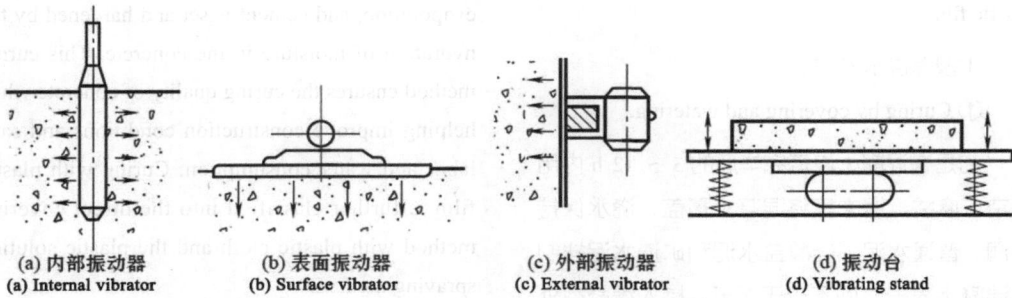

| (a) 内部振动器 | (b) 表面振动器 | (c) 外部振动器 | (d) 振动台 |
| (a) Internal vibrator | (b) Surface vibrator | (c) External vibrator | (d) Vibrating stand |

图4-10 振动机械示意图

Figure 4-10 Schematic Diagram of Vibrator

the impact force of a tamping hammer or rod to compact and shape the concrete, with low efficiency and poor effect. Mechanical vibration is to transmit the vibrating force of a vibrator to the concrete, so that forced vibration will damage the gel structure of the cement slurry, reduce the viscosity of the cement slurry and the friction between the aggregates, improve the fluidity of the mixture, until the concrete is compacted and shaped. With high efficiency, high density and good quality, machinery concrete vibrators are divided into internal vibrators, surface vibrators, external vibrators and vibrating stands according to their working modes, as shown in Figure 4-10.

4.3.2 混凝土养护

4.3.2 Concrete Curing

混凝土浇筑捣实后，逐渐凝固硬化，这个过程主要由水泥的水化作用来实现，而水化作用必须在适当的温度和湿度条件下才能实现。因此，为了保证混凝土有适宜的硬化条件，使其强度不断增长，必须对混凝土进行养护。混凝土养护方法分自然养护和人工养护。

The concrete poured and tamped is gradually set and hardened, which is mainly realized by the hydration effect of cement under appropriate temperature and humidity conditions. Therefore, in order to ensure that the concrete is hardened with increasing strength under suitable conditions, concrete curing is a necessary step. Concrete is cured naturally and manually.

（1）自然养护

(1) Natural curing

自然养护是指在自然气温条件下（平均气温高于 5℃），用适当的材料对混凝土表面进行覆盖、浇水、保温等养护措施，使水泥的水化作用在所需的适当温度和湿度条件下顺利进行。自然养护又可分为覆盖浇水养护和塑料薄膜养护两种。

Under the natural curing mode, concrete surface is covered with an appropriate material, watered, thermally insulated and other curing measures applied under atmospheric temperature conditions(average temperature higher than 5℃), so that the hydration of cement can work normally under the appropriate temperature and humidity conditions. Natural curing is further classified into curing by covering and watering and curing with plastic film.

①覆盖浇水养护。

① Curing by covering and watering.

它是在混凝土浇筑完毕后的 3 ~ 12 h 内用草帘、麻袋、锯末等将混凝土覆盖，浇水保持湿润。普通水泥、硅酸盐水泥和矿渣水泥拌制的混凝土养护时间不少于 7 天，掺入缓凝型外加剂和抗渗混凝土养护时间不少于 14 天。

The concrete is covered with straw mats, sacks, sawdust, etc., within 3-12 hours after the end of concrete pouring, and is watered to keep it moist. The curing duration of the concrete mixed with ordinary cement, portland cement and slag cement shall not be less than 7 days, and that of the concrete mixed with set retarding-type admixtures and impermeable concrete shall not be less than 14 days.

对于地坪、楼面板、屋面板等大面积结构可采用蓄水养护方法；对于贮水池一类工程可在拆除内模后采取注水养护方法；对于地下基础工程可采取覆土养护方法。

For large-area structures such as floor, floorboard and roof slabs, the water storage curing method may be adopted; for works like water reservoirs, water may be injected for curing after the internal formwork is removed; for underground foundation works, curing by soil covering may be adopted.

②塑料薄膜养护。

② Curing with plastic film.

该方法以塑料薄膜为覆盖物，使混凝土与空气隔绝，水分不再蒸发，水泥靠混凝土中的水分完成水化作用而凝结硬化。这种方法能够改善施工条件，节省人工，节约用水，保证混凝土的养护质量。塑料薄膜养护可分为塑料布直接覆盖法和喷涂薄膜养生液法。

Concrete is covered with plastic film for isolation from the air, thus stopping moisture from

evaporation, and cement is set and hardened by the hydration of moisture in the concrete. This curing method ensures the curing quality of concrete while helping improve construction conditions and save labor and water consumption. Curing with plastic film is further classified into the direct covering method with plastic cloth and the plastic solution spraying method.

（2）人工养护

(2) Manual curing

人工养护就是人为控制混凝土的养护温度和湿度，使混凝土强度增长，如蒸汽养护、热水养护、太阳能养护等。人工养护主要用来养护预制构件，现浇构件大多用自然养护。

Under the manual curing mode, the curing temperature and humidity of concrete are controlled manually to increase the concrete strength, and there are steam curing, hot water curing, solar energy curing, etc. This mode is mainly adopted for curing prefabricated members, and most cast-in-situ members are naturally cured.

4.3.3 现浇整体结构混凝土施工

4.3.3 Concrete Construction of Cast-in-situ Monolithic Structure

（1）钢筋混凝土框架结构的浇筑

(1) Pouring of reinforced concrete frame structure

浇筑多层框架按分层分段施工，水平方向以结构平面的伸缩缝分段，垂直方向按结构层次分层。在每层中先浇筑柱和墙，在柱子和墙体浇捣完毕后，停歇 1 ~ 1.5 h，使柱和墙混凝土初步沉实后，再浇筑梁、板。梁、板混凝土应同时浇筑，只有较大尺寸的梁（梁的高度大于 1 m），才可先单独浇筑梁混凝土。水平施工缝设置在板下 20 ~ 30 mm 处。

Multi-storey frames shall be poured by

sections horizontally at the expansion joints of the structural planar layout, and by layers vertically according to the structural floors. Columns and walls shall be poured and vibrated first on each floor, and are kept standstill for 1-1.5 h so that the concrete is initially set before proceeding with the beam and slab pouring. Concrete shall be poured simultaneously for beams and slabs. Only beams with large dimensions(a height greater than 1 m) may be poured separately first. Horizontal construction joints shall be set 20-30 mm under the slab.

柱子浇筑宜在梁、板模板安装后，钢筋未绑扎前进行，浇筑排柱的顺序应从两端同时开始，向中间推进，以免浇筑混凝土后由于模板吸水膨胀、断面增大而产生横向推力，最后使柱发生弯曲变形。

Column pouring should be carried out after the erection of beam and slab formwork and before the binding of reinforcement. Pouring of a row of columns should be started from both ends at the same time and advanced toward the middle, so as to avoid lateral thrust due to expansion and section increase of the formwork after absorbing water from the concrete poured, which finally results in bending and deformation of columns.

（2）基础大体积混凝土浇筑

(2) Foundation mass concrete pouring

混凝土结构实体最小几何尺寸不小于 1 m 的大体积混凝土，或预计会因混凝土中胶凝材料水化引起的温度变化和收缩而导致有害裂缝产生的混凝土，称为大体积混凝土。

Mass concrete is regarded when the minimum physical geometric dimension for concrete structures is not less than 1 m, or it is expected to cause harmful cracks due to temperature changes and shrinkage caused by the hydration of cementitious materials in concrete.

大体积混凝土结构多为工业建筑中的设备基础及高层建筑中厚大的桩基承台或基础底板等。混凝土浇筑面和浇筑量大，整体性要求高，不能留施工缝。浇筑后水泥的水化热量大且聚集在构件内部，形成较大的内外温差，易造成混凝土表面产生收缩裂缝甚至是全截面贯通性裂缝。

Mass concrete structures are mostly used as equipment foundations in industrial buildings and large pile caps or foundation slabs in high-rise buildings. The concrete pouring surface and pouring volume are large, the integrity requirements are high, and construction joints are not allowed. The hydration heat of the poured cement is huge and accumulated inside the members, creating a large temperature difference between inside and outside, and causing shrinkage cracks or even full-section through cracks on the concrete surface.

①基础大体积混凝土浇筑方案。

① Foundation mass concrete pouring scheme.

大体积混凝土结构的浇筑方案，一般分为全面分层、分段分层和斜面分层三种，如图 4-11 所示。

There are three pouring schemes for mass concrete structures, i.e., full layering, sectional layering and sloped layering, as shown in Figure 4-11.

全面分层：第一层全面浇筑完毕，在初凝前浇筑第二层。施工时从短边开始，沿长边逐层进行。适用于平面尺寸不大的构件。

Full layering: On the fully poured first layer, the second layer shall be poured before the initial setting of the first layer. Starting from the short side, pouring shall be carried out layer by layer along the long side. It is suitable for members with small plane dimensions.

分段分层：混凝土从底层开始浇筑，进行 2 ~ 3 m 后再回头浇第二层，同样依次浇筑各层。适用于厚度不大而面积或长度较大的构件。

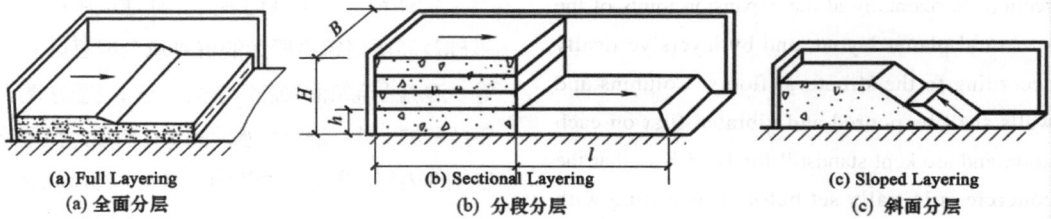

图4-11 大体积混凝土浇筑方案
Figure 4-11 Mass Concrete Pouring Scheme

Sectional layering: Concrete is poured from the bottom layer for 2-3 m before the next upper layer is poured, and such operation is repeated to finish pouring at all layers in turn. It is suitable for members with a small thickness and large area or length.

斜面分层: 浇筑工作从浇筑层的下端开始, 逐渐上移。要求斜坡坡度不大于 1/3, 适用于结构长度超过厚度 3 倍的构件。

Sloped layering: Starting from the lower end of the concrete lift, pouring is conducted gradually upward. With a slope gradient of not more than 1/3, it is suitable for the members whose structure length is more than 3 times the thickness.

②大体积混凝土早期裂缝预防措施:

② Preventive measures against early cracks of mass concrete:

a.宜选用水化热较低的水泥, 如矿渣水泥、火山灰或粉煤灰水泥。

a. Cement with low hydration heat should be used, such as slag cement, puzzolana or flyash cement.

b.掺入缓凝剂或缓凝型减水剂, 也可掺入适量粉煤灰等矿物掺和料。

b. Set-retarder or set-retarding water reducer is added, or an appropriate amount of mineral admixtures such as flyash may also be dosed.

c.降低混凝土入模温度, 可在砂、石堆场,

运输设备上搭设简易遮阳装置或覆盖草包等隔热材料, 采用低温水或冰水拌制混凝土。

c. Aggregates yards and transport equipment are sheltered by simple sunshade devices or thermal insulation materials such as straw bales, and concrete is prepared with low-temperature water or ice water, to reduce the temperature of concrete placed.

d.扩大浇筑面和散热面, 减少浇筑层厚度和降低浇筑速度, 必要时在混凝土内部埋设冷却水管, 用循环水来降低混凝土温度。

d. The pouring surface and heat dissipation surface are expanded, the concrete lift thickness and pouring rate are reduced, and if necessary, cooling water pipes are buried inside the concrete to reduce the concrete temperature with circulating water.

e.加强混凝土保温、保湿养护, 严格控制大体积混凝土的内外温差, 温差不宜超过 25℃, 故可采用草包、炉渣、砂、锯末、油布等不易透风的保温材料或蓄水养护, 以减少混凝土表面的热扩散和延缓混凝土内部水化热的降温速率（混凝土浇筑体在入模温度基础上的温升不宜大于 50℃, 每天降温不宜大于 2℃, 混凝土表面与大气温差不宜大于 20℃）。

e. Measures are taken to ensure thermal and moisture retention during the curing of concrete, by strictly controlling the temperature difference between the inside and outside of mass concrete not exceeding 25℃. Therefore, airtight thermal retention materials such as straw bales, slag, sand, sawdust and tarpaulin may be used or water storage curing may be adopted to reduce the thermal diffusion on the concrete surface and delay the cooling rate of hydration heat inside the concrete(the temperature rise of concrete from pouring into the formwork should be not greater than 50°C; the temperature should not drop more than 2°C a day, and the temperature difference between the concrete surface and the atmosphere should not be greater than 20°C).

4.3.4 混凝土工程质量检查验收

4.3.4 Quality Inspection and Acceptance of Concrete Works

混凝土质量检查包括施工中检查和施工后检查。

Concrete quality inspection includes in-process inspection and post-process inspection.

施工中检查：对混凝土拌制和浇筑过程中所用材料的质量及用量、搅拌及浇筑地点的坍落度的检查，每工作班内至少检查 2 次；对执行混凝土搅拌制度及现场振捣质量也应随时检查。

In-process inspection: The quality and dosage of materials used in the process of concrete mixing and pouring, and the slump of concrete at the mixing and pouring sites shall be inspected at least twice per shift; the implementation of the concrete mixing rules and on-site vibration tamping quality shall also be inspected frequently.

施工后检查：对已完成施工的混凝土工程进行外观质量及强度检查，有抗冻、抗渗要求的混凝土工程应进行抗冻、抗渗性能检查。

Post-process inspection: The visual quality and strength of the completed concrete shall be inspected, and the frost resistance and anti-permeability performance of the concrete with such requirements shall be inspected.

（1）混凝土外观质量检查

(1) Visual quality inspection of concrete

混凝土结构拆模后，应从外观上检查其表面有无麻面、蜂窝、孔洞、露筋、缺棱掉角、缝隙夹层等缺陷，外形尺寸是否超过规范允许偏差。

After formwork is removed, the surface of concrete structures shall be visually inspected for defects such as pits, voids, cavities, exposed reinforcement, missing edges and corners, gaps and interlayers, and whether the overall dimensions exceed the allowable deviation of the specification.

（2）混凝土的强度检验

(2) Strength inspection of concrete

混凝土的强度检验主要是抗压强度检验，它既是评定混凝土是否达到设计强度的依据，是混凝土工程验收的控制性指标，又可为结构构件的拆模、出厂、吊装、张拉、放张提供混凝土实际强度的依据。

The strength inspection of concrete mainly involves compressive strength, which is not only the basis for evaluating whether the concrete reaches the design strength as a critical index for the acceptance of concrete works, but also the basis for the actual strength of concrete for formwork removal, delivery, hoisting, tensioning and releasing of structural members.

4.3.5 混凝土工程的冬期施工

4.3.5 Concrete Works in Winter

当室外日平均气温连续 5 天稳定低于 5°C 时，应采取冬期施工措施；当室外日平均气温连续 5 天稳定高于 5°C 时，可解除冬期施工措施。当混凝土未达到受冻临界强度而气温骤降至 0°C 以下时，应按冬期施工的要求采取应急防护措施。

Winter construction measures shall be taken at an outdoor daily average temperature stable below 5°C for 5 consecutive days; shall be removed at an outdoor daily average temperature stable above 5°C for 5 consecutive days. Emergency protective measures shall be taken according to the requirements of winter construction, when the concrete construction encounters a sudden temperature drop below 0°C without reaching the critical strength in frost resistance.

混凝土工程的冬期施工应根据自然气温条件、结构类型、工期要求，拟定混凝土在硬化过程中防止早期受冻的各种措施，确定混凝土工程冬期施工的养护方法。

For winter construction of concrete works, various measures against the early freezing of concrete in the hardening process shall be proposed and the curing methods shall be determined according to the natural temperature conditions, structure types, and construction period requirements.

混凝土冬期施工的养护方法有两大类：一类是人为地创造一个正温环境，以保证新浇筑的混凝土强度能够正常地不间断地增长，甚至可以加速增长，主要方法有蓄热养护法、综合蓄热养护法、蒸汽加热养护法、电加热养护法和暖棚养护法；第二类为混凝土负温养护法，

是在拌制混凝土时加入适量的外加剂，可以降低水的冰点，使混凝土中的水在负温环境中仍保持液态，能继续与水泥进行水化作用，使得混凝土强度得以在负温环境中持续地增长。选用负温养护法后一般不再对混凝土加热。在选择混凝土冬期施工方法时，应保证混凝土尽快达到冬期施工临界强度，避免遭受冻害。

There are two main types of concrete curing methods in winter construction. One is to artificially create a positive temperature environment to ensure that the newly poured concrete can grow in strength normally and uninterruptedly, or even accelerate the strength increase, mainly by methods including the heat storage curing method, comprehensive heat storage curing method, steam heating curing method, electric heating curing method, and warm shed curing method; the other is to add an appropriate amount of admixtures in concrete preparation to reduce the freezing point of water and keep the water in the concrete in a liquid state at negative temperatures, so that hydration can be continued with cement and the concrete strength can continue to grow at negative temperatures. Concrete heating is generally unnecessary in this type. The winter construction method of concrete shall be selected to ensure that the concrete reaches the critical strength of winter construction as soon as possible to avoid freezing damage.

5

预应力混凝土工程

Prestressed Concrete Works

学习目标：

Learning objectives:

了解预应力混凝土的工作原理；了解先张法台座的类型；掌握张拉程序、张拉力控制和放张方法；掌握先张法、后张法的施工工艺；了解锚固夹具类型及张拉设备；熟悉无黏结预应力筋的施工工艺；掌握预应力混凝土工程质量检验和质量控制的方法；掌握预应力混凝土工程施工的安全技术。

Learn about the working principle of prestressed concrete; learn about the types of pre-tensioned pedestals; master the tensioning procedure, tensioning force control and release methods, the construction processes of pre-tensioning method and post-tensioning method; learn about the types of anchoring, grips and tensioning equipment; get familiar with the construction process of the unbonded prestressing tendon; master the methods of quality inspection and quality control for prestressed concrete works and the safe construction technology of prestressed concrete works.

技能抽查要求：

Skill checking requirements:

能进行预应力混凝土工程的质量检验；能编制预应力混凝土工程的施工方案。

Be able to carry out quality inspection for the prestressed concrete works and to prepare the construction schemes of prestressed concrete works.

建筑八大员岗位资格考试要求：

Qualification examination requirements for eight major posts of construction engineering:

掌握先张法、后张法（包括有黏结与无黏结）施工工艺；掌握预应力混凝土工程施工质量与安全要求。

Master the construction processes of pre-tensioning method and post-tensioning method (including bonded and unbonded) and the construction quality and safety requirements of the prestressed concrete works.

5.1 预应力混凝土及其分类

5.1 Prestressed Concrete and Its Classification

混凝土的极限拉应变很小，在正常使用条件下，若构件的受拉区开裂，刚度下降，变形较大，其使用范围将受到限制。为了控制构件的裂缝和变形，可采取加大构件的截面尺寸、增加钢筋用量以及采用高强混凝土和高强钢筋等措施。但是采用增加截面尺寸和用钢量的方法不经济，并且当荷载及跨度较大时会导致构件自重增大；如提高混凝土的强度等级，由于其抗拉强度提高得很小，对提高构件抗裂性和刚度的效果也不明显；如果提高钢筋的强度，则钢筋达到屈服强度时的抗拉强度很大，与混凝土的极限拉应变相差悬殊。

Due to extremely small ultimate tensile strain, the tensile zone of the member cracks, the stiffness decreases and the deformation is large under normal service conditions, all of which limit the application of the prestressed concrete. To control the cracking and deformation of the member, measures such as increasing the section size of the member, increasing the amount of the reinforcement, and using high-strength concrete and high-strength reinforcement can be taken. However, it is uneconomical to increase the section size or the reinforcement consumption, and the self-weight of the member will increase in case of large load and span. For example, if the strength grade of the concrete is increased, the improvement of the crack resistance and stiffness of the member is not obvious due to quite small tensile strength improvement; if the reinforcement strength is increased, the tensile strain of the reinforcement will be very large when the reinforcement reaches the yield strength, which is quite different from the ultimate tensile strain of concrete.

由此可见，在普通钢筋混凝土构件中，高强混凝土和高强钢筋是不能充分发挥作用的。

Therefore, the high-strength concrete and high-strength reinforcement cannot play a full role in the ordinary reinforced concrete members.

为了充分利用高强混凝土及高强钢筋，可以在混凝土构件受力前，在其使用时的受拉区内预先施加压力，使之产生预压应力，造成人为的应力状态。

In order to make full use of the high-strength concrete and high-strength reinforcement, pressure can be pre-applied in the tensile zone when the concrete member is put into use before such member is stressed to produce compressive pre-stress and have the member in the artificial stress state.

当构件在荷载作用下产生拉应力时，首先要抵消混凝土构件内的预压应力，然后随着荷载的增加，混凝土构件受拉并出现裂缝，此时应减小裂缝的宽度，满足使用要求。这种在构件受荷前预先对混凝土受拉区施加应力的结构称为"预应力混凝土结构"。

When the concrete member produces tensile stress under the load, the compressive pre-stress produced by the member will be offset at first, and then with the increasing load, the member will be tensioned and cracked, thus, the width of the crack can be reduced to meet the service requirements. This structure is referred to as the "prestressed concrete structure", in which the tensile zone of the concrete is stressed in advance before any load is applied on the member.

预应力混凝土构件与普通混凝土构件相比，除能提高构件的抗裂度和刚度外，还具有能增加构件的耐久性、节约材料、减少自重等优点，为建造大跨度结构创造了条件。

Compared with ordinary concrete members, prestressed concrete members can not only improve the crack resistance and stiffness of members, but also improve the durability of the members, save materials and reduce self-weight, thus allowing for the construction of long-span structures.

预应力混凝土根据其预应力施加工艺的不同，可分为先张法和后张法。

For prestressed concrete, there are pre-tensioning method and post-tensioning method according to the different prestress application processes.

5.2 先张法

5.2 Pre-tensioning Method

先张法是在浇筑混凝土之前，先张拉预应力筋，并将预应力筋临时固定在台座或钢模上，待混凝土达到一定强度（一般不低于混凝土设计强度标准值的 75%），混凝土与预应力筋具有一定的黏结力时，放松预应力筋，使混凝土在预应力筋的反弹作用下，令构件受拉区的混凝土承受预压应力，如图 5-1、图 5-2 所示。

In particular, the prestressing tendon is tensioned and fixed onto the pedestal or steel formwork temporarily. When the concrete reaches a certain strength(generally not less than 75% of the standard value of concrete design strength) and there is a certain adhesive force between the concrete and the prestressing tendon, the prestressing tendon is relaxed so that the concrete in the tensile zone of the member bears the compressive prestress due to the rebound of the prestressing tendon, as shown in Figure 5-1 and Figure 5-2.

先张法的主要优点为生产工艺简单、工序少、效率高、质量好、比较经济，适用于预制构

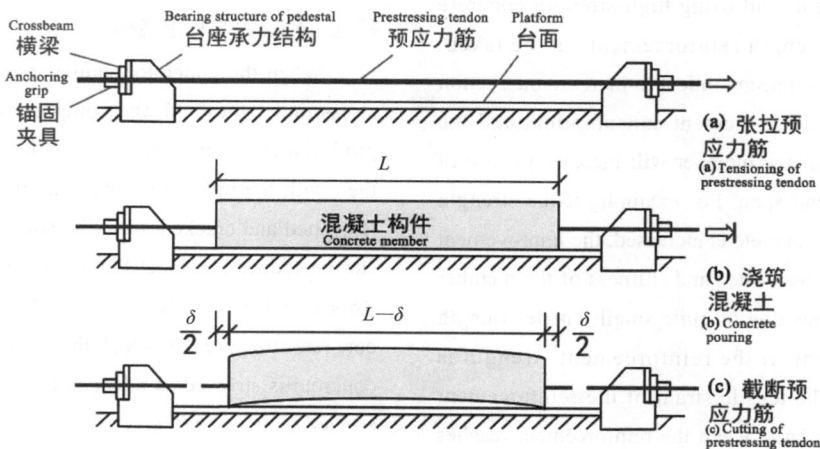

图5-1　先张法施工工艺示意图

Figure 5-1 Diagram of Pre-tensioning Construction Process

台座准备 → 刷隔离剂 → 铺放预应力筋

Preparation of pedestal — Application of release agent — Laying of prestressing tendon

张拉　Tensioning

Installation of side formwork and binding of tranverse reinforcement — 安侧模、绑扎横向筋

Concrete pouring — 浇筑混凝土 ⇨ 制作试块 — Fabrication of test block

Curing — 养护

堆放 ← 出槽 ← 脱模 ← 放松预应力筋 ← 试块试压

Stockpiling — Member lifting — Formwork removal — Relaxation of prestressing tendon — Pressure test for test block

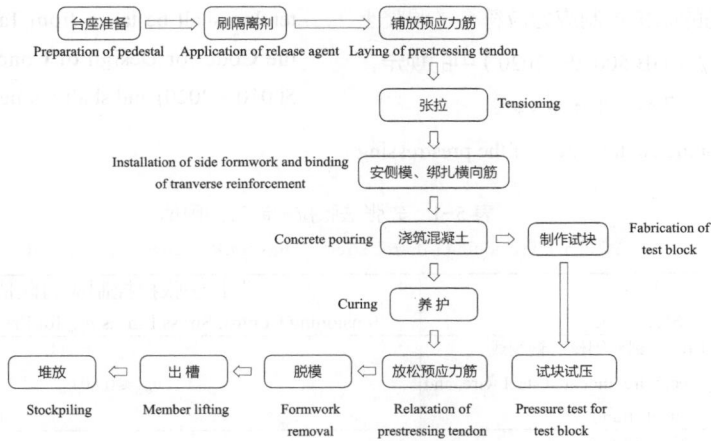

图5-2　先张法施工工艺流程简图

Figure 5-2 Schematic Diagram of Pre-tensioning Construction Process

件厂大批生产定型的中小型预应力混凝土构件。

The main advantages of pre-tensioning method include simple production process, relatively few procedures, high efficiency, good quality, relative economy, and suitable for the prefabricated member plant to fabricate the small- and medium-sized finalized prestressed concrete members on a large scale.

先张法只适用于直线预应力筋，施工时需要台座设施。

This method is only applicable to the linear prestressing tendon, and pedestals are required during construction.

5.2.1 先张法施工要点

5.2.1 Key Construction Points of Pre-tensioning Method

（1）预应力筋的铺设

(1) Laying of prestressing tendon

预应力筋铺设前先做好台面的隔离层。隔离剂不得使钢筋受污，以免影响钢筋与混凝土的黏结。预应力钢丝和钢绞线下料，应采用砂轮切割机，不得采用电弧切割。

Before laying the prestressing tendon, the isolation layer of the platform shall be prepared. The reinforcement shall not be contaminated by the release agent to avoid any impact on the bonding between the reinforcement and the concrete. The prestressed steel wire and steel strand shall be blanked by the abrasive wheel cutter instead of the arc cutting.

（2）预应力筋的张拉

(2) Tensioning of prestressing tendon

预应力筋的张拉可单根张拉或多根同时张拉。当预应力筋数量不多、张拉设备拉力有限时，常采用单根张拉。当预应力筋数量较多且密集布筋，且张拉设备拉力较大时，则可多根同时张拉。

The prestressing tendon may be tensioned one or more at a time. In the case of only a few prestressing tendons and limited tensioning force of the tensioning equipment, the prestressing tendon is usually tensioned one at a time. In the case of a large number of prestressing tendons arranged intensively and a large tensioning force of the tensioning equipment, more than one prestressed reinforcement may be tensioned at the same time.

①张拉控制应力。

① Tensioning control stress.

预应力筋的张拉控制应力应符合《混凝土结构设计规范》（GB 50010—2020）中的规定，按表 5-1 取值，且不应小于 0.4。

The tensioning control stress of the prestressing tendon shall be taken from Table 5-1 according to the Code for Design of Concrete Structures (GB 50010—2020) and shall not be less than 0.4.

表 5-1　先张法张拉控制应力限值

Table 5-1 Tensioning Control Stress Limits for Pre-tensioning Method

钢种 Steel Type	先张法张拉控制应力限值σ_{con} Tensioning Control Stress Limits σ_{con} for Pre-tensioning Method
消除应力钢丝、刻痕钢丝、钢绞线 Stress-eliminating steel wire, indented steel wire, and steel strand	$\sigma_{con} \leqslant 0.80 f_{ptk}$
热处理钢筋 Heat-treated reinforcement	$\sigma_{con} \leqslant 0.75 f_{ptk}$
冷拉钢筋 Cold-drawn reinforcement	$\sigma_{con} \leqslant 0.95 f_{pyk}$

注：σ_{con} 为预应力筋张拉控制应力；f_{ptk} 为预应力筋极限抗拉强度标准值；f_{pyk} 为预应力筋屈服强度标准值。

Note: σ_{con} is the tensioning control stress of the prestressing tendon; f_{ptk} is the standard value of ultimate tensile strength of the prestressing tendon; and f_{pyk} is the standard value of yield strength of the prestressing tendon.

②张拉程序。

② Tensioning procedures.

预应力筋的张拉程序可按下列程序之一进行：

The prestressing tendon can be tensioned according to either of the following processes:

$$0 \rightarrow 105\%\sigma_{con} \xrightarrow{\text{持荷 2 min}} \sigma_{con}；\text{或 } 0 \rightarrow 103\%\sigma_{con}$$

$$0 \rightarrow 105\%\sigma_{con} \xrightarrow{\text{Load holding for 2 min}} \sigma_{con}；\text{ or } 0 \rightarrow 103\%\sigma_{con}$$

预应力筋进行超张拉（103% → 105%σ_{con}）主要目的是减小预应力筋的松弛应力损失。

The over-tensioning (103% → 105%σ_{con}) of the prestressing tendon is mainly aimed at reducing the stress loss caused by the relaxation of such prestressing tendon.

③预应力筋伸长值与应力的测定。

③ Determination of elongation value and stress of prestressing tendon.

预应力筋张拉后，一般应校核预应力筋的伸长值。如实际伸长值与计算伸长值的偏差超过 ±6%，应暂停张拉，查明原因并采取措施予以调整后，方可继续张拉。预应力筋的实际伸长值，宜在初应力约为 10%σ_{con} 时开始测量，但必须加上初应力以下的推算伸长值。预应力筋的位置不允许有过大偏差，对设计位置的偏差不得大于 5 mm，也不得大于构件截面最短边长的 4%。

Generally, the elongation value of the prestressing tendon shall be checked after tensioning. If the difference between the actual and calculated elongation values exceeds ±6%, the tensioning shall be stopped for cause investigation. Only after appropriate adjustment measures are taken, the tensioning can be continued. The actual elongation value of the prestressing tendon should be measured when the initial stress is about 10%σ_{con}, but the calculated elongation value before such initial stress must be added. The deviation of the actual position of the prestressing tendon from the design position shall not be too large, usually not more than 5 mm and not more than 4% of the length of the shortest side of the member section.

5.2.2 混凝土的浇筑与养护

5.2.2 Concrete Pouring and Curing

（1）混凝土浇筑要求

(1) Concrete pouring requirements

预应力筋张拉完毕，应立即浇筑混凝土。混凝土应一次浇筑完，不得留施工缝。新浇筑的混凝土应振捣密实。

The concrete shall be poured immediately after the tensioning of the prestressing tendon. The pouring must be completed at a time without any construction joint left. The newly poured concrete shall be vibrated dense.

混凝土浇筑时，振动器不得碰撞预应力筋。混凝土未达到要求强度前，也不允许碰撞或踩动预应力筋。

During the vibration, the vibrator shall not collide with the prestressing tendon. It is forbidden to collide with or step on the prestressing tendon before the required concrete strength is reached.

（2）混凝土养护要求

(2) Concrete curing

预应力混凝土可采用自然养护或蒸汽湿热养护。当构件在台座上进行湿热养护时，应防止温差引起的预应力损失。先张法在台座上生产混凝土构件，其最高允许的养护温度应根据设计规定的允许温差（张拉与养护时的温度之差）计算确定。当混凝土强度达到 7.5 N/mm^2（粗钢筋配筋）或 10 N/mm^2（钢丝、钢绞线配筋）以上时，则可不受设计规定的温差限制。

Both natural curing and steam curing are applicable to the prestressed concrete. When steam curing is taken for the members on the pedestal, measures shall be taken to prevent the prestress loss caused by the temperature difference. When concrete member is fabricated on the pedestal by the pre-tensioning method, the maximum allowable curing temperature shall be calculated according to the allowable temperature difference (the difference between the tensioning temperature and the curing temperature) specified in the design. When the concrete strength exceeds 7.5 N/mm^2 (for thick reinforcement) or 10 N/mm^2 (for steel wire or steel strand reinforcement), it may not be limited by such temperature difference.

5.2.3 预应力筋的放张

5.2.3 Tension Release of Prestressing Tendon

（1）放张要求

(1) Tension release requirements

预应力筋放张时，预应力混凝土构件的强度必须符合设计要求。设计无要求时，其强度不低于混凝土设计强度标准值的 75%。

When the tension of the prestressing tendon is released, the strength of the prestressed concrete member must meet the design requirements. In case of no such design requirement, the strength shall be not less than 75% of the designed standard value of concrete strength.

（2）放张顺序

(2) Tension release sequence

①轴心受预压构件，所有预应力筋应同时放张。

① For axial compressive prestressed members, the tension of all the prestressing tendons shall be released at the same time.

②偏心受预压构件，应先同时放张预压力较小区域的预应力筋，再同时放张预压力较大区域的预应力筋。

② For eccentric compressive prestressed members, the tension of the prestressing tendon in the area under small precompression shall be released at the same time, followed by those in the area under large precompression.

③不能满足上述要求时，应分阶段、对称、交错地放张，防止构件在放张过程中弯曲、产生裂纹或预应力筋断裂。

③ In case of failure to meet the requirements mentioned above, the tension shall be released in stages, symmetrically and staggeredly to prevent the members from bending, cracking, or prestressing tendon breaking during the tensioning.

④采用湿热养护的预应力混凝土构件宜热态放张，不宜降温后放张。

④ The prestressed concrete members cured by steam should be released in hot state other than after cooling.

5.3 后张法

5.3 Post-tensioning Method

后张法是先制作构件并预留孔道，待构件混凝土达到规定强度后，在孔道内穿入预应力筋，张拉并锚固，然后孔道灌浆。后张法不需台座，构件在张拉过程中完成混凝土的弹性压缩。后张法广泛应用于现场生产的大型预应力构件和现浇混凝土结构中，如图 5-3 所示。

By this method, the members will be fabricated at first, with ducts reserved. When the concrete of the members reaches the specified strength, the prestressing tendons will be inserted into the ducts, tensioned, and anchored, and then the ducts will be grouted. No pedestal is required for this method. The elastic compression of concrete will be completed during the tensioning of the members. It is widely used in large prestressed members fabricated on the site and cast-in-site concrete structures, as shown in Figure 5-3.

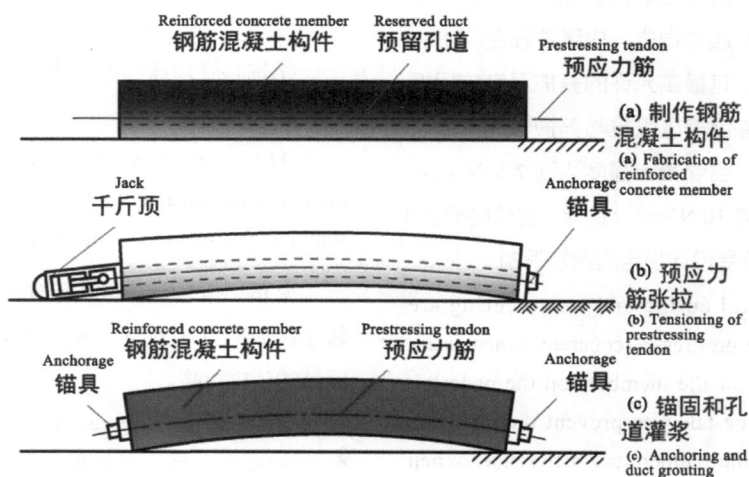

图5-3 后张法主要工序示意图

Figure 5-3 Diagram of Main Processes of Post-tensioning Method

将先张法和后张法对比可以看出，先张法的生产工序少，工艺简单，质量容易保证。同时，先张法不用工作锚具，生产成本较低，台座越长，一条长线上生产的构件数量越多，所以适合于工厂内成批生产中小型预应力构件。但是，先张法生产所用的台座及张拉设备一次性投资费用较大，而且台座一般只能固定在一处，不够灵活。后张法直接在混凝土构件或结构上进行预应力筋的张拉和锚固，故不需要固定的台座设备，不受地点限制，其又是预制构件拼装的一种手段，将构件分成几个小型块体后预制制作，运到工地后，穿入预应力筋，施加预应力拼装成整体。然而，后张法构件只能单一、逐个地施加预应力，工序较多，操作也较麻烦。同时，后张法构件的锚具耗钢量大，锚具加工要求的精度较高，成本较高。因此，后张法适用于运输不便的大中型构件。

By the comparison of the pre-tensioning method and the post-tensioning method, it can be seen that the pre-tensioning method is characterized by few production procedures, simple processes, and easy to be guaranteed quality, as well as no requirement for the working anchorages, resulting in low production costs. In addition, the longer the pedestal is, the more the members are fabricated in a long pedestal. Thus, the pre-tensioning method is suitable for the batch production of small- and medium-sized prestressed members in the plant. However, the pedestal and tensioning equipment used in the pre-tensioning production require a large sum of one-time investment, and the pedestal shall be fixed in one place, impairing flexibility. By the post-tensioning method, the prestressing tendon is tensioned and anchored directly on the concrete members or structures, so it requires no fixed pedestal and equipment and is not limited by location. This method is also a means of assembling prefabricated members, by which the members are divided into several small blocks that are prefabricated and transported to the construction site, and the prestressing tendons are inserted in such blocks to apply prestress, and then these blocks are assembled as a complete member. However, the post-tensioned members can only be prestressed one by one, resulting in many processes and complicated operation. In addition, the anchorages used for the post-tensioned members require a large steel consumption and high accuracy for the anchorage processing, resulting in high costs. Therefore, the post-tensioning method is suitable for medium- and large-sized members which are inconvenient for transportation.

后张法施工步骤是先制作构件，预留孔道。待构件混凝土达到规定强度后，在孔道内穿放预应力筋，对预应力筋进行张拉并锚固；最后孔道灌浆施工、封端，如图 5-4 所示。

By this method, the members will be fabricated at first, with ducts reserved. When the concrete of the members reaches the specified strength, the prestressing tendons will be inserted into the ducts, tensioned, and anchored, and then the ducts will be grouted, and in the end, it will be sealed, as shown in Figure 5-4.

5.3.1 孔道留设

5.3.1 Duct Reservation

孔道留设是后张法构件制作的关键工序之一。预应力筋的孔道形状有直线、曲线和折线三种，其直径与布置根据构件的受力性能、张拉锚固体系特点及尺寸确定。

Duct reservation is one of the key processes in the fabrication of post-tensioned members. For prestressing tendons, there are straight ducts, curved ducts, and polyline ducts, the diameter and arrangement of which depend on the performance under stress, characteristics of the anchoring system and size of the member.

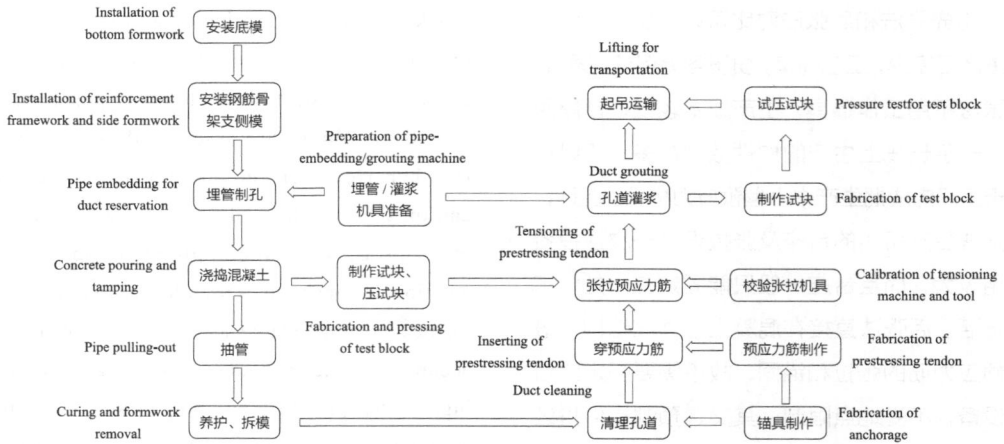

图5-4　后张法施工工艺

Figure 5-4 Post-tensioning Construction Process

（1）钢管抽芯法——用于直线孔道

(1) Steel pipe core-pulling method(for straight ducts)

预先将钢管埋设在模板内的孔道位置处，钢管要平直，表面要光滑，每根长度一般不超过15 m，钢管两端应各伸出构件约500 mm。较长的构件可采用两根钢管，中间用套管连接。钢管的位置一般用钢筋井字架固定，间距为1～2 m。在混凝土浇筑过程中和混凝土初凝后，每间隔一定时间慢慢转动钢管，不让混凝土与钢管黏牢，等到混凝土终凝前抽出钢管。抽管过早，会造成坍孔事故；太晚，则混凝土与钢管黏结牢固，抽管困难。常温下抽管时间约在混凝土浇灌后3～6 h。抽管顺序宜先上后下，抽管可采用人工或用卷扬机，速度必须均匀，边抽边转，与孔道保持直线。抽管后应及时检查孔道情况，做好孔道清理工作。

The steel pipe is embedded in the duct position in the formwork in advance. The steel pipe shall be straight and the surface shall be smooth. The length of each pipe does not exceed 15 m generally, and both ends of the steel pipe shall extend out of the member by about 500 mm. Two steel pipes connected by sleeves may be used for long members. The steel pipe is usually fixed with the #-shaped reinforcement frame, with a spacing of 1-2 m. During concrete pouring and after the initial setting of concrete, the steel pipe shall be slowly rotated at certain intervals to prevent the pipe from bonding with the concrete firmly and pulled out before the final setting of concrete. If the pipe is pulled out early, the duct may collapse; if late, the concrete will be firmly bonded with the steel pipe, making it difficult to pull out the pipe. The pipe is usually pulled out about 3-6 h after concrete pouring at ambient temperature. The pipes should be pulled out from top to bottom manually or with a winch at a constant speed. When pulled out, the pipes shall be rotated all the time and kept in line with the duct. After the pipe is pulled out, the duct shall be checked and cleaned in time.

（2）胶管抽芯法——可用于直线或曲线孔道

(2) Rubber pipe core-pulling method(for straight or curved ducts)

胶管弹性好，便于弯曲，一般有五层或七层夹布胶管和钢丝网橡皮管两种。胶管具有一定弹性，在拉力作用下，其断面能缩小，故在混凝土初凝后即可把胶管抽拔出来。夹布胶管质软，必须在管内充气或充水。浇筑混凝土前，在胶皮管中充入压力为 0.6 ～ 0.8 MPa 的压缩空气或压力水，此时胶皮管直径可增大 3 mm 左右，胶管的位置一般用钢筋井字架固定，间距不宜大于 0.5 m。浇筑混凝土，待混凝土初凝后，放出压缩空气或压力水，胶管孔径变小，并与混凝土脱离，随即抽出胶管，形成孔道。抽管顺序一般应为先上后下，先曲后直。

The rubber pipe has good elasticity and is easy to bend. Generally, it includes the five- or seven-layer cloth-inserted rubber pipe and the steel wire mesh rubber pipe. The rubber pipe has certain elasticity, and its section can be reduced under tension, so it can be pulled out after the initial setting of concrete. The cloth-inserted rubber pipe is very flexible so it must be filled with air or water. Before concrete pouring, the rubber pipe shall be filled with compressed air or pressure water with a pressure of 0.6-0.8 MPa. Then the diameter of the pipe can be increased by about 3 mm. The pipe is usually fixed with the #-shaped reinforcement frame, with a spacing of not more than 0.5 m. After concrete pouring, the compressed air or water will be released after the initial setting of concrete, and the diameter of the pipe will get smaller and separate from the concrete. Then the pipe shall be pulled out and thus a duct is built. Generally, the pipes are pulled out from top to bottom and from the curved ones to the straight ones.

（3）预埋管法

(3) Pipe embedding method

预埋管法是用钢筋井字架（间距不宜大于 0.8 m）将黑铁皮管、薄钢管或镀锌双波纹金属软管固定在设计位置上，在混凝土构件中埋管成型的一种施工方法。预埋管具有质量小、刚度好、弯折方便、连接简单等特点，可制成各种形状的孔道，并省去了抽管工序。适用于预应力筋密集或曲线预应力筋的孔道埋设，但在电热后张法施工中，不得采用波纹管或其他金属管埋设的管道。波纹管安装时，宜先在构件

底模、侧模上弹安装线，并检查波纹管有无渗漏现象，避免漏浆堵塞管道。同时，尽量避免波纹管多次反复弯曲并防止电火花烧伤管壁。

By this method, the black iron sheet pipe, thin steel pipe or galvanized double corrugated metal hose is fixed at the design position with the #-shaped reinforcement frame, with a spacing of not more than 0.8 m, and embedded in the concrete member to form a duct. The embedded pipe is characterized by small weight, good stiffness, easy bending, and simple connection. It can be used to construct various shapes of ducts, without pulling out any pipe. It is suitable for ducts used for intensive prestressing tendons or curved prestressed tendons. However, in the electric heating post-tensioning construction, no corrugated ducts or other metal pipes shall be used. During the installation of corrugated ducts, it is better to draw installation lines on the bottom formwork and side formwork of the member and to check for any leak on the duct to prevent the pipe from blocking by the leakage of grouting. At the same time, the corrugated ducts shall be prevented from repeated bending and the pipe wall shall be prevented from getting burned by the electric sparks, if possible.

5.3.2 预应力筋张拉

5.3.2 Tensioning of Prestressing Tendon

（1）混凝土的张拉强度

(1) Tensioning strength of concrete

预应力筋张拉时混凝土的强度应符合设计要求。当设计无具体要求时，不应低于设计强度标准值的 75%。

When the prestressing tendon is tensioned, the concrete strength shall meet the design requirements. If there is no specific design requirement, the concrete strength shall be not less than 75% of the standard value of the design strength.

（2）预应力筋的张拉控制应力

(2) Tensioning control stress of prestressing tendon

张拉控制应力越高，建立的预应力越大，构件的抗裂性越好。但是张拉控制应力过高，构件使用过程中经常处于高应力状态，构件出现裂缝的荷载与破坏荷载很接近，往往构件破坏前没有明显预兆，且当控制应力过高，构件混凝土预应力过大将导致混凝土的徐变应力损失增加。因此，控制应力应符合设计规定。在施工中预应力筋需要超张拉时，张拉控制应力可比设计要求提高 5%，但最大不得超过表 5-2 的规定。

The higher the tensioning control stress is, the greater the prestress imposed is and the better the crack resistance of the member is. However, in the case of high tensioning control stress, the member is often highly stressed during its use and the cracking load of the member is very close to the failure load. Thus, there is no obvious sign before the damage of the member. The large concrete prestress of the member results in an increase in the creep stress loss of the concrete. From the above, the control stress shall meet the design requirements. When the prestressing tendon needs to be overtensioned during construction, the control stress may be 5% higher than the design requirements, but the maximum tension control stress shall be not more than that specified in Table 5-2.

表 5-2 后张法张拉控制应力限值

Table 5-2 Tensioning Control Stress Limits for Post-tensioning Method

钢种 Steel Type	后张法张拉控制应力限值 σ_{con} Tensioning Control Stress Limits σ_{con} for Post-tensioning Method
消除应力钢丝、刻痕钢丝、钢绞线 Stress-eliminating steel wire, indented steel wire, and steel strand	$\sigma_{con} \leqslant 0.80 f_{ptk}$
热处理钢筋 Heat-treated reinforcement	$\sigma_{con} \leqslant 0.70 f_{ptk}$
冷拉钢筋 Cold-drawn reinforcement	$\sigma_{con} \leqslant 0.90 f_{pyk}$

（3）预应力筋张拉程序

(3) Tensioning procedure of prestressing tendon

为了减少预应力筋的松弛损失，预应力筋的张拉程序可为下列之一：

The prestressing tendon can be tensioned according to either of the following processes to reduce the loss caused by the relaxation of the prestressing tendon:

$$0 \to 105\%\sigma_{con} \xrightarrow{\text{持荷 2 min}} \sigma_{con}; \text{ 或 } 0 \to 103\%\sigma_{con}$$

$$0 \to 105\%\sigma_{con} \xrightarrow{\text{Load holding for 2 min}} \sigma_{con}; \text{ or } 0 \to 103\%\sigma_{con}$$

（4）预应力筋张拉顺序

(4) Tensioning sequence of prestressing tendon

预应力筋张拉顺序应按设计规定进行，如设计无规定时，应分批、分阶段、对称地进行，以免构件受过大的偏心压力而扭转与侧弯。

The prestressing tendons shall be tensioned according to the design sequence. If there is no such sequence, the tendons shall be carried out symmetrically in batches and stages to avoid torsion and lateral bending of the member due to excessive eccentric pressure on the member.

如图 5-5（a）所示为两束预应力筋，能同时张拉，宜将两台千斤顶分别设置在构件两端进行对称张拉。图5-5（b）对称的四束预应力筋，不能同时张拉，应分批、对称张拉，用两台千斤顶分别在两端张拉对角线上的两束，然后张拉另两束。

In Figure 5-5 (a), there are two prestressing tendons that can be tensioned at the same time. Thus, two jacks shall be placed at both ends of the member respectively for symmetrical tensioning. In Figure 5-5 (b), there are four prestressing tendons that shall be tensioned symmetrically in batches, they should not be tensioned at the same time. Thus, two jacks shall be placed at the diagonal ends of the member respectively for symmetrical tensioning of two tendons and then moved to the other diagonal ends for another two tendons.

图5-5（c）所示的预应力混凝土吊车梁，配有多根不对称预应力筋，应分批、分阶段、对称张拉。采用两台千斤顶先张拉上部两束预应力筋，下部四束曲线预应力筋采用两端张拉方法分批进行。为使构件对称受力，每批两束先按一端张拉方法进行张拉，待两批四束均进行一端张拉后，再分批在另一端张拉，以减少先批张拉筋所受的弹性压缩损失。

In Figure 5-5 (c), there is a prestressed concrete crane beam with several asymmetric prestressing tendons that shall be tensioned symmetrically in batches and stages. The two prestressing tendon in the upper part of the beam shall be tensioned by two jacks while the four curved prestressing tendons in the lower part shall be tensioned from both ends in batches. For symmetrical stress on the member, every two tendons are tensioned from one end, the four tendons shall be tensioned at one end in two batches and then at the other end also in two batches to reduce the elastic compression loss suffered by the two tendons tensioned in the first batch.

（5）预应力筋的张拉方法

(5) Tensioning method of prestressing tendon

为了减少预应力筋与预留孔壁摩擦引起的预应力损失，预应力钢筋应根据设计和专项施工方案的要求采用一端或两端进行张拉。采用两端张拉时，宜两端同时张拉，也可一端先张拉锚固，另一端补张拉。当设计对张拉端设置无具体要求时，可按下列规定设置：

To reduce the prestress loss caused by the friction between the prestressing tendon and the wall of the reserved duct, the prestressed reinforcement shall be tensioned at one end or both ends according to the requirements of the design and special construction scheme. For tensioning at both ends, the reinforcement can be tensioned at both ends at the same time or tensioned and anchored at one end and then tensioned at the other end. If there is no specific design requirement for the setting at the tensioning end, the setting can be as follows:

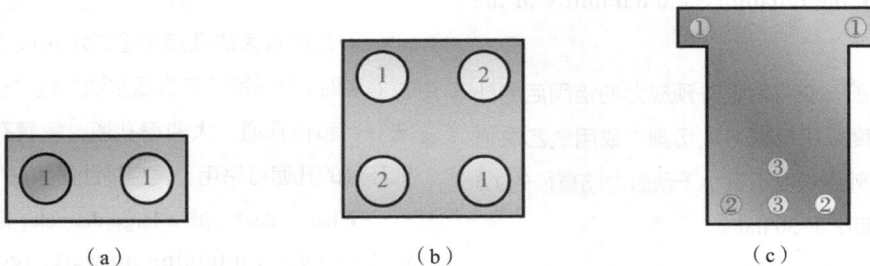

（a）　　　　（b）　　　　（c）

图5-5　预应力筋张拉顺序

Figure 5-5 Tensioning Sequence of Prestressing Tendons

当有黏结预应力筋长度不大于 20 m 时，可一端张拉，大于 20 m 时，宜两端张拉；当预应力筋为直线形时，一端张拉的长度可延长至 35 m；当无黏结预应力筋长度不大于 40 m 时，可一端张拉；当大于 40 m 时，宜两端张拉。

The bounded prestressing tendon with a length of not more than 20 m can be tensioned at one end. Otherwise, it shall be tensioned at both ends. When the prestressing tendon is linear, the length of one end can be extended to 35 m. When the length of unbonded prestressing tendon is not more than 40 m, one end can be tensioned. Otherwise, it shall be tensioned at both ends.

5.3.3 孔道灌浆

5.3.3 Duct Grouting

后张法有黏结预应力筋张拉完毕并经检查合格后，应尽早进行孔道灌浆，防止钢筋锈蚀，增加结构的整体性和耐久性，提高结构抗裂性能和承载力。通过孔道灌浆，使预应力筋与混凝土相互黏结，可减轻锚具传递预应力的负担，提高了锚固的可靠性与耐久性。

The duct grouting shall be carried out as soon as possible after the post-tensioning of the bounded prestressing tendon and applicable inspection to prevent reinforcement corrosion and improve the integrity, durability, crack resistance, and bearing capacity of the structure. This method can bond the prestressing tendon with the concrete to reduce the prestress transferred by the anchorage so as to improve the reliability and durability of the anchorage.

灌浆前，应将后张法预应力筋锚固后的外露多余长度采用机械方法切割（或用氧乙炔焰切割），其外露长度不宜小于预应力筋直径的 1.5 倍，且不应小于 30 mm。

Before grouting, the excessive exposed part of the prestressing tendon that has already been post-tensioned and anchored shall be cut mechanically

(or with the oxy-acetylene flame). The length of the exposed part shall be not less than 1.5 times the diameter of the prestressing tendon, and in any case, not less than 30 mm.

孔道灌浆前应确认孔道、排气兼泌水管及灌浆孔畅通；对预埋管成型孔道，可采用压缩空气清孔，并采用水泥浆、水泥砂浆等材料封闭端部锚具缝隙，也可采用封锚罩封闭外露锚具；采用真空灌浆工艺时，应确认孔道系统的密封性。灌浆时宜先灌下层孔道，后灌上层孔道。灌浆工作应缓慢均匀进行，不得中断，并应排气通畅，在出浆口冒出浓浆并封闭排气口后，继续加压至 0.5 ~ 0.7 N/mm^2 稳压 2 min，再封闭灌浆孔。

Before the duct grouting, it is necessary to make sure that the duct, exhaust and bleeding pipe, and grouting hole are unblocked. If the embedded duct is used, the duct can be purged and cleaned with compressed air and the gaps around the anchorage at the end shall be sealed with cement paste, cement mortar, and other appropriate materials. An anchor seal can also be used to seal the exposed anchor. If vacuum grouting is used, the tightness of the duct system shall be checked. The grouting shall be carried out from the lower ducts to the upper ducts. Grouting shall proceed slowly and evenly without interruption and in good exhaust condition. After thick grout comes out of the grout outlet and the exhaust hole is sealed, it is necessary to pressurize to 0.5-0.7 N/mm^2 and stay for 2 min, and then the grouting hole shall be sealed.

对直径较大的孔道不掺减水剂或膨胀剂进行灌浆时，可采取"二次压浆法"或"重力补浆法"。超长孔道、大曲率孔道、扁管孔道、腐蚀环境的孔道可采用"真空辅助压浆法"。

When a duct with a large diameter is grouted with cement containing no water reducer or expansion agent, the "secondary grouting method" or "gravity grout replenishment method" may be used. For ultra-long ducts, ducts with large curvature, flat pipe ducts, and ducts in a corrosive environment, it is advisable to use the "vacuum-assisted grouting method".

灌浆施工时宜先灌注下层孔道，后灌注上层孔道；灌浆应连续进行，直至排气管排出的浆体稠度与注浆孔处相同且无气泡后，再顺浆体流动方向依次封闭排气孔；全部出浆口封闭后，宜继续加压 0.5 ~ 0.7 MPa，并应稳压 1 ~ 2 min 后封闭灌浆口；孔道内水泥浆应饱满、密实。

During grouting construction, the lower ducts shall be grouted first, and then the upper ones. The grouting shall be carried out continuously until the consistency of the grout coming out of the exhaust pipe is the same as that at the grouting hole and there is no blistering in the grout. And then the exhaust holes shall be sealed in turn along the flow direction of the grout. After all grout outlets are sealed, it is advisable to pressurize to 0.5-0.7 MPa and stay for 1-2 min, and the grout inlet shall be sealed. The cement slurry in the duct shall be full and dense.

5.3.4 后张法施工注意事项

5.3.4 Precautions for Post-tensioning Method

①后张法预应力构件，断裂或滑脱的数量严禁超过同一截面预应力筋总根数的 3%，且每束钢丝不得超过一根（对于多跨双向连续板，其同一截面应按每跨计算）。

① For post-tensioned prestressed members, the number of broken or slipped prestressing tendons must not exceed 3% of the total number of the prestressing tendons on the same section, and the number of broken or slipped wire must not exceed one for each steel tendon(for multi-span two-way continuous slabs, the same section shall be calculated per span).

②预应力筋张拉锚固后实际建立的预应力值与设计规定检验值的相对允许误差为 ±5%。同一检验批内应抽查预应力筋总数的 3%，且不少于 5 束。

② The relative allowable error between the actual prestress after the tensioning and anchoring of the prestressing tendon and the inspection value specified in the design is ±5%. In each inspection lot, 3% of the total number of the prestressing tendons, and in any case, not less than 5 tendons, shall be randomly inspected.

③后张法预应力筋锚固后的外露部分宜用机械方法切割，其外露长度不宜小于预应力筋的 1.5 倍，且不小于 30 mm。长期使用的锚具，可涂刷防锈油漆或用封端混凝土封裹。

③ The exposed part of the post-tensioned and anchored prestressing tendon should be cut mechanically. The length of the exposed part shall be not less than 1.5 times the diameter of the prestressing tendon, and in any case, not less than 30 mm. Permanent anchorages can be coated with antirust paint or sealed and wrapped with end-sealing concrete.

④现浇混凝土构件的侧模板宜在预应力张拉前拆除，底模支架拆除时，孔道灌浆强度不应低于 15 N/mm^2。

④ The side formwork of cast-in-situ concrete members shall be removed before prestressing tendon tensioning. The bottom formwork support can be removed only when the strength of the duct grouting is not less than 15 N/mm^2.

⑤金属波纹管或无黏结预应力筋铺设后，其附近不得进行电焊作业，否则应采取防护措施。

⑤ When the corrugated metal ducts or unbonded prestressing tendons are laid, no electric welding is allowed nearby. If necessary, protective measures shall be taken.

⑥混凝土浇筑时，应防止振动器触碰金属

波纹管、无黏结预应力筋或端部预埋件，不得踏压或撞碰预应力筋、钢筋支架。

⑥ During concrete pouring, the vibrator shall not touch the corrugated metal ducts, unbonded prestressing tendons, or embedded parts at the ends, or collide with or step on the prestressing tendons or the reinforcement supports.

5.4 预应力混凝土施工安全措施

5.4 Safety Measures for Prestressed Concrete Construction

①所用张拉设备仪表由专人负责使用与管理，并定期进行维护与检验，设备的测定期不超过半年，否则必须及时重新测定。施工时，根据预应力筋种类等合理选择张拉设备，预应力筋的张拉力不应大于设备额定张拉力，严禁在负荷时拆换油管或压力表。接电源时，机壳必须接地，经检查绝缘可靠后，才可试运转。

① The tensioning equipment and instrument shall be used and managed by specially-designated personnel, and maintained and inspected regularly. The equipment must be calibrated once at most every six months, otherwise, it must be re-measured in time. During construction, the tensioning equipment shall be reasonably selected according to the type of prestressing tendon. The tensioning force on the prestressing tendon shall be not greater than the rated tensioning force of the equipment. It is strictly prohibited to remove or replace the oil pipe or pressure gauge under load. When energized, the enclosure must be grounded. The commissioning of the equipment can be started only when it is confirmed that the equipment is insulated reliably.

②先张法施工中，张拉机具与预应力应在一条直线上；顶紧锚塞时，用力不要过猛，以防钢丝折断。台座法生产，其两端应设有防护设施，并在张拉预应力筋时，沿台座长度方向每隔 4 ~ 5 m 设置一个防护架，两端严禁站人，更不准进入台座。

② During the pre-tensioning construction, the tensioning machines and tools shall be in a straight line with the prestress. When jacking the anchor plug, the force applied shall be appropriate to prevent the steel wire from breaking due to excessive force. When the pedestal is used for production, there shall be protections at both ends. When the prestressing tendon is tensioned, protective frames shall be set every 4-5 m along the length of the pedestal. No one is allowed to stand at both ends or enter the pedestal.

③后张法施工中，张拉预应力筋时，任何人不得站在预应力筋两端，同时要在千斤顶后面设置防护装置。操作千斤顶的人员应严格遵守操作规程，应站在千斤顶侧面工作。在油泵开动过程中，不得擅自离开岗位，如需离开，应将油阀全部松开或切断电路。

③ During the post-tensioning construction, no one is allowed to stand at both ends of the prestressing tendon when tensioning, and there shall be protections behind the jack. The jack operator shall strictly abide by the operating procedures and shall stand on the side of the jack when operating. During the startup of the oil pump, the operator shall not leave the post without permission. If the operator has to leave, all the oil valves shall be loosened or the circuit shall be open.

6

结构安装与钢结构工程施工

Structure Installation and Construction of Steel Structure Works

学习目标：

Learning objectives:

掌握单层工业厂房结构的吊装施工方法、工艺流程、质量检验要求和安全措施。

Master the lifting construction method, process flow, quality inspection requirements and safety measures of single-storey industrial plant structure.

技能抽查要求：

Skill checking requirements:

能对常见基础的施工质量进行检测，能编制常见基础工程的施工方案。

Be able to inspect the construction quality of common foundation and prepare the construction scheme of common foundation project.

建筑八大员岗位资格考试要求：

Qualification examination requirements for eight major posts of construction engineering:

掌握单层工业厂房结构的吊装工艺、施工方法、质量和安全要求。

Master the lifting process, construction method, quality and safety requirements of single-storey industrial plant structure.

6.1 单层工业厂房安装

6.1 Installation of Single-storey Industrial Plant

单层工业厂房多采用装配式钢筋混凝土结构，主要承重结构除基础在施工现场就地灌注外，其他构件（如柱、吊车梁、屋架、屋面板等）多采用钢筋混凝土预制构件。其中，尺寸大、较重的大型构件一般在施工现场就地预制，中小型构件多集中在预制厂预制，后运到现场吊装。结构安装工程是单层工业厂房施工中的主导工程，其施工过程是将各种预制构件按设计要求采取合理的施工方法在现场进行安装。

Single-storey industrial plants mostly adopt prefabricated reinforced concrete structures. For the main bearing structures, except that the foundation is poured on the construction site, other members (such as columns, crane beams, roof trusses, roof slabs) are mostly reinforced concrete precast members. Among them, large and heavy members are generally prefabricated on site, while small- and medium-sized members are mostly prefabricated in the prefabrication plant and then transported to the site for installation. The structure installation work is the main work in the construction of single-storey industrial plants, and the construction process is to install various prefabricated members on site using reasonable construction methods according to the design requirements.

6.1.1 构件安装前的准备工作

6.1.1 Preparation before Member Installation

为保证单层工业厂房结构安装时的施工质量和进度，在安装前应做好准备工作。安装前的准备工作包括：构件的检查与清理、弹线与编号，混凝土杯型基础的准备，构件的运输、堆放等。

In order to ensure the installation construction quality and progress of single-storey industrial plant structure, preparations shall be made before installation, including check and cleaning of members, line snapping and numbering of members, preparation of concete cup-like foundation, transportation and stacking of members, etc.

（1）构件的检查与清理

(1) Check and cleaning of members

①检查构件的型号和数量。

① Check the model and quantity of members.

②检查构件的截面尺寸。

② Check the section size of members.

③检查构件外观质量（有无变形、缺陷、损坏等）。

③ Check the appearance quality of members (whether there is deformation, defect, damage, etc.).

④检查构件的混凝土强度。构件安装时混凝土强度应不低于设计强度标准值的 75%，对于大跨度构件，如屋架，则应达到 100%。

④ Check the concrete strength of members. The concrete strength shall not be less than 75% of the standard value of design strength. For a large-span member, such as roof trusses, it shall reach 100%.

⑤检查预埋件、预留孔的位置及质量等，并做相应清理工作。

⑤ Check the positions and quality of embedded parts and reserved holes, and clean them accordingly.

（2）构件的弹线与编号

(2) Line snapping and numbering of members

①构件的弹线。

① Line snapping of members.

构件上弹出的安装定位线，可作为构件安装、对位、校正的依据。

A positioning line shall be snapped on the member as a reference for installation, alignment and correction.

a. 柱子。在柱身三面弹出安装中心线，所弹中心线的位置与柱基杯口面上的安装中心线相吻合。此外，在柱顶与牛腿面上还要弹出安装屋架及吊车梁的定位线。

a. Columns. The installation centerlines shall be snapped on three sides of the column body, and the position of the centerline shall coincide with the installation centerline on the cup rim surface of the column foundation. In addition, the positioning lines for installing the roof truss and crane beam shall be snapped on the column top and bracket surface.

b. 屋架。屋架上弦顶面上应弹出几何中心线，并将中心线延至屋架两端下部，再从跨中央向两端分别弹出天窗架、屋面板的安装定位线。

b. Roof trusses. The geometric centerline shall be snapped on the top surface of the upper chord of the roof truss, which shall be extended to the lower parts of both ends of the roof truss, and then the installation positioning lines of the skylight truss and roof slab shall be snapped from the center of the span to both ends respectively.

c. 吊车梁。在吊车梁的两端及顶面弹出安装中心线。

c. Crane beams. The installation centerlines shall be snapped on both ends and top surface of the crane beam.

②构件的编号。

② Numbering of members.

构件编号应编写在构件明显的部位，并在构件上用记号标明不易辨别上下左右的构件。

The member number shall be marked at an obvious position of the member, and the members that are difficult to distinguish UDLR(Up, Down, Left, Right) shall be marked.

（3）混凝土杯型基础的准备

(3) Preparation of concrete cup-like foundation

①杯口弹线。

① Line snapping at cup rim.

先检查杯口的尺寸，再在基础顶面弹出十字交叉的安装中心线，用红油漆画上三角形标志。中心线对定位轴线的允许偏差为 ±10 mm。

First check the size of the cup rim, then snap the cross installation centerline on the top surface of the foundation, and draw a triangle mark with red paint. The allowable deviation of the centerline from the positioning axis is ±10 mm.

②杯底抄平。

② Cup bottom leveling.

浇筑基础时，杯底标高一般比设计标高降低50 mm。具体操作：在杯口内抄上平线，一般此线比杯口设计标高低10 cm。

For foundation grouting, the cup bottom elevation is generally 50 mm lower than the design elevation. Specifically, you need to mark a horizontal line in the cup rim, which is generally 10 cm lower than the design elevation of the cup rim.

（4）构件的运输

(4) Transportation of members

一些质量不大而数量较多的定型构件，如屋面板、连系梁、轻型吊车梁等，宜在预制厂预制，用汽车将构件运至施工现场。起吊及运输时，必须保证构件的强度符合要求，吊点位置符合设计规定。构件支垫的位置要正确，数量要适当，每一构件的支垫数量一般不超过2个支承处，且上下层支垫应在同一垂线上。运输过程中，要确保构件不倾倒、不损坏、不变形。构件的运输顺序、堆放位置应按施工组织设计的要求和规定进行，以免增加构件的二次搬运工作量。

Some small but large number of shaped members, such as roof slabs, connecting beams and light crane beams, should be prefabricated in the prefabrication plant and transported to the construction site by trucks. During lifting and transportation, the member strength must meet the requirements, and the position of the lifting points meet the design requirements. The positions and quantity of the member support pads shall be correct and appropriate, generally not more than 2 support pads for each member, and the upper and lower support pads shall be on the same vertical line. During transportation, it is necessary to ensure that the members are not dumped, damaged or deformed. The transportation sequence and stacking positions of members shall comply with the requirements of the construction organization design to avoid secondary handling of members.

（5）构件的堆放

(5) Stacking of members

堆放构件的地面应平整、坚实，排水良好，构件应按设计的受力情况搁置在垫木或支架上。重叠堆放时，一般梁可堆叠2～3层；大型屋面板不超过6块；空心板不超过8块。构件吊环要向上，标志要向外。

The ground on which members are stacked shall be flat and solid, with good drainage, and members should be placed on skids or supports according to the designed stress conditions.

Generally, the beams can be overlapped for 2-3 layers; there should be not more than 6 large roof slabs and not more than 8 hollow slabs overlapped. Lifting rings of members shall be upward and marks outward.

6.1.2 构件的吊装工艺

6.1.2 Lifting Process of Members

装配式单层工业厂房的结构安装构件有柱子、吊车梁、基础梁、连系梁、屋架、天窗架、屋面及支撑等。构件的吊装工艺包括绑扎、吊升、对位、临时固定、校正、最后固定等工序，对于现场制作的构件需要翻身、扶直，按吊装要求排放后再进行吊装。

The structure installation members of the prefabricated single-storey industrial plant include columns, crane beams, foundation beams, connecting beams, roof trusses, skylight trusses, roofs and supports. The lifting process of members includes binding, lifting, alignment, temporary fixation, correction, final fixation, etc. The members fabricated on site need to be turned over, straightened and arranged as required before lifting.

（1）柱子的吊装

(1) Lifting of columns

单层工业厂房的柱子一般为现场预制，其截面类型有矩形、工字形、双肢形等。当混凝土的强度达到75%混凝土强度标准值以上时方可吊装。

Columns of single-storey industrial plants are generally prefabricated on site, with rectangular, I-shaped and double-limb sections. They can only be lifted when the strength of concrete reaches more than 75% of the standard value of concrete strength.

①绑扎。

① Binding.

柱身绑扎点数和绑扎位置，要保证柱在吊装过程中受力合理，不发生变形、产生裂缝或折断。一般中、小型柱绑扎一点；重型柱或配筋少而细长的柱常绑扎两点甚至两点以上，以减少柱的吊装弯矩。必要时，绑扎点数须经吊装应力和裂缝控制计算后确定。有牛腿的柱，一点绑扎的位置常选在牛腿以下，上柱较长时也可选在牛腿以上；工形断面柱的绑扎点应选在矩形断面处；双肢柱的绑扎点应选在平腹杆处。

The number of binding points and binding positions on the column body shall ensure that the column is reasonably stressed during lifting without deformation, cracks or breakage. Generally, small- and medium-sized columns shall be bound at one point; heavy columns or columns with few and slender reinforcement are often bound at two or more points to reduce the lifting bending moments of columns. If necessary, the bound point shall be determined based on the calculation of the lifting stress and crack control. For columns with brackets, the binding point at one point is often below the brackets, or above the brackets when the upper column is long; the binding point of I-shaped section columns shall be at the rectangular section;

the binding point of double-limb columns shall be at the flat belly rod.

按柱吊起后柱身是否能保持垂直状态，绑扎方法可分为一点绑扎斜吊法和一点绑扎直吊法。

Based on whether the column body can be kept vertical after lifting, there is inclined lifting binding method and vertical lifting binding method.

a. 一点绑扎斜吊法。当柱平卧起吊的抗弯刚度满足要求时可采用斜吊绑扎法，如图 6-1 所示，无须将柱翻身，起重钩低于柱顶，当柱身较长、起重机臂长不够时较为方便。但因柱身倾斜，起吊后柱身与杯底不垂直，对中就位较困难。

a. Inclined lifting with one-point binding. When the bending rigidity of the column lifted horizontally meets the requirements, the inclined lifting binding method is adopted, as shown in Figure 6-1, the column using this method does not need to be turned over, and the lifting hook is lower than the column top. It is preferred when the column is long and the crane boom is not long enough. However, due to the inclination of the column body, the column body is not perpendicular to the bottom of the cup after lifting, and it is difficult to align and position it.

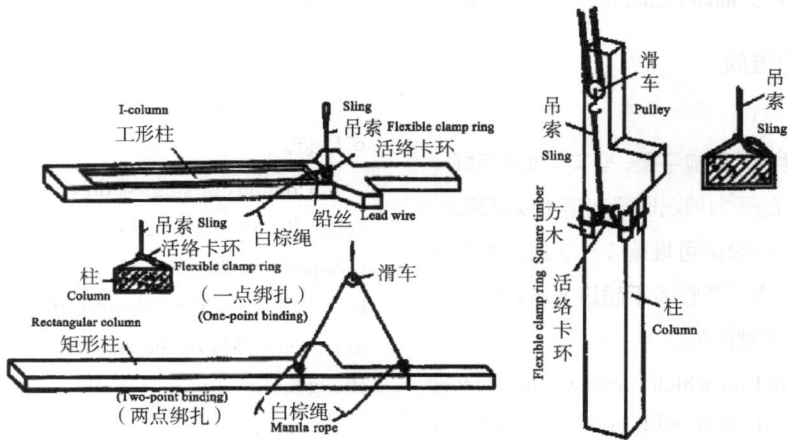

图6-1 一点绑扎斜吊法

Figure 6-1 Inclined Lifting with One-point Binding

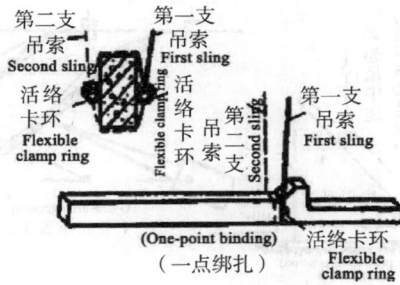

图6-2 一点绑扎直吊法

Figure 6-2 Vertical Lifting with One-point Binding

b. 一点绑扎直吊法。当柱平卧起吊的抗弯刚度不足时，须先将柱翻身后再绑扎起吊，如图 6-2 所示，吊索从柱两侧引出，上端通过卡环或滑轮挂在铁扁担上，柱身成垂直状态，便于插入杯口和对中校正，由于铁扁担高于柱顶，起重臂长度稍长。

b. Vertical lifting with one-point binding. When the bending rigidity of the column for horizontal lifting is insufficient, the column shall be turned over, and then bound and lifted, as shown in Figure 6-2, the slings in this method are led out from both sides of the column, and the upper end is hung on the iron carrying pole by a snap ring or pulley. The column body is vertical to facilitate insertion into the cup and alignment correction. Since the iron carrying pole is higher than the column top, the boom is slightly longer.

②吊升。

② Lifting.

按柱在吊升过程中柱身运动的特点可分为旋转法和滑行法；按使用起重机的数量可分为单机起吊和双机拍吊。

There is rotation method and sliding method according to the characteristics of column movement during the lifting, and single-crane lifting and double-crane lifting according to the number of cranes used.

a. 单机吊装旋转法。

a.Single-crane lifting and rotation method.

柱布置时柱脚靠近杯口，柱的绑扎点、柱脚与杯口中心三者均位于起重半径的圆弧上（即三点共弧），起吊时，起重机边升钩、边回转，使柱绕柱脚旋转而成直立状态，吊离地面插入杯口，如图 6-3 所示。旋转法振动小、效率高，一般中小型柱多采用旋转法吊升，但此法对起重机的回转半径和机动性要求较高，适用于自行杆式（履带式）起重机吊装。

When the column is arranged, the column base is close to the cup rim, and the binding point of the column, the column base and the center of the cup rim are all located on the circular arc of the lifting radius(i.e., the three points on the same arc). When lifting, the crane lifts the hook and rotates at the same time, so that the column rotates around the column base to form an upright state, and is lifted off the ground and inserted into the cup, as shown in Figure 6-3. The rotation method has small vibration and high efficiency. Generally, small- and medium-sized columns are lifted by the rotation method, but this method has high requirements on the turning radius and maneuverability of cranes and is suitable for lifting by self-propelled(crawler) cranes.

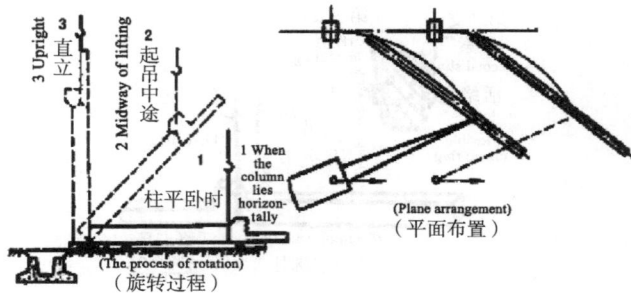

图6-3 单机吊装旋转法

Figure 6-3 Single-crane Lifting and Rotation Method

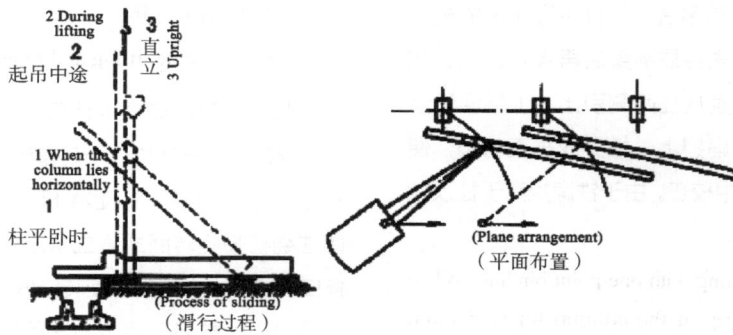

图6-4 单机吊装滑行法

Figure 6-4 Single-crane Lifting and Sliding Method

b. 单机吊装滑行法。

b. Single-crane lifting and sliding method.

柱布置时吊点靠近杯口，柱的绑扎点与杯口中心均位于起重半径的圆弧上（即两点共弧），起吊时，起重机只升钩、不回转，使柱脚沿地面滑行，至柱身直立吊离地面插入杯口，如图6-4所示。此法的特点是柱的布置灵活、起重半径小、起重杆不转动，操作简单，适用于柱子较长较重、现场空间狭窄或使用桅杆式起重机吊装的情况。

When the column is arranged, the lifting point is close to the cup rim, and the binding point and the center of the cup are located on the circular arc of the lifting radius(i.e., the two points on the same arc). When lifting, the crane only lifts the hook and does not rotate, so that the column base slides along the ground until the column body is lifted upright from the ground and inserted into the cup, as shown in Figure 6-4. This method is characterized by flexible arrangement of column, small lifting radius, non-rotation of lifting rods and simple operation, and is suitable for lifting of long and heavy columns, on narrow site or lifting with mast crane.

③对位与临时固定。

③ Alignment and temporary fixation.

柱子对位是将柱子插入杯口并对准安装准线的一道工序。临时固定是用楔子等将已对位的柱子临时固定的一道工序。

Column alignment is a process of inserting

the column into the cup and aligning it with the installation reference line. Temporary fixation is a process of temporarily fixing the aligned columns with wedges.

对位：采用直吊法时，应将柱悬离杯底 30～50 mm 处对位，采用斜吊法时则需将柱送至杯底，在吊索一侧的杯口插入两个楔子，再通过起重机回转使其对位。对位时，在柱四周向杯口内放入 8 只楔子，用撬棍拨动柱脚，使吊装准线对准杯口上的吊装准线。

Alignment: In the vertical lifting method, the column shall be suspended 30-50 mm from the bottom of the cup for alignment. In the inclined lifting method, the column shall be sent to the bottom of the cup, two wedges shall be inserted into the cup on one side of the sling, and then the column shall be rotated by the crane for alignment. During alignment, 8 wedges shall be placed into the cup around the column to move the column base with a crowbar to align the lifting reference line with the lifting reference line on the cup rim.

临时固定：对位后，应将塞入的 8 只楔子逐步打紧对柱子进行临时固定，以防对好线的柱脚移动。细长柱子的临时固定应增设缆风绳。

Temporary fixation: After alignment, the 8 wedges shall be tightened gradually for temporary fixation to prevent the column base aligned from moving. Temporary fixation of slender columns shall be provided with cable wind ropes.

④校正。

④ Correction.

柱子校正是对已临时固定的柱子进行全面检查（平面位置、标高、垂直度等）及校正的一道工序。柱子校正包括平面位置、标高和垂直度的校正。对重型柱或偏斜值较大的柱则用千斤顶、缆风绳、钢管支撑等校正，如图 6-5、图 6-6 和图 6-7 所示。

Column correction is a process of comprehensive inspection(plane position, elevation, verticality, etc.) and correction of temporarily fixed columns. Column correction involves the correction of plane position, elevation and verticality. Heavy columns or columns with large deflections shall be corrected by jacks, cable wind ropes, steel pipe supports, etc., as shown in Figure 6-5, Figure 6-6 and Figure 6-7.

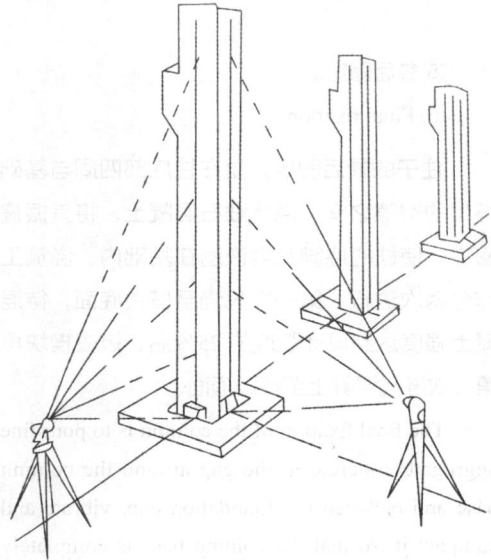

图6-5 柱子垂直度校正

Figure 6-5 Correction of Column Verticality

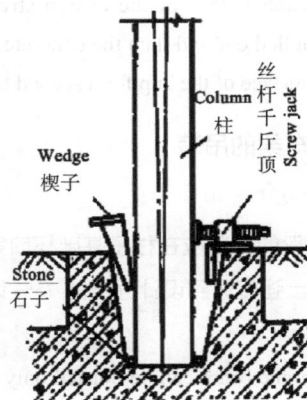

图6-6 立顶法校正柱垂直度

Figure 6-6 Column Verticality Correction by Vertical Top Method

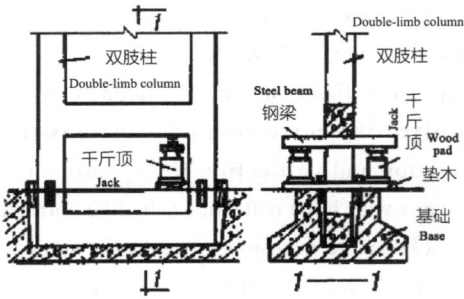

图6-7 平顶法校正柱垂直度

Figure 6-7 Column Verticality Correction by Flat Top Method

⑤最后固定。

⑤ Final fixation.

柱子的最后固定，是在柱底部四周与基础杯口的空隙之间，浇筑细石混凝土，将其振捣密实，使柱的底脚完全嵌固在基础内。浇筑工作分两次进行，第一次先浇至楔块底面，待混凝土强度达到设计强度的25%后，拔去楔块再第二次浇筑混凝土至杯口顶面。

The final fixation of the column is to pour fine aggregate concrete in the gap around the column base and between the foundation cup, vibrate and compact it, so that the column base is completely embedded in the foundation. The pouring is carried out twice. For the first time, it is to pour to the bottom surface of the wedge. After the concrete strength reaches 25% of the design strength, the wedge is pulled out and then the concrete is poured to the top surface of the cup for a second time.

（2）吊车梁的吊装

(2) Lifting of crane beam

吊车梁的吊装应在柱子基础杯口第二次浇筑的混凝土强度达到设计强度的75%以上方可进行。

The crane beam can be hoisted only when the concrete strength of the second pouring at the cup rim of the column foundation reaches more than 75% of the design strength.

由于吊车梁的高度低、长度小，一般采用平吊法。采用平吊法吊装时，吊车梁状态与其使用时工作状态一致。

Due to the low height and small length of the crane beam, the horizontal lifting is generally adopted. Horizontal lifting means that the lifting state of crane beam is same as its working state during use.

①绑扎、吊升、就位与临时固定。

① Binding, lifting, positioning and temporary fixation.

为了便于安装，吊车梁的绑扎应采用两点绑扎，使吊车梁在起吊后能基本保持水平（两根吊索要等长，绑扎点要对称设置），如图6-8所示。

In order to facilitate installation, the crane beam shall be bound at two points, so that the crane beam can be basically kept horizontal after being lifted(two slings shall be of equal length and the binding points shall be symmetrically arranged), as shown in Figure 6-8.

吊车梁的两头须用溜绳控制，以防碰撞柱子。就位时应缓慢落钩，以便对准安装线。在纵轴方向不宜用撬棍撬动吊车梁，因为柱子在纵轴方向刚度很差，如果在此方向撬吊车梁，很容易使柱子弯曲而产生垂直偏差。截面高度 $h \leqslant 600$ mm 的吊车梁在就位时用垫铁垫稳即可；当 $h > 800$ mm 或当梁的高宽比大于 4 时，可用

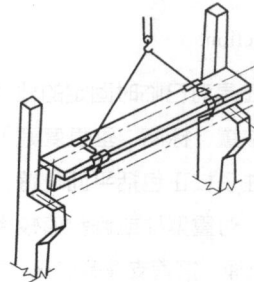

图6-8 吊车梁的吊装

Figure 6-8 Lifting of Crane Beam

铁丝将吊车梁捆于柱上做临时固定。

Both ends of the crane beam must be controlled with slip ropes to prevent collision with the column. During lifting in place, the hook shall be lowered slowly to align with the installation line. It is inadvisable to pry the crane beam with a crowbar in the longitudinal axis direction, because the column in the longitudinal axis direction has very poor rigidity. If the crane beam is pried in this direction, the column is prone to bend to cause vertical deviation. When the crane beam with a section height $h \leqslant 600$ mm is lifted in place, it can be stabilized with pad iron. When $h > 800$ mm or the height-width ratio of the beam is greater than 4, the crane beam can be tied to the column with iron wires for temporary fixation.

②校正及最后固定。

② Correction and final fixation.

吊车梁校正包括垂直度校正和平面位置校正，两者应同时进行。

Crane beam correction includes verticality correction and plane position correction, which shall be carried out simultaneously.

a. 垂直度校正。

a. Verticality correction.

吊车梁垂直度用靠尺、线锤检查，允许偏差应在规范规定的 5 mm 以内。T 形吊车梁测其两端垂直度，鱼腹式吊车梁测其跨中两侧垂直度。校正吊车梁的垂直度时，需在吊车梁底端与柱牛腿面之间垫斜垫块，为此，可根据吊车梁的轻重使用撬棍或千斤顶等工具将吊车梁抬起，也可在柱上或屋架上悬挂倒链，将吊车梁须垫斜垫块的一端吊起。

The verticality of the crane beam shall be checked with a guiding rule and a plumb bob, and the allowable deviation shall be within 5 mm as specified in the specification. The verticality at both ends of the T-shaped crane beam, and the verticality at both sides of the midspan of the fish-belly crane beam shall be measured. For correction of the verticality of the crane beam, it is necessary to add an inclined cushion block between the crane beam bottom and the column bracket surface, so that the crane beam can be lifted with a crowbar or jack depending on the weight of the crane beam, or a chain block can be hung on the column or roof truss to lift the end of the crane beam that needs an inclined cushion block.

b. 平面位置校正。

b. Plane position correction.

吊车梁的平面位置校正常用通线法和平移轴线法。

The straight-through line method and axis translation method are commonly used for correction of plane position of crane beam.

通线法：根据柱的定位轴线，在车间两端地面用木桩定出吊车梁定位轴线的位置，并设置经纬仪。先用经纬仪将车间两端的四根吊车梁位置校正准确，用钢尺检查两列吊车梁之间的跨距是否符合要求。然后在四根已校正的吊车梁端部设置支架（或垫块），垫高 200 mm，并根据吊车梁的定位轴线拉钢丝通线，然后根据通线来逐根用撬棍拨正吊车梁。

Straight-through line method: According to the positioning axis of the column, the position of the positioning axis of the crane beam shall be marked with wooden pegs on the ground at both ends of the workshop, and a theodolite shall be set up. First, a theodolite shall be used to correct the positions of the four crane beams at both ends of the workshop, and a steel ruler to check whether the span between the two crane beams meets the requirements. Then, supports(or cushion blocks) shall be provided at the ends of the four corrected crane beams to raise them by 200 mm, steel wires pulled through based on the positioning axes of the crane beams, and finally a crowbar used to set right the crane beams one by one based on the wires.

平移轴线法：首先，在柱两边设置经纬仪；然后，逐根将杯口上柱的吊装中心线投影到吊车梁顶面处的柱身上，并进行标记。

Axis translation method: First, theodolites shall be placed on both sides of the column, and then the lifting centerline of the column on the cup rim shall be projected onto the column body at the top surface of the crane beam one by one, and marks shall be made, too.

边吊边校法：对于较重的吊车梁，由于脱钩后校正比较困难，所以一般采取边吊边校法。

Correction while lifting: For heavy crane beams, it is difficult to correct them after unhooking, so they are usually corrected while being lifted.

吊车梁的最后固定：在吊车梁校正完毕后，将梁与柱子上的预埋铁件进行焊接，并在接头处支模，浇灌细石混凝土。

Final fixation of crane beam: After the crane beam is corrected, the beam is welded with the embedded iron piece on the column, and the formwork is erected at the joint and fine aggregate concrete is poured.

（3）屋架的吊装

(3) Lifting of roof truss

单层工业厂房的钢筋混凝土预应力屋架一般在施工现场平卧叠浇生产。屋架安装施工的主要工序有：绑扎，扶直与就位，吊升、对位与临时固定，校正及最后固定等。

Reinforced concrete prestressed roof trusses of single-storey industrial plants are usually poured horizontally on the construction site. The main procedures of roof truss installation construction include binding, straightening and positioning, lifting, alignment, temporary fixation, correction, final fixation, etc.

①绑扎。

① Binding.

屋架的绑扎点与绑扎方式与屋架的形式和跨度有关，其绑扎的位置及吊点的数目一般参考设计方案。屋架绑扎时吊索与水平面的夹角不宜小于45°，以免屋架上弦杆承受过大的压力使构件受损，如加大夹角，则吊索过长，起重机的起重高度不够时，可采用横吊梁。当屋架跨度不大于18 m时采用两点绑扎；屋架跨度为18～24 m时采用四点绑扎；跨度为30～36 m时采用9 m横吊梁、四点绑扎，如图6-9、图6-10所示。

The binding point and binding method of the roof truss are related to the form and span of the roof truss, and the binding positions and the number of lifting points are generally referred to the design scheme. During binding of roof truss, the included angle between the sling and the horizontal plane should not be less than 45°, so as to avoid damage to the member due to excessive pressure on the upper chord of the roof truss. If the included angle is increased, the sling is too long. When the lifting height of the crane is not enough, the cross hanging beam can be used. When the span of the roof truss is no more than 18 m, two-point binding is adopted; when the span of the roof truss is 18-24 m, four-point binding is adopted; when the span is 30-36 m, 9 m cross hanging beam and four-point binding are adopted, as shown in Figure 6-9 and Figure 6-10.

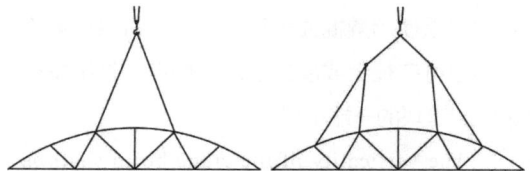

图6-9　屋架两点绑扎和屋架四点绑扎

Figure 6-9　Two-point Binding of Roof Truss and Four-point Binding of Roof Truss

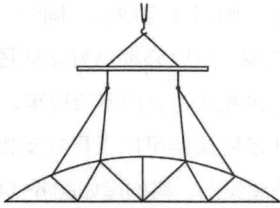

图6-10　用横吊梁四点绑扎

Figure 6-10 Four-point Binding with Cross Hanging Beam

②扶直与就位。

② Straightening and positioning.

屋架都是平卧生产，吊装前必须先翻身扶直。由于屋架平面刚度差，翻身中易损坏，18 m以上的屋架应在屋架两端用方木搭设井字架，高度与下一榀屋架的上平面相同，以便屋架扶直后搁置其上。24 m以上的屋架当验算得出抗裂度不够时，可在屋架下弦中节点处设置垫点，使屋架在翻身过程中下弦中节点始终着实。扶直后，下弦的两端应着实、中部则悬空，因此中垫点的厚度应适中。屋架高度大于 1.7 m 时，应加绑木、竹或钢管横杆，以加强屋架平面刚度，如图 6-11、图 6-12 所示。

Roof trusses are produced in a horizontal position, and must be turned over and kept upright before lifting. Due to the poor plane rigidity of the roof truss, it is likely to be damaged during turning over. For a roof truss above 18 m, square timbers shall be used to erect well-shape frames at both ends of the roof truss, with a height same as the upper plane of the next roof truss, so that the roof truss can be placed on it after being straightened. When the caculated crack resistance of the roof truss above 24 m is not enough, cushion points can be set at the middle node of the lower chord of the roof truss, so that the middle node of the lower chord is always attached solidly during the turning of the roof truss. After straightening, both ends of the lower chord shall be attached solidly, and the middle part shall be suspended, so the thickness of the middle cushion point shall be moderate. When

the height of the roof truss is greater than 1.7 m, timber, bamboo or steel tube cross bars shall be added to strengthen the plane rigidity of the roof truss, as shown in Figure 6-11 and Figure 6-12.

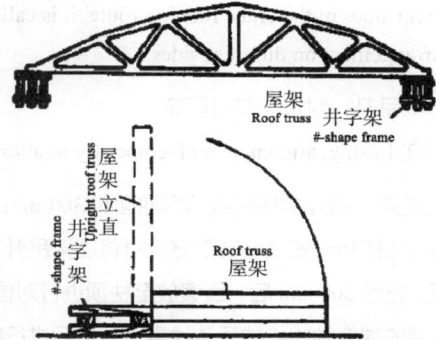

图6-11　屋架重叠生产的翻身扶直

Figure 6-11 Turnover and Straightening of Roof Trusses Produced in an Overlapped Manner

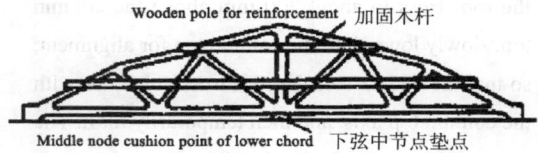

图6-12　屋架设置中垫点的翻身扶直

Figure 6-12 Turnover and Straightening of Cushion Points in Roof Truss Setting

屋架扶直后应排列、摆放在规定位置，排列、摆放位置与起重机的性能和吊装方法有关，应少占场地，便于吊装，并考虑屋架的吊装顺序、两端朝向等问题。当屋架排列、摆放位置与屋架的预制位置在起重机开行路线同一侧时，称同侧排放。当屋架排列、摆放位置与屋架预制位置分别在起重机开行路线不同侧时，叫异侧排放。

After the roof truss is straightened, it shall be arranged at the specified position which is related to the performance and lifting method of the crane. It shall occupy less site area for easy lifting, and the lifting sequence and the orientation of both ends of the roof truss shall be considered. When the roof truss storage position and the roof truss prefabrication position are on the same side of the crane running route, it is called the arrangement on the same side. When the roof truss storage position and the roof truss prefabrication position are on different sides of the crane running route, it is called the arrangement on different sides.

③吊升、对位与临时固定。

③ Lifting, alignment and temporary fixation.

屋架的吊升是将屋架吊离地面约 300 mm，然后将屋架转至安装位置下方，再将屋架吊升至柱顶上方约 300 mm 后，缓缓放至柱顶进行对位，使屋架的端头轴线与柱顶轴线重合，然后进行临时固定的过程。屋架固定稳妥后起重机才能脱钩。

The lifting of roof truss is to lift the roof truss about 300 mm above the ground, then turn the roof truss below the installation position, and next lift the roof truss to about 300 mm above the column top, slowly lower it to the column top for alignment, so that the end axis of the roof truss coincides with the column top axis, and then temporarily fix it. The crane can be unhooked only after the roof truss is fixed properly.

第一榀屋架的临时固定必须可靠，因为它

是单片结构，侧向稳定性差；同时，它是第二榀屋架的支撑，所以必须做好临时固定。一般采用四根缆风绳从两边把屋架拉牢，如图 6-13 所示。其他各榀屋架可用工具式支撑临时固定在前面一榀屋架上，其构造如图 6-14 所示。工具式支撑的构造，如图 6-15 所示。

The temporary fixation of the first roof truss must be reliable, because it is a monolithic structure with poor lateral stability and it is also the support of the second roof truss. Generally, four cable wind ropes are used to fasten the roof truss at both sides, as shown in Figure 6-13. Other roof trusses can be temporarily fixed on the previous roof truss with tool supports, and the structure is as shown in Figure 6-14. The structure of the tool support is shown in Figure 6-15.

④校正及最后固定。

④ Correction and final fixation.

屋架主要校正垂直度偏差。屋架的垂直度偏差可用锤球或经纬仪检查，在屋架的中间和两端设置三处卡尺。挑出屋架中心线 50 cm，观

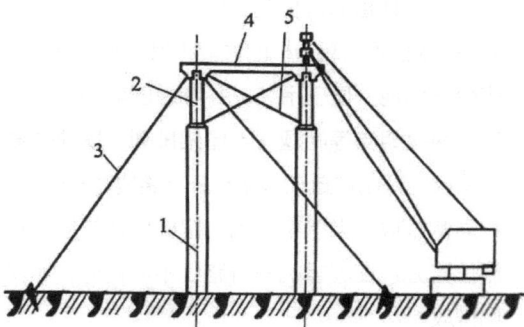

1.柱子；2.屋架；3.缆风绳；4.工具式支撑；5.屋架垂直支撑

1. column; 2. roof truss; 3. cable wind rope; 4. tool support; 5. roof truss vertical support

图6-13　屋架的临时固定

Figure 6-13 Temporary Fixation of Roof Truss

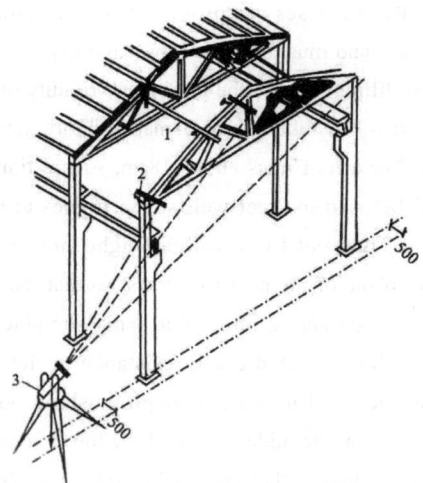

1.工具式支撑；2.卡尺；3.经纬仪

1. tool support; 2. caliper; 3. theodolite

图6-14　屋架的临时固定与校正（单位：mm）

Figure 6-14 Temporary Fixation and Correction of Roof Truss

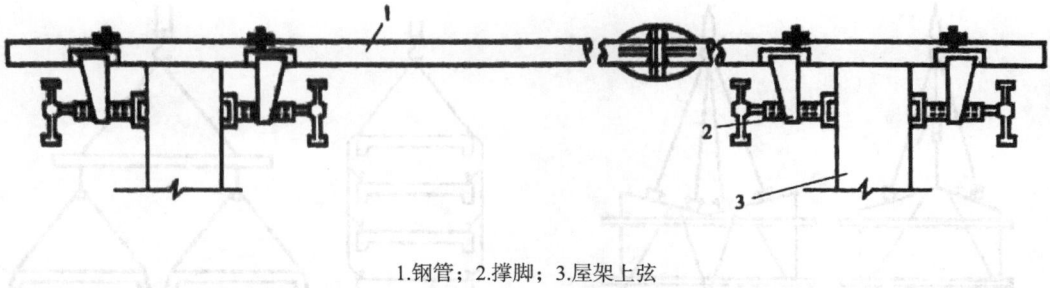

1.钢管；2.撑脚；3.屋架上弦

1. steel tube; 2. brace; 3. upper chord of roof truss

图6-15 工具式支撑的构造

Figure 6-15 Structure of Tool Support

测三个卡尺的标志是否在同一垂直面上，存在误差时，转动工具式屋架校正器上的螺栓加以校正，在屋架两端的柱底上嵌入斜垫铁，如图6-16所示。

The verticality deviation of the roof truss is mainly corrected. The vertical deviation of the roof truss can be checked with a plumb bob or theodolite. First, place three calipers respectively in the middle and at both ends of the roof truss, overhanging from the centerline of the roof truss for 50 cm. Then, observe whether the marks of the three calipers are on the same vertical plane. In case of any error, rotate the bolt on the tool-type roof truss corrector for correction; next, insert inclined shim plates into the column bases at both ends of the roof truss, as shown in Figure 6-16.

校正无误后立即用电焊固定，焊接时应在屋架的两侧同时对角施焊，不得同侧同时施焊。

After that, immediately fix it by electric welding by diagonal welding on both sides of the roof truss at the same time, instead of welding on the same side.

（4）天窗架及屋面板的吊装

(4) Lifting of Skylight Truss and Roof Slab

天窗架常单独吊装，也可与屋架拼装成整体同时吊装。单独吊装时，应待屋架两侧屋面板吊装完成后进行，采用两点或四点绑扎，并用工具式夹具或圆木进行临时加固，如图6-17所示。

The skylight truss is usually lifted separately or assembled with the roof truss and lifted together. If it is lifted separately, it shall be lifted after the roof slabs on both sides of the roof truss are lifted, bound at two or four points, and temporarily reinforced with tool clamps or logs, as shown in Figure 6-17.

屋面板多采用一钩多块叠吊或平吊法，以

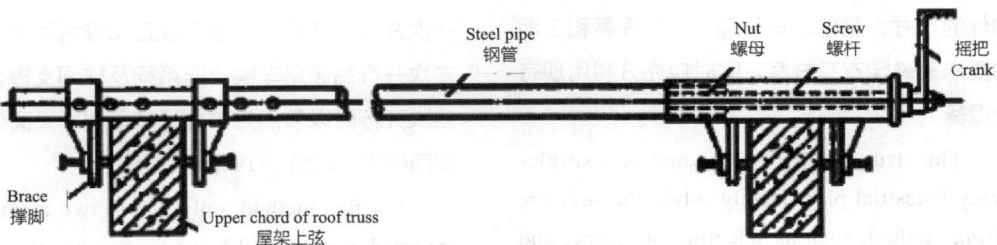

图6-16 屋架校正器

Figure 6-16 Roof Truss Corrector

– 161 –

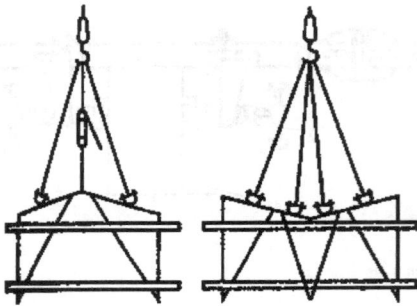

图6-17 天窗架的绑扎、吊装

Figure 6-17 Binding and Lifting of Skylight Truss

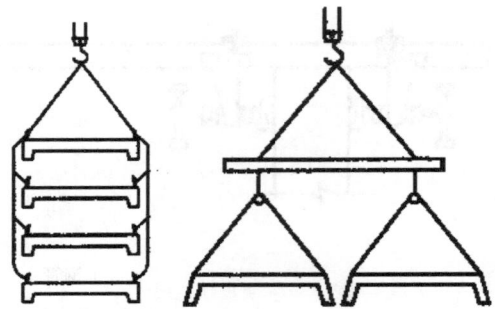

图6-18 多块叠吊与多块平吊

Figure 6-18 Multi-piece Overlapped Lifting and Multi-piece Horizontal Lifting

发挥起重机的效能。吊装顺序: 由两边檐口开始, 左右对称逐块向屋脊安装, 避免屋架承受半跨荷载。屋面板对位后应立即焊接牢固, 每块板角点焊接不少于三个, 如图 6-18 所示。

Roof slabs are mostly lifted horizontally or after being overlapped by one hook to give full play to the efficiency of cranes. Lifting shall start from the cornices on both sides, and install the roof truss symmetrically on the left and right one by one to avoid half-span load on the roof truss. After alignment, roof slabs shall be welded firmly immediately at least three corner points for each slab, as shown in Figure 6-18.

6.1.3 结构安装方案

6.1.3 Structure Installation Scheme

单层工业厂房结构吊装方案主要解决结构吊装方法, 如选择起重机, 确定起重机的开行路线和平面布置等内容。应根据厂房结构形式, 构件的尺寸、重量、安装高度, 工程量和工期的要求来确定安装方案, 同时应充分利用现有的设备。

The structure lifting scheme of a single-storey industrial plant mainly solves the structure lifting method, such as selecting the crane, and determining the running route and plane layout of the crane. The installation scheme shall be determined based on the structural form of the plant, the member size, weight, installation height, as well as quantities and construction period, while the existing equipment therein shall be fully utilized.

（1）结构吊装方法

(1) Structure lifting method

单层工业厂房结构的结构安装方法有分件吊装法和综合吊装法。

There is piece-by-piece lifting and integrated lifting for the structure installation of a single-storey industrial plant.

①分件吊装法。

① Piece-by-piece lifting.

分件吊装法是在厂房结构吊装时, 起重机每开行一次仅吊装一种或两种构件。例如, 第一次开行吊装柱, 并进行校正和最后固定, 第二次开行吊装吊车梁、连系梁及柱间支撑; 第三次开行时以节间为单位吊装屋架、天窗架及屋面板等, 如图 6-19 所示。

For this method, only one or two members are lifted every time the crane runs when the plant structure is lifted. For example, the columns are lifted first, corrected and finally fixed; second, the crane beams, connecting beams and column bracings are lifted; third, the roof trusses, skylight trusses and roof slabs are lifted in sections, as shown in Figure 6-19.

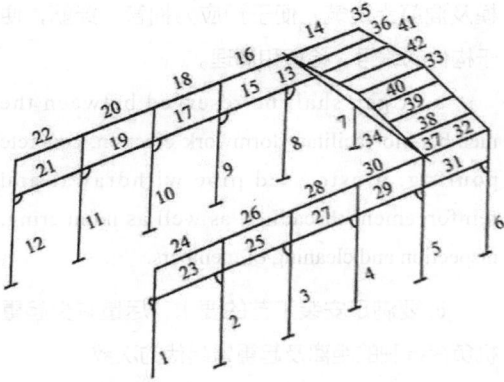

1~12.柱；13~32.单数是吊车梁，双数是连系梁；33、34.屋架；35~42.屋面板

1-12. columns; 13-32. singular for crane beams, even numbers for connecting beams; 33, 34. roof trusses; 35-42. roof slabs

图6-19　分件安装时的构件吊装顺序

Figure 6-19 Lifting Sequence of Members During Piece-by-piece Lifting

②综合吊装法。

② Integrated lifting.

综合吊装法是在厂房结构安装过程中，起重机一次开行，以节间为单位安装所有的结构构件。这种吊装方法具有起重机开行路线短，停机次数少的优点。但是由于综合吊装法要同时吊装各种类型的构件，起重机的性能不能充分发挥；索具更换频繁，影响生产率的提高；构件校正要配合构件吊装工作进行，校正时间短，给校正工作带来困难；构件的供应及平面布置也比较复杂。所以，在一般情况下，不宜采用这种吊装方法，只有在吊装轻型车间（结构构件种类相差不大）时，或采用移动困难的起重机（如桅杆式起重机）吊装时才采用综合吊装法。

The crane runs at one time to install all structural members by sections during the installation of plant structures. In this way, the crane has a short running route and less shutdown times. However, various types of members need to be lifted at the same time with this method, so the performance of the crane cannot be fully exerted; besides, the frequent replacement of rigging will affect the improvement of productivity; the correction of members needs to be carried out in coordination with the lifting of members, and the correction time is short, which makes it difficult to correct; the supply and plane layout of members are also complex. Therefore, in general, this lifting method should not be adopted, except that when the light workshop(with little difference in types of structural members) are lifted, or when it is difficult for the crane(such as mast crane) to move.

（2）起重机的选择

(2) Selection of crane

起重机的选择主要包括选择起重机的类型、型号、臂长及起重机的数量。一般中小型厂房多选择履带式等自行式起重机；当厂房的高度和跨度较大时，可选择塔式起重机吊装屋盖结构；当一台起重机无法吊装时，可采用两台起重机抬吊；遇到缺乏自行式起重机或受到地形的限制起重机难以移动的情况，或在自行式起重机难以到达的地方，可选择桅杆式起重机。

The selection of crane mainly considers the crane type, model, boom length and number. Self-propelled cranes such as crawler cranes are generally selected for small- and medium-sized plants; when the plant has a large height and span, the tower crane can be selected to lift the roof structure, and two cranes can be used if one is not enough; when self-propelled cranes are absent or there are terrain constraints, or it is difficult for self-propelled cranes to reach, mast cranes may be selected.

起重机类型确定后，要根据构件的重量、尺寸和安装高度确定起重机型号，使所选起重机的三个工作参数（起重量、起重高度、起重半径）满足结构吊装的要求。

After the crane type is determined, the crane model shall be determined based on the weight, size and installation height of the member, so that the three working parameters of the selected crane(lifting capacity, lifting height and lifting radius) shall meet the requirements of structure lifting.

吊装屋架、屋面板等屋面构件时，起重机宜跨中开行；吊装柱子时，则视跨度大小、构件尺寸、质量及起重机性能，可沿跨中开行或跨边开行。

When lifting roof members such as roof truss and roof slab, the crane should run along the midspan; when lifting columns, the crane can run along the midspan or sidespan depending on the span size, member size, mass and crane performance.

（3）构件的平面布置和吊装前的堆放

(3) Plane layout and stacking before lifting of members

构件的平面布置与吊装方法、起重机性能、构件制作方法有关。在选定起重机型号，确定施工方案后，可根据施工现场实际情况确定如何进行构件的平面布置。

The plane layout of member is related to the lifting method, crane performance and member fabrication method. After the crane model and the construction scheme are selected, it can be determined according to the actual situation on the construction site.

①构件的平面布置原则。

① Principles for plane layout of members.

a. 每跨的构件可布置在本跨内，也可布置在跨外便于吊装的地方。

a. The members of each span are either arranged in this span, or in a place easy for lifting outside the span.

b. 构件之间应预留有一定空隙，应便于支模及混凝土浇筑，便于预应力抽管、穿筋，便于构件的编号、检查和清理。

b. A gap shall be reserved between the members to facilitate formwork erection, concrete pouring, prestressed pipe withdrawal and reinforcement threading, as well as numbering, inspection and cleaning of members.

c. 要满足安装工艺的要求，尽量减少起重机负荷行驶的距离及起重臂起伏的次数。

c. It is necessary to reduce the running distance of the crane under load and the lifting times of the boom in order to meet the requirements of the installation process.

d. 要为起重机械、运输车辆预留开行道路以及为起重机预留回转工作面。

d. Roads for lifting machinery and transport vehicles, slewing working face for crane shall be reserved.

e. 要注意布置时构件的朝向，特别是屋架等构件，以免在安装时需要空中调头，影响安装进度，也不安全。

e. Attention should be paid to the orientation of members during plane layout, especially roof trusses, so as not to turn around in the air during installation, which will affect the installation progress and cause potential safety hazards.

f. 构件要布置在坚实的地基上，在新填土的地基上布置构件时，必须采取一定的措施，防止地基下沉，影响构件质量。

f. The members shall be arranged on a solid foundation. If on a newly filled foundation, measures must be taken to prevent foundation subsidence that will affect the quality of the members.

②柱子的布置。

② Layout of columns.

a. 斜向布置：预制柱子与厂房纵轴线成一斜角。这种布置主要是为了配合旋转起吊。

a. Oblique layout: The prefabricated column forms an oblique angle with the longitudinal axis of the plant. This layout is mainly to adapt to rotary lifting.

旋转法吊装柱子时最好按图6-20（a）所示的三点（杯型基础中心、柱脚、绑扎点）共弧斜向布置。当场地受限制或柱子较长，柱的平面布置按三点共弧有困难时，可采用两点（杯型基础中心、柱脚）共弧，如图6-20（b）所示。

When the column is lifted by rotating, it is better to arrange the column obliquely in the same arc at three points(cup foundation center, column base and binding point), as shown in Figure 6-20(a). If it is difficult to do this due to restrictions on the site or a long column, it may be arranged obliquely at two points(cup foundation center and column base) in the same arc, as shown in Figure 6-20(b).

b. 纵向布置：柱子预制与厂房轴线平行排列，主要是配合滑行法起吊柱子。

b. Longitudinal layout: The prefabricated columns are arranged in parallel with the axis of the plant, mainly for lifting of columns by sliding.

布置时可考虑让起重机停于两柱之间，每停机一次安装两根柱子，柱子的绑扎点应布置在起重机吊装该柱时的起重半径上。

During layout, it can be considered that the crane stops between two columns. For each stop, two columns shall be installed. The binding point of the column shall be arranged on the lifting radius when the crane is lifting the column.

③屋架的布置。

③ Layout of roof trusses.

钢筋混凝土或预应力混凝土屋架多采用在跨内平卧叠层预制，每叠3～4榀，布置方式有斜向布置、正反斜向布置和正反纵向布置三种，多采用斜向布置，因其便于扶直和排放，只有在场地受到限制时，才考虑其他两种形式。

Reinforced concrete or prestressed concrete roof trusses are mostly prefabricated horizontally and overlapped in the span, with 3-4 trusses for each lamination. There are three types of layout: oblique layout, forward/reverse oblique layout, and forward/reverse longitudinal layout, and oblique layout is mostly adopted because it is convenient for straightening and placement, but other two layouts are considered only when the site is restricted.

④吊车梁的布置。

④ Layout of crane beam.

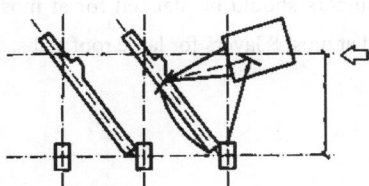

（a）三点共弧斜向布置

(a) Obliquely arranged in the same arc at three points

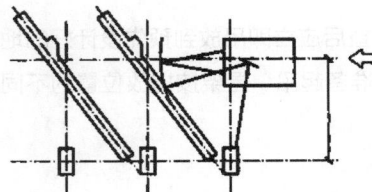

（b）两点共弧斜向布置

(b) Obliquely arranged in the same arc at two points

图6-20　旋转法柱子的平面布置

Figure 6-20 Plane Layout of Columns with Rotation Method

当吊车梁在现场预制时，可靠近柱基础纵向轴线或略倾斜布置，也可插在柱子之间预制。如具备运输条件，也可在场外集中预制。

When the crane beam is prefabricated on site, it can be arranged near the longitudinal axis of the foundation or slightly inclined, or inserted between the columns. If the transportation conditions are available, it can also be prefabricated off-site in a centralized manner.

⑤吊装阶段构件的布置、运输及堆放。

⑤ Layout, transportation and stacking of members during lifting.

由于柱在预制阶段已按吊装阶段的就位要求进行布置，当柱的混凝土强度达到设计要求的等级后，即可现行吊装，以便空出场地布置其他构件。吊装阶段的就位布置是指柱已吊装完毕后，屋架的扶直排放，吊车梁、屋面板的运输、排放等。

Since the column has been arranged according to the positioning requirement for lifting in the prefabrication stage, when the concrete strength of the column reaches the design strength, it can be lifted immediately, so as to leave the site for arrangement of other members. The arrangement in place in the lifting stage refers to the vertical placement of the roof truss, the transportation and placement of the crane beam and roof slab, etc. after the lifting of the column.

a. 屋架的扶直排放。

a. Vertical placement of roof trusses.

屋架扶直后应立即吊放到预先设计好的地面位置上，准备起吊。屋架按排放位置的不同可分为同侧排放和异侧排放。同侧排放时，屋架的预制位置与排放位置均在起重机开行路线的同一边；异侧排放时，需将屋架由预制的一边转至起重机外行路线的另一边排放。

After the roof truss is straightened, it shall be immediately placed on the ground position pre-designated for lifting. The roof truss can placed on the same side or on different sides. When it is placed on the same side, it shall be prefabricated and placed on the same side of the crane running route; when it is placed on different sides, it shall be turned from the prefabrication side to the other side of the crane running route for placement.

b. 吊车梁、连系梁及屋面板的排放。

b. Placement of crane beams, connecting beams and roof slabs.

单层厂房除了柱和屋架一般在施工现场制作外，其他构件如吊车梁、连系梁、屋面板等均在预制厂制作，然后运到现场按施工组织设计所规定的位置就位或集中堆放。梁式构件的叠放不宜超过2层，大型屋面板的叠放不宜超过8层。

Except that the columns and roof trusses of the single-storey workshop are generally fabricated on the construction site, other members such as crane beams, connecting beams and roof slabs are fabricated in the prefabrication plant, and then transported to the site and placed or stacked in a centralized manner in positions specified in the construction organization design. The beam members should be stacked for at most 2 layers, and at most 8 layers for large roof slabs.

6.2 钢结构安装

6.2 Installation of Steel Structure

6.2.1 钢结构施工概述

6.2.1 Overview of Steel Structure Construction

钢结构是用钢板、型钢和圆钢等通过焊接连接、螺栓连接等方式制造的结构。目前常用的钢结构类型主要有钢梁、钢柱、钢桁架、钢网架结构等。

A steel structure is made of steel plate, section steel and round steel by welding connection, bolt connection and so on. At present, the commonly used steel structures mainly include the structure of steel beams, steel columns, steel trusses and steel grids.

（1）钢结构的特点

(1) Characteristics of steel structure

钢结构房屋是用钢板或各种型钢，通过焊接、螺栓连接组装成结构骨架的房屋。

When it comes to the steel structure building, steel plates or various section steels are assembled by welding and bolting to form the structural framework of the building.

钢结构房屋与其他结构房屋相比优点很多：钢材近似匀质，强度高、韧性好，有利于抗震，理论计算与实际最接近；用钢材轧制的各种理想断面可使构件轻巧；构件连接构造简单，可建造大跨度、大空间建筑；结构占用面积小；符合绿色、环保的理念；便于机械化、工厂化制作后再现场安装，施工周期短，质量较高。但钢结构房屋的缺点也很明显：薄壁杆件多，容易发生整体或局部失稳；构件相互连接的节点多，容易产生次应力；易锈蚀，不耐火，需

要定期维护；高层立柱有较大的压缩变形。

Compared with other structure buildings, steel structure building has many advantages such as approximately uniform steel, high strength, good toughness and seismic resistance. Besides, the theoretical calculation is closest to the reality; various ideal sections rolled with steel allow lightweight members; the member connection is simple, which allows to build large-span and large-space buildings; this structure occupies a small area, in line with the concept of green and environmental protection; it is easy for on-site installation after mechanized factory fabrication, with short construction period and high quality. However, the steel structure building also has obvious disadvantages. For example, there are many thin-walled members and bars, which may cause overall or local instability. Moreover, there are many connection nodes of members, which is likely to cause secondary stress; the members are easy to rust, not fireproof and need regular maintenance; high-rise columns have large compressive deformation.

（2）钢结构的材料

(2) Materials of steel structure

①钢板和钢带。

① Steel plates and strips.

建筑钢结构使用的钢板和钢带的轧制方法通常为冷轧和热轧两种。

The rolling methods of steel plates and strips used in building steel structures are usually cold rolling and hot rolling.

钢板是指平板状、矩形的，可直接轧制或

由宽钢剪切而成的板材；而钢带一般成卷交货，宽度不小于 600 mm 的为宽钢带，宽度小于 600 mm 的为窄钢带，窄钢带可直接轧制也可由宽钢带纵剪而成。板厚 4 mm 以下的为薄钢板，4 ~ 60 mm 的为厚钢板，厚度大于 60 mm 的称为特厚钢板。

Steel plates refer to flat, rectangular plates that can be directly rolled or cut from wide steel. Steel strips are generally delivered in coils, with a width of not less than 600 mm for wide steel strips and less than 600 mm for narrow steel strips. Narrow steel strips can be directly rolled or longitudinally cut from wide steel strips. Steel plates with a thickness of less than 4 mm are thin steel plates, those with a thickness of 4-60 mm are thick steel plates, and those with a thickness of more than 60 mm are extra-thick steel plates.

②结构用钢管。

② Steel tubes for structures.

结构用钢管有热轧无缝钢管和焊接钢管两大类。

There are mainly two types of structural steel tubes: Hot-rolled seamless steel tubes and welded steel tubes.

其中，焊接钢管由钢带卷焊而成，依据管径大小，又分为直缝焊和螺旋焊两种。

Welded steel tubes are made of steel strips by rolling welding, and there are straight welding tubes and spiral welding tubes depending on the pipe diameter.

③型钢。

③ Section steel.

工字钢、槽钢和角钢是工程结构中最早使用的传统型钢，随着轧制技术的发展，更多截面性能优良的型钢相继问世，如圆钢管、方钢管、H 型钢。

I-steel, channel steel and angle steel are

the earliest traditional section steels used in construction structures. With the development of rolling technology, more and more section steels with excellent section properties have been produced successively, such as round steel tubes, square steel tubes and H-steel.

（3）钢结构构件的加工

(3) Processing of steel structure members

钢结构构件一般在工厂加工制作，然后运至工地进行安装。钢结构制作工序较多，加工流程要安排周密，避免工件倒流，减少往返运输时间。

Steel structure members are generally processed and fabricated in the plant, and then transported to the construction site for installation. There are many fabrication procedures for steel structures, and the processing sequence shall be carefully arranged to avoid return of workpieces and reduce round-trip transportation time.

钢结构构件加工包括放样、画线、切割、矫正、弯制成型、制孔、涂装等工序。

The fabrication procedures of steel structure members include setting-out, marking, cutting, correction, bending and forming, hole making, painting, etc.

（4）钢结构构件的验收、运输、堆放

(4) Acceptance, transportation and stacking of steel structure members

①钢结构构件的验收。

① Acceptance of steel structure members.

钢结构构件加工制作完成后，应按照施工图和验收规范进行验收，钢结构构件出厂时，应提供齐备的技术资料。

The steel structure members shall be checked and accepted in accordance with the construction drawings and acceptance specifications after the processing and fabrication, and the complete technical data shall be furnished when the steel structure members leave the plant.

②钢结构构件的运输。

② Transportation of steel structure members.

a. 钢结构构件的运输可采用公路、铁路或水路运输。运输构件时，应根据构件的长度、重量、断面性质、运输形式的要求选用合理的运输方式。

a. Steel structure members can be transported by road, railway or water. Reasonable transportation modes shall be selected according to the requirements member length, weight, section property and transportation form.

b. 大型或重型构件的运输宜编制运输方案。

b. A transportation plan should be prepared for the transportation of large or heavy members.

c. 运输的构件单件质量超过 3 t 的，宜在易见部位用油漆标上重量及重心位置。

c. If the mass of a single member transported is more than 3 t, the weight and center-of-gravity position should be marked with paint visibly.

d. 构件运输时，构件在运输车辆上的支点应符合要求。两端伸长的长度及绑扎方法均应保证构件不产生永久性变形、不损伤涂层。构件起吊必须按设计吊点起吊。

d. When the members are transported, the pivots of the members on the transport vehicles shall meet the requirements. The extension length at both ends and the binding method shall ensure that the members will not be subject to permanent deformation and the coating will not be damaged. The members must be lifted at the design lifting points.

③钢结构构件的堆放。

③ Stacking of steel structure members.

a. 构件的堆放场地应平整、坚实，无水坑、冰层，地面干燥，并应排水通畅，有较好的排

水设施，同时有便于车辆进出的回路。

a. The stacking area of members shall be flat and solid, free of pools and ice cover. The ground shall be dry, with smooth drainage, good drainage facilities. In addition, there shall be a passage for access of vehicles.

b. 构件应按种类、型号、安装顺序划分区域保存，插竖标志牌。构件底层垫块要有足够的支承面，钢结构产品不得直接置于地上，要垫高 200 mm。

b. Members shall be stored in different zones according to the type, model and installation sequence, with sign boards erected. The cushion block at the bottom of the member shall have sufficient bearing surface, and the steel structure products shall not be directly placed on the ground, but shall be cushioned to raise by 200 mm.

c. 对于已堆放好的构件，要派专人进行管理，严禁乱翻、乱移。要对已堆放好的构件进行保护，避免风吹雨打、日晒夜露等对其造成不良影响。

c. Special personnel shall be assigned to take care of members stacked, and it is strictly prohibited to turn over or move them at will. In addition, the stacked members shall be properly protected from weather exposure.

d. 不同类型的构件不要堆放在一起。

d. Members of different types shall not be stacked together.

（5）钢结构构件的连接

(5) Connection of steel structure members

钢结构构件的连接方法有焊接、紧固件连接（螺栓连接、射钉、自攻钉、拉铆钉）及铆接三种。

There are three connection methods for steel members: Welding, fastening(bolts, shoot nails, self-tapping screws and rivets) and riveting.

（6）钢结构的防腐与涂装

(6) Anti-corrosion and coating of steel structures

①钢结构构件防腐涂料的种类。

① Types of anticorrosive paints for steel structure members.

涂料按其基料中的成膜物质可分为 17 类，施工中按其作用及施涂先后顺序可分为底涂料和饰面涂料两种。

There are 17 categories of paints according to the film-forming materials in the binder, and primer and finish paint according to the function and sequences in construction.

②钢结构构件涂装前的表面处理。

② Surface treatment of steel structure members before coating.

涂装前钢材表面的处理是保证涂料防腐效果和钢结构构件使用寿命的关键。

The treatment of steel surface before coating is the key to ensure the anti-corrosion effect of coating and the service life of steel structure members.

钢结构的除锈是构件在施涂之前的一道关键工序，除锈干净可提高底涂料的附着力，确保构件的防腐质量。

The rust removal of steel structure is a key process before coating. It can improve the adhesion of primer and thus guarantee the good corrosion resistance of members.

③涂装施工。

③ Coating construction.

涂装施工前，应确保钢结构制作、安装、校正已完成并验收合格。

Before coating, it should be ensured that the fabrication, installation and correction of steel structures have been completed and accepted.

涂装施工环境温度宜为 15 ~ 30℃，施工环境相对湿度不大于 85%，钢材表面的温度应高于空气露点温度 3℃ 以上。

The ambient temperature during coating construction should be 15-30°C, and the relative humidity of the construction environment should not be more than 85%. The temperature of the steel surface shall be more than 3°C higher than the air dew point temperature.

a. 涂装工序及施工方法。

a. Coating process and methods.

涂装工序：涂刷防锈漆、局部刮腻子、涂装施工、漆膜质量检查。

Coating process: Application of antirust paint, local puttying, coating and checking paint film quality.

施工方法：刷涂法、滚涂法、浸涂法、空气喷涂法、无气喷涂法、粉末涂装法等。

Coating methods: Brush coating, roller coating, dip coating, air spraying, airless spraying, powder coating, etc.

b. 涂装遍数及涂层厚度。

b.Passes and thickness of coating.

涂装遍数、涂层厚度应符合设计要求。涂层干漆膜总厚度一般为：室外 150 μm，室内 125 μm，其允许偏差为 −25 μm。

The coating passes and thickness shall meet the design requirements. The total thickness of the dry paint film of the coating is generally 150 μm outdoors and 125 μm indoors, with an allowable deviation of -25 μm.

c. 钢结构防火涂料涂装施工。

c. Steel structure fireproof coating construction.

钢结构防火涂料类型：按所用黏结剂的不同分为有机类、无机类；按涂层的厚度分为薄涂型（2 ~ 7 mm）、厚涂型（8 ~ 50 mm）两类；按施工环境不同分为室内、露天两类；按涂层受热后的状态分为膨胀型和非膨胀型两类。

Types of fireproof coating for steel structure: There are organic and inorganic coatings based on different binders; thin coating(2-7 mm) and thick coating(8-50 mm) based on the coating thickness; indoor and outdoor types according to the different construction environment; intumescent coating and non-intumescent coating according to the state of the coating after heating.

施工要求：钢结构涂装应在安装就位完毕并经验收合格后进行。钢结构防火涂料施涂前应搅拌均匀，方可施涂。

Construction requirements: Steel structure coating shall be applied after installation and acceptance. Steel structure fireproof coating shall be mixed evenly before construction.

双组分涂料应按说明书规定的配比配制，随用随配。配制的涂料应在规定的时间内用完。

The two-component coating shall be prepared according to the proportion in the specification immediately before use. The coating prepared shall be used up within the specified time.

d. 薄涂型钢结构防火涂料涂装施工。

d. Thin-coated steel structrue fireproof coating construction.

喷涂方法：底层涂料宜喷涂；面层涂料可采用刷涂、喷涂或滚涂；局部修补及小面积施工可采用抹灰刀等工具手工抹涂。

Coating methods: The primer should be sprayed; the surface coating can be brushed, sprayed or rolled; local repair and small-area application can be done manually with trowel and the other tools.

施工要点：底层涂料一般喷 2 ~ 3 遍，待前一遍干燥后再喷后一遍，第 2、3 遍每遍喷涂厚度不宜超过 2.5 mm；底层涂料厚度应符合设计规定，基本干燥后再涂面层，面层涂料一般涂饰 1 ~ 2 遍，第 1 遍从左至右，第 2 遍则从右至左，保证覆盖全部底涂层。喷涂时，喷

枪要持稳，枪嘴与构件宜垂直或呈 70°，喷口距构件宜为 400 ~ 600 mm。涂层应厚薄均匀，不漏喷、不流淌，接槎平整，颜色均匀一致。

Key points: The primer is generally sprayed for 2-3 times, every time it shall be dried. The thickness of coating for the second and third time should not exceed 2.5 mm. The thickness of the primer coat shall meet the design requirements. The surface course shall not be constructed until the primer coat is basically dry. The surface course is generally painted for 1-2 times, from left to right for the first time and from right to left for the second time, so as to ensure that the primer coat is fully covered. During spraying, the spray gun shall be held stably, the nozzle shall be vertical or 70° to the member, with a distance of 400-600 mm from the member. The coating shall be uniform in thickness, without missing spraying or flowing, with flat joint and uniform color.

6.2.2 钢结构安装基础知识

6.2.2 Basic Knowledge of Steel Structure Installation

钢结构安装前应进行图纸会审，对施工的场地条件、钢构件核查等相关作业条件进行准备布置，以便于钢结构施工安装工作的顺利开展。

Before the installation of steel structures, the drawings shall be jointly reviewed, and the relevant operation conditions such as construction site conditions and steel member verification shall be prepared and arranged, so as to facilitate the smooth progress of steel structure construction and installation.

钢结构安装施工中除了起重设备外，还需运用到校正构件安装偏差的千斤顶、用于垂直水平运输的卷扬机、用于固定缆风绳的地锚、用于起吊轻型构件的倒链等索具设备。

In addition to lifting equipment, jacks for correcting member installation deviation, winches for vertical and horizontal transportation, ground anchors for fixing cable wind ropes, chain blocks and rigging for lifting light members will also be used for the installation of steel structures.

（1）钢结构工程安装方法选择

(1) Selection of installation methods for steel structure works

钢结构工程安装方法有分件安装法、节间安装法和综合安装法。

The piece-by-piece installation, inter-section installation and integrated installation may be adopted for steel structure works.

①分件安装法。

① Piece-by-piece installation.

分件安装法是指起重机在节间内每开行一次仅吊装一种或两种构件。例如，起重机第一次开行中先吊装全部柱子，并进行校正和最后固定，然后依次吊装地梁、柱间支撑、墙梁、吊车梁、托架、屋架、天窗架、屋面支撑和墙板等构件，直至整个建筑物吊装完成。有时屋面板的吊装也可在屋面上单独用桅杆或层面小吊车来进行。

The piece-by-piece installation means that the crane only lifts and installs one or two members once in every operation in the section. For example, during the first operation of the crane, all columns will be lifted, corrected and finally fixed, followed by members such as ground beams, column bracings, wall beams, crane beams, brackets, roof trusses, skylight trusses, roof supports and wall slabs in turn, until all members of the building are lifted. Sometimes, the roof slabs can also be lifted on the roof with masts or small cranes separately.

分件吊装法一般适用于中、小型厂房的吊装。

The piece-by-piece installation is generally suitable for small- and medium-sized plants.

②节间安装法。

② Inter-section installation.

节间安装法是指起重机在厂房内一次开行中，分节间依次安装所有构件，即先吊装一个节间柱子，并立即加以校正和最后固定，然后接着吊装地梁、柱间支撑、墙梁（连系梁）、吊车梁、走道板、柱头系统、托架（托梁）、屋架、天窗架、屋面支撑系统、屋面板和墙板等构件。一个（或几个）节间的全部构件吊装完毕后，起重机行进至下一个（或几个）节间，再进行下一个（或几个）节间全部构件的吊装，直至所有构件吊装完毕。

When it comes to inter-section installation, the crane in a single operation in the plant lifts and installs all types of members in turn, that is, columns for one section are first lifted and installed, corrected immediately and finally fixed, followed by the ground beams, column bracings, wall beams(connecting beams), crane beams, walkway slabs, capitals, brackets(joists), roof trusses, skylight trusses, roof supports, roof slabs, wall slabs and other members. After all members for one(or several) section(s) are lifted, the crane will move to the next(or several) section(s) for lifting of all members.

该方法适合采用回转式桅杆进行吊装，在吊装有特殊要求的结构或须原因局部特殊施工时采用。

It is suitable for lifting with slewing masts, or for structures with special requirements or local special construction for some reason.

③综合安装法。

③ Integrated installation.

综合安装法是将全部或一个区段的柱头以下部分的构件用分件安装法吊装，即柱子吊装完毕并校正固定后，再按顺序吊装地梁、柱间支撑、吊车梁、走道板、墙梁、托架，接着按节间综合吊装屋架、天窗架、屋面支撑系统和屋面板等屋面结构构件。整个吊装过程可按三次流水进行，根据结构特性，有时也可采用两次流水，即先吊装柱子，然后分节间吊装其他构件。

The integrated installation is to lift the members below the column capitals of a or all sections piece by piece, that is, columns are first lifted, corrected and fixed, followed by the ground beams, column bracings, crane beams, walkway slabs, wall beams and brackets in sequence, and then roof trusses, skylight trusses, roof supports, roof slabs and other roof members using the inter-section integrated lifting method. The whole lifting process can be carried out in three flow operations, sometimes in two flow operations depending on the structural characteristics, that is, the columns are lifted first, and then other members are lifted in sections.

综合安装法结合了分件安装法和节间安装法的优点，能最大限度地发挥起重机的能力从而提高效率，缩短工期，是广泛采用的一种安装方法。

In combination with the advantages of the piece-by-piece installation and the inter-section installation, the integrated installation can maximize the capacity and efficiency of the crane and shorten the construction period. It is a widely used installation method.

6.3 单层钢结构安装质量要求与安全技术

6.3 Quality Requirements and Safety Technology for Installation of Single-storey Steel Structures

6.3.1 单层钢结构安装质量要求

6.3.1 Quality Requirements for Installation of Single-storey Steel Structures

钢结构基础施工时，应注意保证基础顶面标高及地脚螺栓位置的准确，其偏差应在允许偏差范围内。

During the construction of steel structure foundation, attention shall be paid to the accuracy of foundation top elevation and anchor bolt position, and the deviation shall be within the allowable range.

钢结构安装应按施工组织设计进行。安装程序必须保持结构的稳定性且不导致永久性变形。

The steel structure shall be installed according to the construction organization design. The installation procedures must guarantee the stability of the structure and not cause permanent deformation.

钢结构安装前，应按构件明细表核对进场的构件，查验产品合格证和设计文件。工厂预拼装过的构件在现场拼装时，应根据预拼装记录进行。

The operators shall check the members on site against the member list, the manufacturer certificates and design documents of products before installation of the steel structure. The pre-assembled members in the factory shall be assembled on site based on the pre-assembly records.

钢结构安装偏差的检测，应在结构形成空间刚度单元并连接固定后进行，其偏差应在允许偏差范围内。

The installation deviation of steel structure

shall be checked after the structure forms a spatial rigid unit and is connected and fixed, and the deviation shall be within the allowable range.

6.3.2 单层钢结构安装安全技术

6.3.2 Safety Technology for Installation of Single-storey Steel Structure

（1）操作人员方面

(1) Operators

从事安装工作的人员要经过体检，心脏病或高血压患者，不能高空作业；不准酒后作业；新工人要经过培训才能上岗。

Personnel involved in installation shall undergo physical examination, and those with heart trouble or hypertension are not allowed to work at heights; do not work after drinking; new workers must be trained before taking up their posts.

操作人员进入现场，必须戴安全帽、手套；高空作业时，必须系好安全带；所用的工具，要用绳子扎好或放入工具包内。

Operators must wear safety helmets and gloves when entering the site; when working at heights, they must wear safety belts; the tools used shall be tied with ropes or put into tool kits.

电焊工在高空焊接时，应系安全带、戴防护面罩；在潮湿地点工作时要穿胶靴。

Welders shall wear safety belts and protective masks when working at heights, and wear rubber boots when working in wet places.

在高空安装构件时，用撬杠校正构件位置时必须防止发生因撬杠滑脱而引起的高空坠物事件。撬构件时人要站稳，最好一只手扶脚手架或构件，另一只手操作，撬杠插进的深度要适宜，循序渐进。

During installation of members at high altitude, be sure to prevent falling from high altitude due to release of the crowbar when correcting the positions of members. When prying the members, stand firmly. It is better to hold the scaffold or members with one hand and operate with the other hand. The crowbar shall be inserted step by step to an appropriate depth.

在冬、雨季施工时，为防止构件因潮湿或有积雪等使操作人员滑倒，必须采取防滑措施。

During construction in winter and rainy season, anti-slip measures must be taken to prevent operators from slipping due to moisture or snow.

登高用的梯子必须牢固，梯子与地面的夹角一般为 65°～75°。

The ladder for climbing must be secure, and the angle between the ladder and the ground is generally 65°-75°.

结构安装时，要听从统一号令，统一行动。

During structure installation, follow the unified command.

（2）起重机械与索具方面

(2) Lifting machinery and rigging

吊装所用的钢丝绳，事先必须认真进行检查，表面磨损、腐蚀达钢丝绳直径的 10% 时，不允许使用。

The steel wire rope used for lifting must be carefully checked in advance, and it shall not be used when the surface wear and corrosion reach 10% of the diameter of the steel wire rope.

吊钩和卡环如有永久性变形或裂纹时，不能使用。

The lifting hook and snap ring cannot be used if there is a permanent deformation or crack.

起重机的行驶道路，必须坚实可靠，如地面为松软土层则要进行压实处理，必要时还须对土层进行加固。

The road for the crane must be solid and reliable. If the ground is of soft soil, it shall be compacted and reinforced if necessary.

履带式起重机必须负荷行走时，重物应在履带的正前方，并用绳索绑住构件，缓慢行驶，构件离地不得超过 50 cm。起重机在接近满荷时，不得同时进行两种操作动作。

When the crawler crane must travel with load, the weight shall be directly in front of the crawler and the member shall be tied with a rope, and the crane shall travel slowly. The member shall not be more than 50 cm above the ground. When the crane is with almost full load, two operations shall not be carried out at the same time.

起重机工作时，严格注意勿碰撞高压架空电线，起重臂、钢丝绳、重物等与架空电线要保持一定的安全距离。

When the crane is working, do not collide with high voltage overhead wires, and maintain a safe distance between the boom, wire rope, heavy object, etc. and the overhead wires.

新到、修复或改装的起重机在使用前必须进行检查、试吊，并要进行静、动负荷试验。

A new, repaired or modified crane must be inspected and tested before use, and subject to static and dynamic load tests.

起吊构件时，升降吊钩要平稳，避免紧急制动和冲击。

When lifting members, it is required to lift or descend smoothly to avoid emergency braking and impact.

起重机停止工作时，起重装置要关闭上锁。吊钩必须升高，防止摆动伤人，并不得在其上悬挂物件。

When the crane is shut down, the lifting device shall be closed and locked. The hook must be raised to prevent swinging and hurting people, without hanging objects above it.

（3）安全设施方面

(3) Safety Facilities

吊装现场周围应设置临时栏杆，禁止非工作人员入内。

The lifting site shall be surrounded by temporary railings, and no unauthorized personnel are allowed in.

工人如需要在高空作业时，应尽可能搭设临时操作平台。

If workers have to work at height, a temporary operation platform shall be put up where possible.

要配备悬挂式斜靠的轻便爬梯，供人上下。

A portable, suspended and leaning ladder shall be available.

如需要在悬空的屋架上弦行走时，应设置安全防护栏杆。

When people need to walk on the upper chord of the overhead roof truss, guard rails shall be mounted.

遇到六级以上大风和雷雨天气时，一般不得进行高空作业；必须进行时，要采取妥善的安全措施。雷雨季节，起重设备在 15 m 以上高度时，必须装设避雷设施。

In case of strong winds above scale six and thunderstorms, high-altitude operations are generally not allowed; if necessary, appropriate safety measures shall be taken. In the thunderstorm season, lightning protection facilities must be installed when the lifting device is at a height of more than 15 m.

7

构件预制工厂规划与建设

Planning and Construction of Member Prefabrication Plant

学习目标:

Learning objectives:

掌握构件预制工厂的选址、预制构件常用的生产模式、构件预制工厂的总体规划、构件预制工厂设备的选择和厂区的布置。

Master the site selection of member prefabrication plant, common production modes of prefabricated members, overall planning of member prefabrication plant, selection of equipment in member prefabrication plant and layout of plant area.

技能抽查要求:

Skill checking requirements:

能进行构件预制生产厂房的基本规划并参与其基本建设。

Be able to make basic planning of member prefabrication plant and participate in its basic construction.

建筑八大员岗位资格考试要求：

Qualification examination requirements for eight major posts of construction engineering:

掌握构件预制生产厂房的规划内容。

Master the planning contents of member prefabrication plant.

装配式建筑指的是将建筑的部分或全部构件在工厂预制完成，然后运至施工现场，将构件通过可靠的连接方式组装而成的建筑。装配式建筑区别于传统建筑的特征之一就是工厂化生产。工厂化生产的优点是标准化程度高（工艺设置标准化、工序操作标准化）、机械化程度高（生产效率高、用工量减少）、产品质量有保证（内控体系）、受气候影响小（室内作业）。构件预制工厂建设的重点问题，首先是要确定生产模式，选择基地化工厂生产还是游牧式工厂生产；其次是确定构件预制工厂规划内容，主要包括厂址选择、工厂总平面布置、生产线布局及设备选型等。

Prefabricated building refers to a building in which some or all members of the building are prefabricated in the plant and then transported to the construction site to be assembled by reliable connection. One of the distinctions of prefabricated buildings from traditional buildings lies in plant production. The plant production has such advantages as high standardization level(standardization of process setting and process operation), high level of mechanization(high production efficiency and reduction of labor consumption), guaranteed product quality(internal control system) and low impact by climate(indoor operation). The key points of construction of a member prefabrication plant is to first determine the production mode, i.e., base plant production or nomadic plant production; secondly, it is to determine the planning contents of member prefabrication plant, mainly including site selection, plant general layout, production line layout and equipment selection, etc.

7.1 PC（混凝土预制）构件生产模式选择与构件预制工厂选址

7.1 Selection of PC(Precast Concrete) Member Production Mode and Member Prefabrication Plant Site

PC 构件生产有基地化工厂生产和游牧式工厂生产两种生产方式。工厂建设应根据市场需求、主要产品类型、生产规模和投资能力等因素首先确定采用什么生产工艺，再根据选定的生产工艺进行工厂布置。

There are two production modes for PC members: Base plant production and nomadic plant production. For plant construction, the production process shall be determined first according to market demand, main product types, production scale and investment capacity, and then the plant layout shall be determined according to the selected production process.

7.1.1 PC 构件生产规模的确定

7.1.1 Determination of PC member production scale

构件预制工厂的生产规模通常以年产预制构件混凝土立方量计，生产板式构件的工厂也可以以平方米计。从内部角度来看，生产规模由生产能力决定，并与工作平台数量和堆场面积有关；从外部角度来看，服务半径内建筑规模、装配式混凝土结构建筑的比例以及其他构件预制厂家的情况是确定工厂生产规模的主要依据。

The production scale of a member prefabrication plant is usually calculated based on cubic meters of precast member concrete per year, or square meters for a plate type member plant. From an internal point of view, the production scale is determined by the production capacity and has something to do with the number of working platforms and the area of the storage yard; from an external point of view, the production scale of the plant mainly depends on the size of the building within the service radius, the proportion of prefabricated concrete structure buildings and the current situation of other member prefabricated plants.

（1）生产能力的确定

(1) Determination of production capacity

可采用表 7-1，通过统计近年各类型预制构件年需求量，推算发展趋势，根据产品结构、周边工厂数量及产能等因素确定构件预制工厂的生产能力。

Table 7-1 can be used to calculate the annual demand for various types of prefabricated members in recent years to learn the development trend, and determine the production capacity of the member

prefabrication plant based on the product structure, the number and production capacity of surrounding plants.

（2）工作平台数量的确定

(2) Determination of working platforms number

预制构件是在工作平台上制作的，因此工作平台的数量决定了构件的产量。例如，假定每件构件的混凝土用量平均为 1.3 m³，根据日产量和工作平台周转次数，可计算所需要的工作平台数量。在实际生产中，为保证工厂的正常生产，工作平台通常要有 20% 的富余量。

Prefabricated members are made on working platforms, so the number of working platforms determines the output of members. For example, the number of working platforms required can be calculated according to the daily output and the turnover times of working platforms, assuming that the average concrete consumption of each member is 1.3 m³. In practice, in order to ensure the normal production of the plant, a margin of 20% is usually considered for the number of working platform.

表 7-1 市场规模预测

Table 7-1 Market Size Forecast

序号 S/N	预测项目 Forecast Item	单位 Unit	数量（按年计） Quantity (count by year)				
			2022	2023	2024	2025	2026
1	有把握参与建设的面积（内部市场） Area determined for construction (internal market)	万㎡ 10 000 ㎡					
2	可以争取参与建设的面积 Area available for construction	万㎡ 10 000 ㎡					
3	（1+2）小计 Subtotal (1+2)	万㎡ 10 000 ㎡					
4	按建筑面积折算成结构体积 Converted to structural volume according to building area	万m³ 10 000 m³					
5	对应建设面积房屋的预制率 Prefabrication rate of building of corresponding construction area	%					

序号 S/N	预测项目 Forecast Item	单位 Unit	数量（按年计） Quantity (count by year)				
			2022	2023	2024	2025	2026
6	钢筋混凝土预制构件体积 Volume of reinforced concrete precast members	万m³ 10 000 m³					

（3）储存区面积的确定

(3) Determination of storage area

构件储存区的面积须从工厂日生产能力，生产后28天的存放能力，预制墙体、叠合板、阳台等主要构件所需要的存放面积，以及必要的检修运输空间等方面考虑。

The member storage area shall be considered in terms of the daily production capacity of the plant, the storage capacity for 28 days after production, the storage area required for main members such as prefabricated walls, laminated slabs and balconies, and the necessary maintenance and transportation space.

7.1.2 PC 构件生产模式的选择

7.1.2 Selection of PC Member Production Mode

装配式混凝土建筑的一大特点就是工厂化生产，也就是预制。工厂化生产是一种生产方式，不等同于在工厂里生产。预制混凝土构件大多在构件预制工厂生产，也可以在现场生产，后者一般被称为游牧式生产。游牧式构件预制厂的所有设备（包括垫层），都可搬迁、可移动，在成本、物流等方面具有基地化PC构件预制厂不可比拟的优势，搬迁后还不影响现场原有土地。基地化工厂与游牧式工厂生产对比见表7-2。

One of the features of prefabricated concrete buildings is plant production, that is, prefabrication. Plant production is merely a mode of production, not production in a plant. Precast concrete members are mostly produced in a member prefabrication plant or on site, and the latter generally refers to nomadic production. All equipment in the nomadic plant, including cushion, can be relocated and moved without affecting the original land on site, which has incomparable advantages over base PC member prefabrication plant in terms of cost and logistics. See Table 7-2 for the comparison between base plant production and nomadic plant production.

表 7-2 基地化工厂生产与游牧式工厂生产对比

Table 7-2 Comparison of Base Plant Production and Nomadic Plant Production

基地化工厂生产 Base Plant Production	游牧式工厂生产 Nomadic Plant Production
投资额大 Large investment	投资额小 Small investment
机械化程度高 High level of mechanization	机械化程度较低 Low level of mechanization
自动流水线+固定模台 Automatic assembly line + fixed die table	固定模台 Fixed die table
蒸汽养护+自然养护 Steam curing + natural curing	自然养护 Natural curing
劳动力需求小 Small labor demand	劳动力需求大 Large labor demand

基地化工厂生产 Base Plant Production	游牧式工厂生产 Nomadic Plant Production
受天气影响小 Less affected by weather	受天气影响大 Greatly affected by weather
运输成本高 High transportation cost	运输成本低 Low transportation cost
固定式，不够灵活 Less flexible	灵活多变，可跟随项目移动 Flexible and relocated as required

对于 PC 构件的生产，没有哪种模式是绝对好的，无论是高度机械化、自动化的生产模式，还是露天作业的生产模式，适用于工程项目的才是最好的。

As for the production of PC members, there is no model that is absolutely good. Whether it is a highly mechanized and automated production mode or an open-air production mode, what works for the project is the best.

在住宅产业化发展相对滞后的地区，游牧式构件预制厂较常规构件预制厂预制构件具有明显优势，具体体现在以下方面：

In areas with relatively backward development of residential industrialization, a nomadic plant has obvious advantages over a conventional plant in prefabrication of members, which are as follows:

①游牧式构件预制厂布置灵活多样，实用性强，投资较小，建设周期短，有利于中小城市或特殊项目建筑工业化的推广应用，且能够有效避免产能不匹配的问题。

① The nomadic plant has flexible and diversified layout, strong practicability, small investment and short construction cycle, which can promote the popularization and application of building industrialization in small- and medium-sized cities or special projects, and effectively avoid the problem of mismatch of production capacity.

②在游牧式构件预制厂预制构件，构件运输距离短、运输设备小型化，减少了构件运输损耗，降低了综合成本，相比大型构件预制生产厂家，在运距、价格等方面具有一定优势。

② In a nomadic plant, short-distance transportation and miniaturization of transportation equipment reduce the transportation loss of members and the comprehensive cost. Compared with large member prefabrication plants, it has certain advantages in transportation distance, freight, etc.

③利于构件尺寸大型化、多样化，通过优化节点避免了裂缝等质量通病。

③ It is beneficial to the large and diversified sizes of members, and avoids the common quality problems such as cracks by optimized nodes.

④能与现场紧密配合，及时发现并快速解决施工中的问题。

④ It can work closely with the site operation to find and solve problems in construction in a timely manner.

⑤产业工人固定化，利于总包单位组织管理，不受第三方因素影响。

⑤ The fixation of industrial workers facilitate the organization and management of the general contractor, which is not affected by third-party factors.

游牧式工厂，也就是设在建筑工地上的 PC 构件预制厂。由于靠近建筑工地，具有运距短、投资少、布置灵活等诸多优点。在目前的建筑施工中，一些超宽、超高的大型 PC 构件，通常在游牧式工厂预制生产。

Nomadic plant is just a PC member precast plant on construction site. As it is close to the construction site, it has many advantages such as short transportation distance, low investment and flexible layout. At present, some ultra-wide and ultra-high large PC members are usually prefabricated in a nomadic plant on site.

7.1.3 构件预制工厂选址

7.1.3 Site Selection of Member Prefabrication Plant

选择在什么样的地方建厂及投资成本的大小，都关系到建筑工程能获得多少经济和社会效益，这也是构件预制工厂建设工作中的重点。

As the focus in the construction of a member prefabrication plant, the plant site selection and the size of investment cost will have a direct bearing on economic and social benefits that can be obtained.

（1）影响构件预制工厂选址的因素

(1) Factors affecting site selection of member prefabrication plant

合理的工厂布置是保证整个生产系统能够高效、安全和经济运行的基础。因此，工厂布置与选址建设时，需结合区域总体规划，综合考虑周边环境、生产内容、产能需求、工艺流程、物流运输等因素。

Reasonable plant layout is the basis for the efficient, safe and economical operation of the whole production system. Therefore, the layout and site selection of the plant shall comprehensively take into account the overall regional planning, the surrounding environment, products, required capacity, process flow, logistics transportation, etc.

①项目选址的区域位置现状。

① Current situation of project site.

a. 场地竖向标高。

a. Vertical elevation of the site.

合理地进行竖向设计，在满足使用功能的前提下尽可能节省土方工程量，可以有效降低土地平整的成本；同时，必须要考虑竖向设计是否会影响部分功能的实现。

Reasonable vertical design can save earthwork as much as possible while meeting the use function, and can effectively reduce the cost of site leveling. At the same time, it is necessary to consider whether the vertical design will affect the realization of some functions.

b. 地块尺寸及形状。

b. Plot size and shape.

项目所在地用于规划区域的尺寸，在规划过程中会直接影响整个厂区生产、物流及配套的布局，进而影响产能规划。项目所在地用于规划的地块形状直接影响着各建筑物间的相对位置关系，合理的区域形状可提高土地利用率，同时，可为高效生产提供各种便利条件。

The size of the planned area on the project site will directly affect the layout of the entire plant's production, logistics and supporting facilities in the planning process, thus affecting the capacity planning. The planned shape of plot on the project site directly affects the relative position relationship between buildings. A reasonable region shape can improve the land use rate and provide various convenient conditions for efficient production.

如图 7-1 所示，两个厂区红线图内地域形状不同，将形成两种不同的规划风格。将上述两个方案进行对比，在投入成本、物流规划以及工艺方案上左侧的方案优势明显：车间与堆场共用起重机，减少了设备投入，物料及成品输送通道更为顺畅。

As shown in Figure 7-1, the shapes in the red line maps of the two plant areas are different, that is, there will be two different options. By comparing the two options, the left one has obvious advantages in input cost, logistics and process. The crane shared by the workshop and the storage yard will reduce equipment input, and the material and finished products transportation channels are more available.

②项目选址现有的建（构）筑物。

图7-1 构件预制工厂规划方案

Figure 7-1 Planning Options of Member Prefabrication Plant

② Existing structures and buildings in project site.

某项目限制条件为一条航油管线斜穿厂区，则在管线两侧各 50 m 范围内不得设置永久性建筑，如龙门吊堆场、生产厂房等。

If an aviation oil pipeline obliquely crosses the plant area, no permanent buildings shall be built within 50 m on both sides of the pipeline, such as gantry crane storage yard and workshop.

③项目选址区域配套设施。

③ Supporting facilities in the project site.

a. 水。

a. Water.

生活用水：必须符合饮用水标准，多采用市政供水管网供水。

Domestic water: It must meet the drinking water standard, and the municipal water supply network is mostly adopted.

生产用水通常指混凝土搅拌用水、蒸汽用水、构件冲洗用水等。

Production water refers to concrete mixing water, steam water, member flushing water, etc.

混凝土搅拌用水必须符合搅拌用水标准，构件冲洗用水可以使用处理后的污水。因生产

用水量较大，大多数工厂采用自打井，通过水源地泵审批即可使用。

Concrete mixing water must meet the mixing water standard, and the member flushing water can be treated sewage. Due to the large production water consumption, most plants adopt self-drilled wells, which can be used after approval of the water source pump.

b. 电——办公、生活用电，生产用电。

b. Electricity—office, living and production electricity.

一般 PC 厂用电总功率不低于 800 kV·A，所以在建厂选址过程中应注意是否需要单独增容设线，这对建设过程中的实际操作及投资额均有影响。

Generally, the total electricity power of PC plant is not less than 800 kV·A, so attention should be paid to whether separate capacity expansion lines are required during the site selection process, which has an impact on the actual operation and investment in the construction process.

c. 气。

c. Gas.

目前，环保审批在项目立项及手续办理过程中较为严谨，涉及锅炉项目，一般燃烧介质

必须使用清洁能源，从综合成本上考虑，燃气锅炉较为经济。同时，燃气作为锅炉燃烧介质也是目前相关管理部门推荐的。

At present, there are more stringent requirements for environmental protection approval for project initiation and formalities handling. For boiler projects, clean energy must be used as the combustion medium. From the perspective of comprehensive cost, gas boilers are more economical. At the same time, gas as the boiler combustion medium is currently recommended by the relevant management department.

d. 供暖。

d. Heating.

目前，按相关政策要求，办公及生活供暖多采用联网集中供暖。车间是否采暖将与生产相关，所以推荐使用蒸汽锅炉自供暖。

At present, office and domestic heating are mostly network central heating according to relevant policies. The workshop heating depends on production, so it is recommended to use steam boiler for self-heating.

e. 蒸汽。

e.Steam.

集中供暖无法满足生产用蒸汽供给要求，一般自建蒸汽锅炉。

Central heating cannot meet the requirements of steam supply for production, so steam boilers are generally built by the plant itself.

上述各种配套设施若单独建设既影响工期，又增大了资本投入，所以建议选址时提前考虑。

If the above supporting facilities are built separately, it will not only affect the construction period, but also increase the capital investment, so it is recommended to consider them during the site selection.

④工业用地指标规定。

④ Industrial land indicators.

《工业项目建设用地控制指标》对投资强度、容积率、建筑系数、行政办公及生活服务设施用地所占比重做了规定。

The Controlling Index for Industrial Project Construction Land stipulates the investment intensity, plot ratio, building coefficient, and the proportion of land for administrative office and living service facilities.

a. 投资强度。

a. Investment intensity.

以柳州为例，PC 构件业属于工艺品及其他制造业行业，行业代码为 3122，依据地区分类，柳州土地等别为六等，城市分类为 2 类，则其投资强度应不低于 1245 万元每公顷。

Taking Liuzhou as an example, the PC member industry belongs to crafts and other manufacturing industry, industry code: 3122, area classification: Liuzhou land grading VI, and city classification: 2, so its investment intensity shall be ≥ RMB 12.45 million/ha.

b. 容积率。

b. Plot ratio.

按行业要求，容积率不低于 0.7。

The plot ratio is not less than 0.7 according to the industry requirements.

c. 建筑系数。

c. Building coefficient.

项目用地范围内各种建筑物、用于生产和直接为生产服务的构筑物占地面积总和占总用地面积的比例，不应低于 30%。

The total floor area of various buildings, structures used for production and directly serving production within the project land area shall not be less than 30% of the total land area.

d.行政办公及生活服务设施用地所占比重。

d. Proportion of land for administrative office and living service facilities.

工业项目所需行政办公及生活服务设施用地面积不得超过工业项目总用地面积的 7%。

The land area for administrative office and living service facilities required for industrial projects shall not exceed 7% of the total land area for industrial projects.

e.绿地率。

e. Ratio of green space.

工业企业内部一般不得安排绿地，但因生产工艺等特殊要求需要安排一定比例绿地的，绿地率也不得超过 20%。

Green spaces are generally not allowed in industrial plants. However, due to production technology or other special requirements, the green space ratio shall not exceed 20%.

⑤工厂生产运营管理。

⑤ Plant production and operation management.

a.厂区面积。

a. Plant area.

根据客户信息输入如项目数量、年产量、生产节拍、供货方式等确认厂区面积基本需求，再根据物流方式、生产方式、功能区域面积、办公区域大小、辅助用房占地、厂区道路及绿化面积对布局进行优化，最终提出实际厂区面积使用需求。

Confirm the basic requirements of the plant area according to the customer information input, such as the number of projects, annual output, production take time, supply mode, etc., and then optimize the layout according to the logistics mode, production mode, functional area, office area, auxiliary room land occupation, plant roads and green space area, and finally determine the

actual plant area.

b.成本。

b.Cost.

如政府优惠政策、土建投资、水暖电气接入厂房的成本及今后使用成本、物流成本及人员交通成本等。

Government preferential policies, civil engineering investment, cost of water heating and electrical access to the plant, future use cost, logistics cost, personnel transportation cost, etc.

c.离项目距离。

c. Distance from the project.

选址时须考虑项目节拍及生产周期是否足够，考虑正常使用和备用供货物流路线。

Consider whether the project take time and production cycle are sufficient during the site selection, and consider normal use and alternate supply logistics routes.

d.人员招聘的便利性。

d. Convenience of personnel recruitment.

周边劳动力资源是否能满足用工需求。

Whether the surrounding labor resources can meet the labor demand.

e.医疗救护的便利性。

e. Convenience of medical aid.

一旦有工伤事故或员工身体不适的情况，厂址附近是否能提供医疗服务。

Medical aid should take into consideration in case of industrial accident or physical discomfort of employees.

（2）构件预制工厂选址的要求

(2) Requirements for member prefabrication plant site selection

工厂选址的五大要求：合法、经济、安全、方便、合理。

Five requirements for plant site selection:

Legality, economy, safety, convenience and rationality.

①合法。

① Legality.

不侵占、不使用国家划定的永久基本农田，选择非永久基本农田并且已办理合法出让手续或手续齐备的工业用地。要取得建设用地规划许可证，并通过建设项目环境影响评价文件审批许可。

It shall not occupy or use permanent basic farmland designated by the state, and shall select industrial land that is not permanent basic farmland and has gone through legal transfer procedures. It shall obtain the construction land use permit and pass the environmental impact assessment of the construction project.

②经济。

② Economy.

首先，要确定所选择的厂址是否在可行性研究报告中所划定的 PC 构件有效经济供应半径以内，工厂与原材料供应地、产品销售地的距离是否超出有效经济供应半径。其次，选择的地块要尽量平整，确保场地整平时填挖平衡，不产生大量的借土和弃土。在一般情况下，尽量不在软基和起伏过大的丘陵山区建厂，以减少工厂建设过程中的软基处理和土石方开挖爆破的工程量，降低工程造价。最后，要考虑地面以上的房屋等建筑拆迁量，庄稼、树木等砍伐量，青苗补偿要在经济、合理的承受范围以内。

Firstly, the plant site shall be within the effective economic supply radius of PC members defined in the feasibility study report, and the distance between the plant and the raw material supply place and the product sales place shall be within the effective economic supply radius. Secondly, the site selected shall be leveled as far as possible to ensure the balance of filling and excavation during site leveling and avoid to

generate a large amount of borrow soil and spoil. Normally, it is not allowed to build plants in soft foundation and hilly and mountainous areas with excessive fluctuations as far as possible, so as to reduce the quantities of soft foundation treatment and earthwork excavation and blasting in the process of plant construction and reduce the project cost. Finally, the demolition of houses and other buildings above the ground, the felling of crops and trees shall be considered, and the crop compensation shall be economically and reasonably affordable.

③安全。

③ Safety.

工厂的地理位置和环境，要满足相关法律法规规定的防洪、防雷要求，避开滑坡、泥石流等地质灾害频发地带，远离危险化学品、易燃易爆品等危险源。

The geographical location and environment of the plant shall meet the requirements of flood control and lightning protection stipulated by relevant laws and regulations, avoid geological disaster zones such as landslides and debris flows, and keep away from dangerous chemicals, flammable and explosive sources.

构件预制工厂建成后也不得对周围环境和常住人群的生活环境造成破坏和污染。

After the member prefabrication plant is completed, it shall not cause damage and pollution to the surrounding environment and the living environment of the permanent residents.

④方便。

④ Convenience.

工厂选址要考虑方便，主要是为了优化物流运输、降低运营成本、提高生产效率。

The primary consideration for plant location selection is convenience, aiming to optimize logistics and transportation, reduce operational costs, and enhance production effciency.

⑤合理。

⑤ Rationality.

首先，要考虑工厂附近和经济运距范围内是否有可靠的资源供应和能源供给，例如砂石料的供应，附近是否有电、水、天然气、通信的接入条件，周围的交通条件能否满足各种原材料和产品及时顺利地进出工厂的需求。其次，也要考虑工人日后生活的方便性。最后，还要关注工厂周围的民风民俗，施工团队能否与周围的居民和谐共处，这也是以后构件预制工厂能否顺利开展生产活动的一个重要影响因素。

First of all, it is necessary to consider whether there is a reliable resource supply and energy supply near the plant and within the length of economic haul, such as the supply of sand and gravel, whether there are access conditions for electricity, water, natural gas and communication nearby, and whether the surrounding traffic can meet the needs of various raw materials and products entering and leaving the plant in a timely and smooth manner. Secondly, the convenience of workers' future life should also be considered. Finally, we should know the folk customs to live in harmony with the surrounding residents. This is also an important consideration for the smooth production of member prefabrication plant in the future.

因此，工厂选址应多考察几个地块，在综合考虑以上因素，进行比对分析后，再从中选取一个优良的厂址。严禁随意、仓促选址。

For this reason, more areas shall be investigated for the site selection of the plant. After comprehensive consideration of the above factors and comparative analysis, the right one shall be selected. Do not select the site at will or in a hurry.

下文以广西建工轨道装配预制混凝土有限公司（简称"PC 生产基地"）为例展示工厂选址实践。

The following takes Guangxi Construction Engineering Track Assembly Precast Concrete Co., Ltd. (referred to as PC production base) as an example to display plant site selection practice.

PC 生产基地位于柳州市柳北区白露片区A-1-1 号柳州装配式建筑现代化产业园，总用地 169 亩（约合 112 667 m²），基地主厂房占地约 35 000 m²，2018 年 3 月开工建设，5 月底已实现主厂房屋面封顶，6 月初设备进场安装，8 月份完成设备安装调试，9 月份已实现正式投产。PC 生产基地采用 6 跨单层轻钢结构，单跨跨度为 27 ~ 30 m，配套设施包括室外堆场、配电房、锅炉房、道路、管网、围墙、门卫室及绿化用地等工程。工艺生产按照"4+1"生产线设计（1条综合环形生产线、1 条双皮墙生产线、1 条叠合板生产线、1 条固定模台生产线及 1 条钢筋加工线）。投产后产品涵盖装配式建筑所需的外墙板、内墙板、叠合板、楼梯、阳台板、空调板、梁柱等全类型构件，设计年产 12 万 m³ 预制构件，可供住宅项目约 400 万 m²。如图 7-2 所示。

The PC production base is located in Liuzhou Prefabricated Building Modernization Industrial Park, Bailu No. A-1-1, Liubei District, Liuzhou City, with a total land area of 169 mu(about 112 667 m²) and a main plant building of about 35 000 m². The construction started in March 2018, the roof of the main plant building was capped at the end of May, the equipment was mobilized and installed in early June, and the equipment installation and commissioning were completed in August. It was officially put into operation in September. The PC production base is of 6-span single-storey light steel structure with a single span of 27-30 m. The supporting facilities include outdoor storage yard, power distribution room, boiler room, roads, pipe network, enclosure, guard room, greening facilities, etc. The process production is designed based on "4+1" production line(1 integrated circular production line, 1 double-faced laminated shear wall production line, 1 laminated slab production line, 1 fixed die table production line and 1 reinforcement processing line). After it is put into production, the products cover all types of members required for prefabricated buildings, including exterior wall panels, interior wall panels, laminated slabs, stairs, balcony slabs, air conditioner boards, beams and columns, etc. The designed annual output of prefabricated members is 120 000 m³, which can be used for residential projects of about 4 million m², as shown in Figure 7-2.

图7-2　广西建工轨道装配预制混凝土有限公司

Figure 7-2 Guangxi Construction Engineering Track Assembly Precast Concrete Co., Ltd.

7.2　构件预制工厂总体规划

7.2 Overall Planning of Member Prefabrication Plant

构件预制工厂规划时，首先应根据构件预制工厂可行性研究报告、企业的经济技术状况、对构件预制工厂的预期等，进行工厂的总体规划；其次，还应考虑厂址所在地允许扩展的空间、PC 产品定位和产量需求、PC 构件生产线主要设备的性能参数以及堆场面积的需求等因素。构件预制工厂建设的主要内容是生产车间（钢结构厂房），成品堆场，办公及生活配套设施（如办公研发楼、宿舍餐饮楼），锅炉房、搅拌站等生产配套设施，园区综合管网，成品展示区等。

In the planning of member prefabrication plant, the overall planning of the plant shall be carried out first according to the feasibility study report of member prefabrication plant, the economic and technological conditions of the enterprise and the expectation of the plant; secondly, factors such as allowable expansion space of the plant site, PC product positioning and output demand, performance parameters of main equipment of PC member production line and the requirement of storage yard area shall also be considered. The main construction contents of the member prefabrication plant include production workshop(steel structure workshop), finished product storage yard, office and living supporting facilities(such as office and R&D building, dormitory catering building), boiler room, mixing station and other production supporting facilities, integrated pipe network in the park, finished product exhibition area, etc.

7.2.1 构件预制工厂总体规划原则

7.2.1 Principles for Overall Planning of Member Prefabrication Plant

在选址上，因地制宜，充分利用现有条件，做到交通便利、物流畅通。

In terms of site selection, it is necessary to adjust measures to local conditions, make full use of existing conditions, and achieve convenient transportation and smooth logistics.

在技术上，生产线适用性强，设备性能稳定可靠、运转安全、操作维修方便。

In terms of technocogy, strong applicability of production line, stable and reliable equipment performance, safe and convenient operation and maintenance.

在经济上，建设成本可控，后期运行维护成本低，生产线可塑性强。

In terms of economy, controllable construction cost, low operation and maintenance cost at the later stage, and highly flexible production line.

在环境上，环境绿化与空间组合协调，努力改善工厂和工作环境，符合环保要求。

In terms of environment, environmental greening is coordinated with space combination, and efforts shall be made to improve the plant and working environment to meet the environmental protection requirements.

此外，还需考虑如何对待规划区域及将来扩建的可能性。

In addition, it is necessary to consider how to treat the planned area and the possibility of future expansion.

7.2.2 构件预制工厂总体规划内容

7.2.2 Contents of Overall Planning of Member Prefabrication Plant

构件预制工厂规划设计包括对构件预制工厂建设进行工厂总平面设计（包含厂区规划，生产线工艺规划，厂内物流系统、厂外物流系统、厂内人流系统、厂外人流系统、垂直起吊系统、安全防护系统、生产工艺系统、人员配置、给排水系统、蒸汽养护系统规划等）、工艺设计、生产系统规划、设备选型及经济测算分析等内容。

The planning and design of the member prefabrication plant includes the general layout design of the plant construction(including plant planning, production line process planning, in-plant logistics system, off-plant logistics system, in-plant personnel flow system, off-plant personnel flow system, vertical lifting system, safety protection system, production process system, personnel allocation, water supply and drainage system, steam curing system planning, etc.), process design, production system planning, equipment selection, economic calculation and analysis, etc.

（1）总平面设计原则和工厂总平面布置

(1) Principles for general layout design and general layout of the plant

①总平面设计原则。

① Principles for general layout design.

a. 总平面设计必须执行国家及地方规范标准，依循相关方针政策，按设计任务书进行。简单地列举几个规范，如《建筑设计防火规范》

《民用建筑设计通则》《工业企业总平面设计规范》等。

a. The general layout design must follow the national and local standards and policies and the design specification. Including Code for Fire Protection Design of Buildings, Code for Design of Civil Buildings, Code for General Layout Design Plan of Industrial Enterprises, etc.

b.总平面设计必须以所在城市的总体规划、区域规划为依据，符合总体布局规划要求。如场地出入口位置、建筑体形、层数、高度、公建布置、绿化等都应满足规划要求，与周围环境协调统一。同时，建设项目内的道路、管网应与市政道路与管网合理衔接，以满足生产、方便生活。

b. The general layout design must be based on the overall planning and regional planning of the city, and meet the requirements of the overall layout planning. For example, the location of entrance and exit on site, building shape, floors, height, layout of public buildings, greening shall meet the planning requirements and be harmonious with the surrounding environment. At the same time, the roads and pipe network in the construction project shall be reasonably connected with the municipal roads and pipe network to meet the requirements of production and convenient life.

c.总平面设计应结合当前选址的自然条件，如地形、地势、地质、方位、水文、气象等，依山就势，因地制宜。

c. The general layout design shall be combined with the natural conditions of the current site, such as topography, terrain, geology, orientation, hydrology, meteorology, and shall be in harmony with mountain and local conditions.

d.总平面设计应结合地形，合理地进行用地范围内的建筑物、构筑物、道路及其他工程设施间的平面布置，设计要素如建筑物面积、

高度及出入口的位置等。

d. During the general layout design, the layout of buildings, structures, roads and other engineering facilities within the scope of land use shall be reasonably arranged in combination with the terrain, the building area and height, the locations of entrances and exits, etc.,these design elements should be taken into account.

e.总平面设计应考虑建筑物之间的距离，其距离应满足生产、安全、日照、通风、抗震及管线布置等各方面要求。

e. The distance between buildings shall be considered in the general layout design, which shall meet the requirements of production, safety, sunlight, ventilation, seismic resistance, pipeline layout, etc.

f.总平面设计应对原有的不可拆除的建筑物、构筑物或设施等进行重新的布局考虑，重新评估其供电能力、给排水能力及通风照明能力等。

f. The general layout design shall reconsider the layout, power supply, water supply and drainage, ventilation and lighting of the existing non-removable buildings, structures or facilities.

g.总平面设计应尽力降低成本，对资源、设备、空间、能源等有效地进行最大化利用。

g. The general layout design shall try to reduce the cost and maximize the utilization of resources, equipment, space and energy.

②工厂总平面布置。

② General layout of the plant.

构件预制工厂应按照可行性研究报告中的规划进行设计和布局，同时兼顾整个工厂内各生产项目的投资顺序和预制构件生产线日后提能扩产的要求。构件预制工厂整体由构件生产区、构件成品堆放区、办公区、生活区、相应配套设施等组成，具体可分为生产厂房、办公

研发楼、成品堆场、混凝土原材库、成品展示区、实验室、锅炉房、钢筋及其他辅材库房、配电室、宿舍楼、餐饮楼等。构件生产车间由预制构件生产线、钢筋加工生产线、车间内预制构件临时堆放区、混凝土搅拌运输系统、高压锅炉蒸汽系统、桥式门吊系统、动力系统等组成。

The member prefabrication plant is designed and arranged according to the planning in the feasibility study report after taking into account the investment sequence of each production project in the whole plant and the requirements for future production capacity expansion of the production line. The member prefabrication plant is generally composed of member production area, finished products storage area, office area, living area and other supporting facilities, specifically the production workshop, office and R&D building, finished products storage yard, concrete raw materials warehouse, finished products exhibition area, laboratory, boiler room, steel bars and other auxiliary materials warehouse, power distribution room, dormitory building, catering building, etc. The member production workshop consists of prefabricated member production line, reinforcement processing production line, temporary prefabricated member stacking area in the workshop, concrete mixing and transportation system, high-pressure boiler steam system, bridge gantry crane system, power system, etc.

构件预制工厂的基本设置大体上都一样，按功能分为三大区域：生活办公区、生产区、存放区，如图7-3、图7-4所示。总平面布置根据项目各单项工程规划，工艺流程，物料投入产出，废弃物排出及原料贮存，内外交通运输情况，场地的自然条件，生产要求与功能以及行业、专业的设计规范进行安排。

All member prefabrication plants are

图7-3　某构件预制工厂总平面布置方案一

Figure 7-3 General Layout Plan I of a Member Prefabrication Plant

图7-4 某构件预制工厂总平面布置方案二

Figure 7-4 General Layout Plan II of a Member Prefabrication Plant

basically the same in configuration, including three functional areas: Living & office area, production area and storage area, as shown in Figure 7-3 and Figure 7-4. The general layout shall be arranged according to the individual works, process flow, material input and output, waste discharge and raw material storage, internal and external transportation of the project, as well as local natural conditions, production requirements and functions, and industrial and professional design specifications.

a. 生产区。

a. Production area.

生产区一般为大跨度单层钢结构厂房，车间设计 2 ~ 4 跨不等，生产区长度 120 ~ 180 m，单跨宽 24 ~ 27 m，每跨车间内需配桁车至少 2 台、起吊高度不小于 7 ~ 9 m。地面硬化处理，硬化层不低于 20 mm 厚。生产线振动系统工位及蒸养房工位地面需做地基处理，主要布置部

品部件生产线和钢筋加工线、混凝土搅拌站等，部品部件生产线可布置综合环形生产线（可生产叠合板、内外墙板），固定模台生产线，楼梯、阳台、空调板生产区等。

The production area is 120-180 m long and generally equipped with a large-span single-story steel structure workshop, with 2-4 spans and a width of a single span of 24-27 m. At least 2 truss cranes shall be equipped for each span of the workshop, with a lifting height of not less than 7-9 m. The ground shall be hardened, with a hardened layer thickness of not less than 20 mm. Foundation treatment is required for the ground of the vibration system station of the production line and the station of the steam curing room. The production line of components and parts, reinforcement processing line, concrete mixing station, etc. are mainly arranged. The production line of components and parts can be arranged into an integrated circular production line(which can produce laminated slabs, interior and exterior wall panels), a fixed die table production line, and a production area of stairs, balconies, air conditioner boards, etc.

综合环形生产线主要生产外墙保温板，也可兼顾生产内墙板（含内承重墙和内隔墙）和叠合板。综合环形生产线采用高精度、高结构强度的成型模具，布料机把混凝土浇筑在模具内，振动台振捣后并不立即脱模，而是经预养护和蒸汽养护，使构件强度满足设计强度后方进行拆模处理的。拆模后的 PC 构件成品运输至成品暂存区或室外成品堆放区，而空模台沿输送线自动返回，形成了自动化环形流水作业，如图 7-5 所示。

The integrated circular production line is mainly used to produce exterior wall panels with thermal insulation, but also interior wall panels(including internal load bearing walls and internal partition walls) and laminated slabs. The integrated circular production line adopts the forming die of high precision and high structural strength. The concrete is poured into the die by the distributor and vibrated on the vibrating stand; then, it is subject to pre-curing and steam curing until the strength of the member meets the design strength; next, demolding is done. After that, the finished PC members are transported to the temporary storage area or outdoor stacking area of finished products, and the empty die table automatically returns along the conveyor line, thus forming an automatic circular flow operation, as shown in Figure 7-5.

固定模台生产线：顾名思义，在固定位置放置模台，制作构件的所有操作均在模台上进行，材料、人员相对于模台流动。固定模台生产线是平面预制构件生产线中常用的一种生产模式。模台一般是一块平整度很高的钢结构平台。常用模台尺寸为预制墙板模台（4 m × 9 m），预制叠合楼板（3 m × 12 m），预制柱、梁构件（3 m × 9 m）。在生产时，模台作为构件的底模，与四周可拆卸侧模组成完整的模具。固定模台生产线自动化程度较低，需要更多工人，但是该模式具有所需设备少、投资少、灵活方便等优点，适合制作墙板、楼梯、阳台、飘窗等异

图7-5　自动流水生产区

Figure 7-5 Automatic Flow Production Area

型复杂构件，如图 7-6 所示。

The fixed die table production line is to place the die table at a fixed position, and members are produced completely on the die table, and materials and personnel flow relative to the die table. It is a common production mode for plane prefabricated members production line. The die table is generally a steel structure platform with high flatness. Usually, the die table for prefabricated wall panels is 4 m×9 m, 3 m×12 m for prefabricated laminated slabs, and 3 m×9 m for prefabricated columns and beam members. During production, the die table, as the bottom die of the member, forms a complete mold with the removable side molds around. The fixed die table production line is of low level of automation and requires more workers. Even so, it has such advantages as a small amount of equipment, small investment, flexibility and convenience, and is suitable for making special-shaped complex members such as wallboards, stairs, balconies and bay windows, as shown in Figure 7-6.

固定模台生产线布置时应考虑下列因素：

The following factors shall be considered in the arrangement of the fixed die table production line:

车间面积应满足模台摆放、作业空间和布设安全通道的需要。

The workshop area shall cater for die table placement, working space and layout of safe passage.

每个固定模台要配有蒸汽管道和自动控温装置，可定做移动式覆盖缝来保温覆盖。

Each fixed die table shall be equipped with steam pipe and automatic temperature control device, for which mobile coverage seam can be customized for thermal insulation.

当采用运料罐车运送混凝土时，固定模台处应方便运料罐车进出。

When the concrete is transported by tank

图7-6　固定模台生产区

Figure 7-6　Fixed Die Table Production Area

truck, it shall be easy for the entry and exit of the tank truck.

加工好的钢筋可通过起重机或运输车吊至或运输到模台处。

The processed steel bars can be lifted or transported to the die table by crane or transport vehicle.

混凝土的振捣多采用振动棒，板类构件可以在固定模台上安放附着式振捣器。

Vibrators are mostly used for concrete vibration, and attached vibrators can be placed on the fixed die table for plate type members.

异型构件生产区主要生产预制楼梯、预制阳台和预制空调板等PC构件。楼梯、阳台、空调板的生产采用定制模具的生产工艺，即通过定制与楼梯、阳台、空调板相配套的模具进行生产。其中，楼梯、阳台构件生产台座需特殊处理。

The special-shaped member production area is mainly used to produce PC members such as prefabricated stairs, prefabricated balconies and prefabricated air conditioner boards. The stairs, balconies and air conditioner boards are produced by customized dies. Special treatment is required for the member fabrication table of stairs and balconies.

钢筋加工区是构件预制工厂的重要组成部分，除了预应力板外，各构件工艺的钢筋加工都

设置钢筋加工区，人工在这里完成钢筋原材的调直、切断、成型、绑扎、成品储存等工序，为各生产区供应钢筋半成品及成品。很多工厂将钢筋加工区与构件制作布置在一个厂房内，钢筋加工生产线宜采用计算机数控设备，可选用数控钢筋调直切断机、数控钢筋弯曲机、钢筋网成型机、钢筋连接接头加工机械、钢筋冷加工设备等，或启用自动桁架钢筋生产线，具体设备可根据实际生产需求配置，如图7-7所示。

The reinforcement processing area is an important part of member prefabrication plant. In addition to prestressed slabs, a reinforcement processing workshop is set up for each member process to complete manually the straightening, cutting, forming and binding of raw steel bars, and finished products storage, so as to supply semi-finished and finished steel bar products to PC production areas. In most of plants, the reinforcement workshop and member fabrication workshop are arranged in the same building, and the reinforcement processing production line should adopt automatic CNC(computerized numerical control) equipment, that is, CNC steel bar straightener and shear, CNC steel bar bender, steel mesh former, reinforcement joint processing machine, reinforcement cold processing equipment and automatic truss reinforcement production line can be selected depending on the actual production requirements, as shown in Figure 7-7.

图7-7　钢筋加工区
Figure 7-7 Reinforcement Processing Area

就目前国内建筑结构体系而言，钢筋加工可以采用自动化的预制构件，包括叠合楼板、女儿墙、非承重内隔墙、夹芯保温板、外叶板和非承重外挂墙板等。尚无法做到钢筋加工自动化的构件包括楼梯、阳台板、柱、梁、三明治外墙板、剪力墙板等其他造型复杂的构件。

As far as the current domestic building structure system is concerned, reinforcement processing can be realized for automatic prefabricated members including laminated floor slabs, parapets, non-load-bearing inner partition walls, sandwich insulation boards, exterior wall panels and non-load-bearing external wall panels. While, it is not feasible for stairs, balcony slabs, columns, beams, sandwich exterior wall panels, shear wall panels, and other special-shaped complex members.

构件预制工厂搅拌站有两种类型，工厂专用搅拌站和商品混凝土搅拌站（兼给工厂供应混凝土）。需要注意商品混凝土与构件混凝土的不同，最好是单独设置搅拌机系统。

There are two types of mixing stations in the member prefabrication plant, i.e., special mixing station for plant and commercial concrete mixing station(which also supplies concrete to the plant). It's important to know the difference between commercial concrete and concrete for structural elements. A separate mixer system is preferred.

混凝土搅拌站是构件预制工厂的主要生产设施，搅拌站位置最好布置在距生产线布料点近的地方，以减少路途运输时间，一般布置在车间端部或端部侧面，通过轨道运料系统将混凝土运到布料区。对于固定模台生产线，搅拌站的布设应能满足罐车运输混凝土的需求。

As the main production facility of a PC plant, the concrete mixing station shall be preferentially arranged close to the pouring point of the production line to reduce the transportation time, generally at the end or side of the end of the workshop; the concrete is transported to the pouring area by the rail transportation system. For the fixed die table production line, the mixing station system should consider the conditions available for concrete transportation by tank trucks.

混凝土搅拌站包括原材料储存区、混凝土生产区与中控室等，一般采用全封闭车间生产模式，以减少粉尘和噪声的污染；砂石原材料堆放于指定储存区；水泥、粉煤灰等掺和料采用筒仓储存。车间生产线应根据实际生产规模合理确定水泥筒仓、粉煤灰筒仓、特殊添加剂筒仓等设施，如图7-8所示。

The concrete mixing station consists of raw material storage area, concrete production area, central control room, etc., and the fully enclosed workshop is adopted to reduce dust and noise pollution. Sand and gravel raw materials are stacked in the designated storage area, while cement, fly ash and other admixtures are stored in silo. The production line of the workshop shall be reasonably provided with the cement silo, fly ash silo, special additive silo and other facilities according to the actual production scale, as shown in Figure 7-8.

材料存放区用于存放车间生产用材料，例如钢筋、保温板、铝窗、瓷砖、黏合剂、预埋

图7-8　混凝土搅拌站
Figure 7-8 Concrete Mixing Station

件及其他生产辅助材料等，存放区宜设置在生产工位附近且便于管理的位置。

The material storage area is used to store materials for workshop production, including steel bars, insulation boards, aluminum windows, ceramic tiles, adhesives, embedded parts and other auxiliary production materials, and it should be located near the production station to facilitate management.

b. 构件储存区。

b. Members storage area.

构件预制工厂构件储存区不仅是构件存储的场地，也是构件质量检查、修补、粗糙面处理、表面装饰处理的场所，如图 7-9 所示。构件储存区分为车间内和车间外两种。室外场地面积一般为制作车间的1.5 ～ 2倍。地面尽可能硬化，至少要铺碎石，排水要通畅。室外场地需要配置16 ～ 20 t 龙门式起重机，场地内有构件运输车辆的专用道路。

The members storage area of the member prefabrication plant is not only for member storage, but also for member quality inspection, repair, rough surface treatment and surface finishing, as shown in Figure 7-9. There are members storage areas inside and outside the workshop. The outdoor site area is generally 1.5-2 times that of the production workshop. The ground shall be hardened as much as possible, at least gravel shall be paved, and drainage shall be smooth. The outdoor site needs to be equipped with a 16-20 t gantry crane, and there is a special road for members transportation vehicles within the site.

预制构件储存区应与生产车间相邻，以方便运输，减少运输距离。检验合格的半成品可以通过同一轨道的吊车转入构件储存区，形成流水作业。

The storage areas of prefabricated members shall be arranged adjacent to the production workshop to facilitate transportation and reduce transportation distance. The semi-finished products passing the inspection can be transferred to the members storage areas by the crane on the same track to form a flow operation.

图7-9　构件储存区
Figure 7-9　Members Storage Area

c. 实验室。

c. Laboratory.

实验室一般设在办公楼一楼或车间内，实现主要原材料及生产过程的检验与记录，主要由材料室、混凝土室、力学室、标养室及留样室等组成，其功能主要是做砂石等材料的一般物理测试，进行混凝土试配、强度检验、试块养护、原材留样，配合产品研发。

The laboratory is generally located on the first floor of the office building or in the workshop to realize the inspection and recording of main raw materials and production processes. It is mainly composed of material room, concrete room, mechanics room, standard curing room and sample room. Its main job is to carry out general physical tests of sand and gravel and other materials, concrete trial mixing, strength inspection, test block curing and raw material sample retention, so as to cooperate with product research and development, etc.

d. 锅炉房。

d. Boiler room.

预制构件生产用蒸汽主要用来进行构件的蒸汽养护。如果有市政集中供蒸汽，应采用市政供汽，但要设置自己的换热站。没有集中供蒸汽时须自建锅炉生产蒸汽，一般采用清洁能源（天然气、秸秆等）作为燃料。锅炉房宜就近配备，内设燃气锅炉，通过蒸汽管道为养护仓提供适宜的温度及湿度条件，有效缩短构件养护时间；如果锅炉房距离养护仓较远，蒸汽管道应采取保温措施，以减少热能的损失。

The steam for the production of prefabricated members is mainly used for steam curing of members. If there is municipal centralized steam supply, it shall be adopted, but with our own heat exchange station. If not, a boiler must be built independently to produce steam, generally with clean energy(natural gas, straw, etc.) as fuel. The boiler room should be located nearby, with a gas boiler inside, to provide suitable temperature and humidity conditions for the curing chamber through the steam pipeline, thus effectively shortening the curing time of members. If the boiler room is far away from the curing chamber, the steam pipeline shall be provided with thermal insulation measures to reduce the loss of heat energy.

e. 污水及废弃物处理设施。

e. Sewage and waste treatment facilities.

在车间成品区尾部须设置三级污水处理池，混凝土搅拌站须设置污水处理循环利用系统。为能够进行水循环利用，应根据现场条件设置雨水、废水收集循环系统。厂区内应设置专门的固体废弃物回收处，主要用于临时存放混凝土废渣等废弃物。

A stage III sewage treatment tank shall be set up at the end of the finished product area of the workshop, and a sewage treatment and recycling system shall be set up at the concrete mixing station. For water recycling, rainwater and wastewater collection and circulation systems shall be set up according to site conditions. A solid waste recycling point shall be specially set up within the plant area for temporary storage of concrete waste residue and other wastes.

f. 配电室。

f. Power distribution room.

工厂用电应根据设备负荷合理规划设置配电系统，配电室宜靠近生产车间。比如，年产 10 万 m³ 预制构件的工厂年用电量一般为 80 万 ~ 100 万 kw·h。

The power distribution system shall be reasonably planned according to the equipment load, and the power distribution room should be close to the production workshop. For example, the power consumption of the plant with an annual output of 100 000 m³ of prefabricated members is generally 800 000~1 000 000 kw·h.

g. 配套设施。

g. Supporting facilities.

配套设施主要指生产、生活辅助设施，包括办公楼、职工宿舍、食堂及工作休闲区等。配套设施所在区域一般应该与生产园区有明显分界，使工人工作之余能够更好地休息。应根据生产能力估算所需生产人员和管理人员数量，再根据人数设定办公室、食堂与宿舍的建筑规模和类型。此外，最好能配有工作休闲区，为工人及管理人员提供良好的生活环境。

Supporting facilities mainly refer to auxiliary facilities for production and living, including office buildings, staff dormitories, canteens and work and leisure areas. Generally, the supporting facilities area shall be clearly separated from the production park, so that workers can get better rest after work. It is necessary to estimate the number of production personnel and management personnel required according to the production capacity, and determine the sizes and types of offices, canteens and dormitory buildings according to the number of people. What's more, leisure areas should be taken into consideration, providing a good living environment for workers and management personnel.

h. 厂区道路。

h. Roads in the plant area.

合理的厂区道路规划在满足生产需求的同时，也能使物流顺畅有序。厂区主干道构成环状路网，各路相通。厂区物流出入口应紧邻市政道路。厂区交通要做到人、物分流。厂区道路一般采用水泥混凝土路面。主干道路面宽度不小于 7 m，次干道为 5 m。对厂前区、道路两侧及新建建筑物、构筑物周围皆应予以绿化，种植花草和树木，以达到减少空气中的灰尘、降低噪声、调节空气温度和湿度及美化环境的目的，为工作人员创造一个良好的户外活动场所，如图 7-10 所示。

The roads in the plant area shall meet the living needs and ensure smooth and orderly logistics. The main roads within the plant area form a ring road network, and all roads are connected. The logistics entrances/exits in the plant area shall be close to the municipal roads. The people flow and material flow shall be separated in the plant area. All roads in the plant area are generally of cement concrete pavement. The main road is not less than 7 m wide, and the secondary road is 5 m wide. Greening shall be arranged in the front area, on both sides of roads, and around new buildings and structures. Flowers and trees shall be planted to reduce dust in the air and noise, regulate air temperature and humidity and beautify the environment, so as to create a good place for outdoor activities for the staff, as shown in Figure 7-10.

图7-10　厂区道路

Figure 7-10　Roads in the Plant Area

（2）工艺设计

(2) Process design

预制混凝土构件产品生产的全部工艺设计内容包括外墙板生产工艺（包括正、反打工艺），内墙板生产工艺，叠合楼板生产工艺，空调板、女儿墙生产工艺，楼梯、阳台、PCF 板（预制外挂墙板）等异型构件生产工艺等，以及构件养护工艺。

The production process design of precast concrete members includes production process of exterior wall panels(including front and reverse driving process), production process of interior wall panels, production process of laminated floor slabs, production process of air conditioner boards and parapets, production process of special-shaped members such as stairs, balconies and PCF(precast concrete facade) slabs, and member curing process.

（3）生产系统规划

(3) Production system planning

根据产能需求及生产工艺特点提供生产系统规划，具体内容包括：依据产品种类及生产工艺，规划 PC 构件生产线布局方式，包括混合式生产线、外墙板生产线、内墙板生产线、叠合板生产线、固定模台生产线；混凝土拌合及运输方式的布局规划；钢筋加工系统布局规划及周转方式的确定；工厂及厂区内垂直起吊系统的规划；生产过程物料周转方式的规划；生产车间内的辅助功能区域布局与规划；生产车间内安全通道及人行通道等的规划；构件存储及运输方式的规划。

Production system planning based on capacity requirements and production process characteristics, specific contents include: Layout planning of PC member production lines according to the product category and production process, including hybrid production line, exterior wall panel production line, interior wall panel production line, laminated slab production line and fixed die table production line; layout planning of concrete mixing and transportation modes; layout planning of reinforcement processing system and determination of turnover mode; planning of vertical lifting system in the plant and plant area; planning of material turnover mode in production process; layout and planning of auxiliary functional areas in the production workshop; planning of exit passageway and pedestrian passageway in production workshop; planning of member storage and transportation modes.

（4）生产线布局、辅助设备选型及经济测算

(4) Production line layout, selection of auxiliary equipment and economic estimate

①生产线布局。

① Production line layout.

根据生产工艺特点规划生产线布局及相关配置，比如混合式生产线布局及配置规划、外墙板生产线布局及配置规划、内墙板生产线布局及配置规划、叠合板生产线布局及配置规划、异型构件生产线布局及配置规划。

Plan the production line layout and related configuration according to the characteristics of the production process, such as the layout and configuration planning of the hybrid production line, the layout and configuration planning of the exterior wall panel production line, the layout and configuration planning of the interior wall panel production line, the layout and configuration planning of the laminated slab production line, and the layout and configuration planning of the special-shaped member production line.

②辅助设备选型。

② Selection of auxiliary equipment.

根据生产线布局配置的特点确定相关的辅助设备及设施，包括起重设备选型及配置规划，

搅拌站设备选型及配置规划，钢筋加工设备选型及配置规划，锅炉、空压机设备选型及配置规划，供电供水设施的选型规划、机修设施的选型规划，实验室设施的选型规划，工装系统的配置规划，安全防护系统的配置规划等。

Determine related auxiliary equipment and facilities according to the characteristics of production line layout and configuration, including lifting equipment selection and configuration planning, mixing station equipment selection and configuration planning, reinforcement processing equipment selection and configuration planning, boiler and air compressor equipment selection and configuration planning, selection planning of power supply and water supply facilities, repair facilities selection planning, laboratory facilities selection planning, tooling system configuration planning, and safety protection system configuration planning, etc.

③经济测算。

③ Economic estimate.

根据产能规划及投资规模提供构件成本、建厂投资测算等的数据分析，包括预制构件产品成本分析、盈亏平衡分析、利润测算等。

Provide member cost analysis and plant construction investment estimate data analysis according to capacity planning and investment scale, including cost analysis, break-even analysis

and profit calculation of prefabricated member products, etc.

总而言之，预制构件有不同的制作工艺，采用何种工艺与构件类型和复杂程度有关，与构件品种有关，与投资者的偏好也有关。合理的工厂布置是保证整个生产系统能够高效、安全和经济运行的基础。

In a word, there are different processes for prefabrication members, depending on the types and complexity of members, the varieties of members, and the preference of investors. Reasonable plant layout is the basis for the efficient, safe and economical operation of the whole production system.

7.2.3 工厂基本配置

7.2.3 Basic configuration of Plant

构件预制工厂基本配置见表 7-3。其中，人员数量根据产能和生产工艺灵活调整。在工厂达到满产及正常生产的情况下，辅助人员（保安、厨师、勤杂人员）根据公司运营状况配置。

See Table 7-3 for the basic configuration of plant. Personnel are flexibly adjusted according to production capacity and production process. When the plant produces in a full capacity and normal production, the auxiliary staff(security guards, cooks, servants) will be assigned according to the company's operation condition.

类别 Type	项目 Item	单位 Unit	生产规模/m³ Production scale/m³			
			5万 50 000		10万 100 000	
			固定模台 Fixed die table	流水线 Assembly line	固定模台 Fixed die table	流水线 Assembly line
人员 Personnel	管理技术人员 Technical management Personnel	人 Person	15 ~ 20	15 ~ 20	20 ~ 30	20 ~ 30
	生产工人 Production workers	人 Person	75 ~ 80	25 ~ 40	120 ~ 150	70 ~ 90
	人员合计 Total	人 Person	80 ~ 100	40 ~ 60	130 ~ 170	90 ~ 120
建筑 Building	预制构件制作车间 Prefabrication member workshop	m²	6 000 ~ 8 000	4 000 ~ 6 000	12 000 ~ 16 000	10 000 ~ 12 000
	钢筋加工车间 Reinforcement processing workshop	m²	2 000 ~ 3 000	2 000 ~ 3 000	3 000 ~ 4 000	3 000 ~ 4 000
	仓库 Warehouse	m²	100 ~ 200	100 ~ 200	200 ~ 300	200 ~ 300
	实验室 Laboratory	m²	200 ~ 300	200 ~ 300	200 ~ 300	200 ~ 300
	工人休息室 Workers' break room	m²	50 ~ 100	50 ~ 100	100 ~ 200	100 ~ 200
	办公室 Office	m²	1 000 ~ 2 000	1 000 ~ 2 000	1 000 ~ 2 000	1 000 ~ 2 000
	食堂 Canteen	m²	300 ~ 400	200 ~ 300	400 ~ 500	400 ~ 500
	模具修理车间 Die repair workshop	m²	500 ~ 700	500 ~ 700	800 ~ 1000	800 ~ 1000
	建筑合计 Total	m²	10 150 ~ 14 700	6 050 ~ 10 600	1 700 ~ 24 300	15 700 ~ 20 300
场地、道路 Site and road	构件存放场地 Members storage site	m²	10 000 ~ 15 000	10 000 ~ 15 000	20 000 ~ 25 000	20 000 ~ 25 000
	材料库场 Material warehouse	m²	2 000 ~ 3 000	2 000 ~ 3 000	3 000 ~ 4 000	3 000 ~ 4 000
	产品展示区 Product exhibition area	m²	500 ~ 800	500 ~ 800	500 ~ 800	500 ~ 800
	停车场 Parking lot	m²	500 ~ 800	500 ~ 800	800 ~ 1 000	800 ~ 1 000
	道路 Roads	m²	5 000 ~ 6 000	5 000 ~ 6 000	6 000 ~ 8 000	6 000 ~ 8 000
	绿地 Green space	m²	3 400 ~ 4 600	3 400 ~ 4 600	4 500 ~ 5 500	4 500 ~ 5 500
	场地合计 Total	m²	21 400 ~ 29 200	21 400 ~ 29 200	30 300 ~ 44 300	30 300 ~ 44 300

类别 Type	项目 Item	单位 Unit	生产规模/m³ Production scale/m³			
			5万 50 000		10万 100 000	
			固定模台 Fixed die table	流水线 Assembly line	固定模台 Fixed die table	流水线 Assembly line
设备、能源 Equipment and energy	混凝土搅拌站 Concrete mixing station	m³	1 ~ 1.5	1 ~ 1.5	2 ~ 3	2 ~ 3
	钢筋加工设备 Reinforcement processing equipment	t/h	1 ~ 2	1 ~ 2	2 ~ 4	2 ~ 4
	电容量 Electric capacity	kV·A	400 ~ 500	600 ~ 800	800 ~ 1 000	1 000 ~ 1 200
	水 Water	t/h	4 ~ 5	4 ~ 5	5 ~ 6	5 ~ 6
	蒸汽 Steam	t/h	2 ~ 4	2 ~ 4	4 ~ 6	4 ~ 6
	场地龙门式起重机（20 t） Site gantry crane (20 t)	台 Set	2（16 t、20 t）	2（16 t、20 t）	2 ~ 4（16 t、20 t）	2 ~ 4（16 t、20 t）
	车间行式起重机（5 t、10 t、16 t） Workshop traveling crane (5 t, 10 t, 16 t)	台 Set	8 ~ 12	4 ~ 8	10 ~ 16	4 ~ 8
	叉车（3 t、8 t） forklift(3 t, 8 t)	辆 Nr.	1 ~ 2	1 ~ 2	2 ~ 3	2 ~ 3

7.2.4 PC 构件生产线布置

7.2.4 Production Line Layout for PC Members

（1）生产工艺对比分析

(1) Comparative analysis of production process

PC 生产系统由 PC 生产线、钢筋生产线、混凝土拌合运输、蒸汽生产输送、车间门吊起运等五大生产系统组成。其中，PC 生产线为主线生产系统，钢筋生产线、混凝土拌合运输、蒸汽生产输送和门吊起运系统为辅助生产系统。

The PC production system consists of PC production line, reinforcement production line, concrete mixing and transportation system, steam production and transportation system, and workshop gantry crane system. The PC production line is the main line, and the reinforcement production line, concrete mixing and transportation, steam production and transportation and gantry crane systems are auxiliary.

根据模台的运动与否，PC 构件生产工艺分为平模传送流水线法和固定模台生产线法，如图 7-11 所示。

There are flat-die conveying assembly line and fixed die table production line for the production process of PC members based on the movement of the die table, as shown in Figure 7-11.

固定模台生产线和环形流水生产线是目前构件预制工厂的主流工艺，两种工艺形式对设备、人员等方面的要求各有不同，两种工艺的比较和优缺点见表 7-4、表 7-5。

The fixed die table production line and annular flow production line are the mainstream

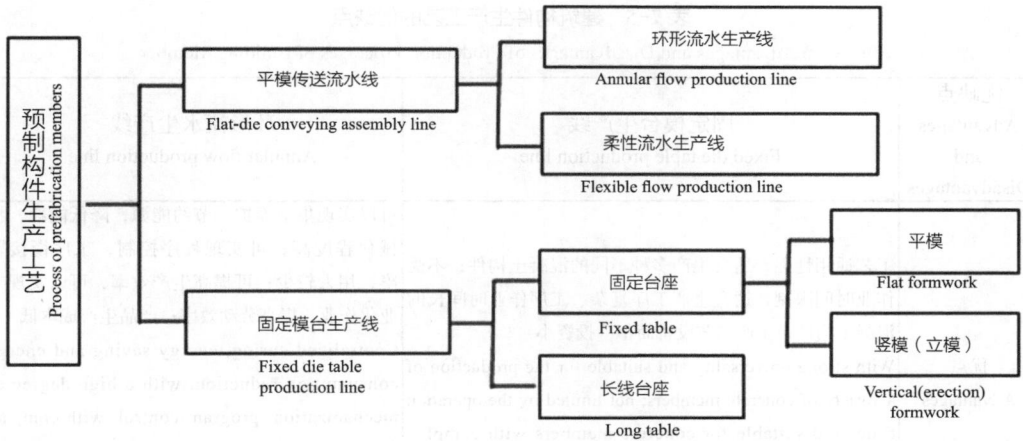

图7-11 PC构件生产工艺分类

Figure 7-11 Classification of PC Members Production Process

processes adopted by the member prefabrication plants at present. The two processes have different requirements on equipment and personnel, etc. The advantages and disadvantages of the two processes are shown in Table 7-4 and Table 7-5.

表 7-4 建筑构件生产工艺形式的比较

Table 7-4 Comparison of Production Processes of Building Member

序号 S/N	项目名称 Description	固定模台生产线 Fixed die table production line	环形流水生产线 Annular flow production line
1	适应范围 Application scope	墙板 Wallboards	板类构件 Plate-type members
2	浇筑设备 Pouring equipment	浇筑机、吊斗 Pouring machine and bucket	浇筑机 Pouring machine
3	成型设备 Forming equipment	台座振动、振捣棒 Table vibrator and vibrating rod	振动台 Vibrating stand
4	养护方法 Curing method	分散蒸养、自然 Decentralized steam curing, natural	集中蒸养 Centralized steam curing
5	通用性 Universality	好 Good	受限制 Restricted
6	作业条件 Operation conditions	一般 Ordinary	好 Good
7	机械化程度 Level of mechanization	低 Low	高 High
8	能耗 Energy consumption	较高 Relatively high	较低 Relatively low
9	产品质量 Product quality	好 Good	较好 Fair
10	劳动效率 Labor efficiency	一般 Ordinary	高 High
11	投资 Investment	较小 Relatively small	大 Large

优缺点 Advantages and Disadvantages	固定模台生产线 Fixed die table production line	环形流水生产线 Annular flow production line
优点 Advantages	工艺通用性强，适合生产多种不同的混凝土构件；不受作业时间限制，适合生产工序复杂、工序作业时间长的混凝土构件；生产工艺设备简单，投资小 With strong universality and suitable for the production of a variety of concrete members; not limited by the operation time, and suitable for concrete members with complex procedures and long operation time; simple production process equipment and small investment	可以实现集中养护，节约能源，降低能耗，机械化程度高；可实现程序控制，工序衔接紧凑，用人较少；可提高生产效率，可以实现专业化作业，提高劳动效率；产品生产成本低 Centralized curing, energy saving and energy consumption reduction, with a high degree of mechanization; program control, with compact connection of procedures, and less labor; production efficiency and labor efficiency improvement, specialized operation; production cost saving
缺点 Disadvantages	构件分散养护，保温设施简单，能耗高；台座分散布置，占地面积大，机械化程度低，用人数量比较多，劳动效率低；由于作业分散，作业环境整洁不好保证；生产成本高 Decentralized curing of members, simple thermal insulation facilities and high energy consumption; large floor area for decentralized arrangement of tables, low degree of mechanization, large number of employees and low labor efficiency; due to decentralized operation, the clean and tidy working environment is not guaranteed; high production cost	受工序作业时间限制，需要在限定时间内完成工序作业内容；一次性投资大 Limited by the process operation time, that is, the process operation needs to be completed within the limited time; large one-time investment

通过两种生产工艺的对比，循环流水线生产工艺可以集中蒸氧，节能降耗，工序设计紧凑，工艺布局科学合理，安全性强，生产效率、机械化程度高，可以根据市场对不同构件的需求比例，灵活安排生产。

Through the comparison of the two production processes, the annular flow production line has such advantages as centralized steam curing, energy saving and consumption reduction, compact process design, scientific and reasonable process layout, high safety, high production efficiency and high degree of mechanization. The production can be flexibly arranged according to the market demand proportion for different members, and it satisfies the principle of process selection.

（2）生产线选型说明

(2) Production line selection instructions

环形流水生产线可生产的产品种类由两个因素决定，一是钢底模的尺寸规格（9 000 mm×4 000 mm），二是蒸养窑的层高限制（层间空隙 450 mm），由以上两个因素可以确定产品的种类为长、宽、高不大于 9 000 mm×4 000 mm×450 mm 的板类及梁、柱构件。比如未来市场潜力比较大的公用建筑装饰外挂板，一般宽度不超过 6 米，本生产线完全适用。

The types of products that can be produced by the annular flow production line are determined by two factors, size and specification of the steel bottom die (9 000 mm×4 000 mm), and layer height limit of the steam curing kiln (interlayer gap 450 mm). Based on this, it can be determined that the products are plates, beams and columns of no more than L×W×H = 9 000 mm×4 000 mm×450 mm. For example, the width of decorative cladding panels for public buildings with great market potential in the future is generally not more than 6 m, so this production line is a right choice.

对于尺寸符合一定规则（长宽 3.5 m×4.5 m，厚度不超过 300 mm）的平板类混凝土预制构件，如果数量较大（墙板、楼板等），适合采用流水线连续生产。此种流水线国外已有成熟技术，国内也已经开发出适宜的工艺生产线。未来工厂的工艺规划和工程建设重点要满足此种流水线的要求。

For slab concrete precast members with certain dimensions (L×W=3.5 m×4.5 m, thickness not more than 300 mm), if the quantity is large(wall panels, floor slabs, etc.), the assembly line is preferred for continuous production. For this, mature technology has been developed abroad, and suitable process production line has been developed in China. The process planning and project construction of the plant in the future shall focus on meeting the requirements of this assembly line.

对于大型建筑预制构件（大型立体墙板、屋面板、工业厂房屋架等），最好采用固定模台方式生产。工厂建设阶段预留车间或者露天场地，但要做好设备选型工作，以适应构件最大尺寸和重量需要。

For prefabricated members of large buildings(large three-dimensional wall panels, roof panels, industrial plant roof trusses, etc.), it is better to adopt the fixed die table for production. Workshops or open sites shall be reserved during the plant construction stage, but equipment selection shall be done well to adapt to the maximum sizes and requirements of weights of members.

（3）生产线布置

(3) Production line layout

目前，主流 PC 自动化流水生产线均采用环形布置，充分考虑各个生产单元功能的不同、所占流水线节拍的长短、与拌合站混凝土运输线路的衔接位置、与钢筋生产线的相对关系等，进行合理布置，如图 7-12 所示。

At present, the mainstream PC automatic assembly line is reasonably arranged in an annular shape, with full consideration of the different functions of each production unit, the length of takt time, the connection position with the concrete transportation line of the mixing station, the relative relationship with the reinforcement production line, etc, as shown in Figure 7-12.

固定模台生产线一般采用线性布置。生产区与存放区相邻，且位于车间门吊的行走范围内。

The fixed die table production line is generally arranged linearly. The production area is adjacent to the storage area and located within the traveling range of the gantry crane in the workshop.

①生产线布置的原则。

① Principles for production line layout.

a. 方便流畅原则：各工序要有机结合，相关联工序集中放置，流水化布局。

a. Convenience and smoothness: Organic combination of all processes, centralized placement of related processes and streamlined layout.

b. 最短距离原则：减少搬运距离，避免流程交叉，直线运行。

b. Minimum distance: Reduction of handling distance, avoidance of cross-process, and straight-line operation.

c. 平衡均匀原则：工位之间资源配置、速率配置平衡。

c. Balance and uniformity: Balanced resource and rate configurations between stations.

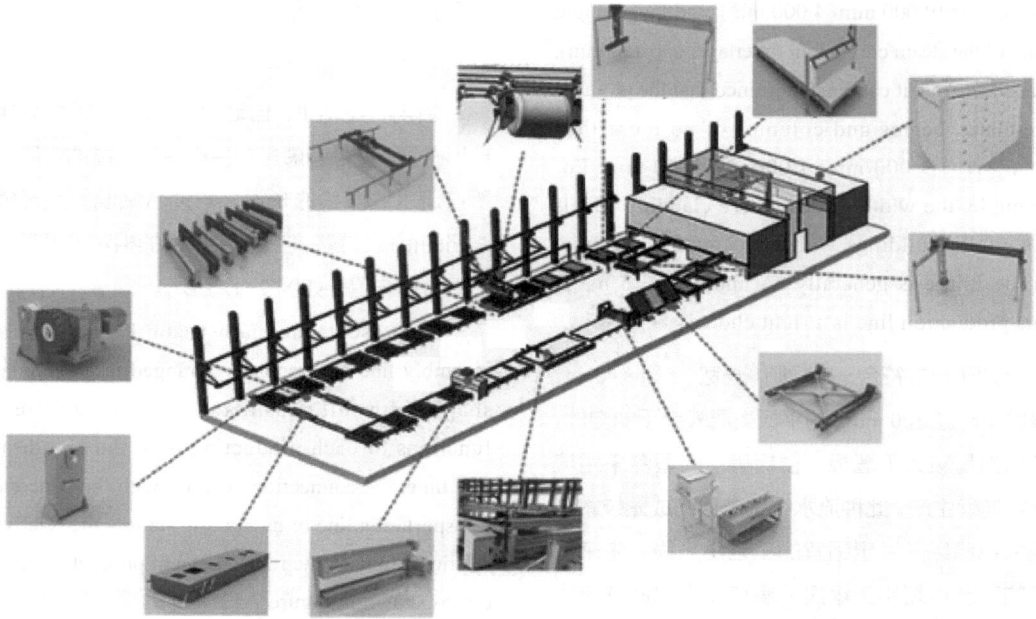

图7-12 环形流水线示意图

Figure 7-12 Schematic Diagram of Annular Flow Production Line

d. 固定循环原则：固定工位，减少诸如搬运等无价值活动。

d. Fixed cycle: Fixed stations to reduce worthless activities such as handling.

e. 安全合规原则：电气设备的安装、高压蒸汽条件下元件的保护，要符合相关法规、规程，进行合规的安装布局。模台运行、物体起运要设安全保险装置。

e. Safety and compliance: The installation of electrical equipment and the protection of elements at high-pressure steam shall comply with relevant laws and regulations, and the installation layout shall be compliant. Safety devices shall be provided for die table operation and objects lifting and transportation.

f. 经济产量原则：生产线布置要能适应最小批量生产，尽可能利用车间空间。

f. Economic yield: Adapt to minimum batch production and make use of workshop space as much as possible for production line layout.

g. 柔韧性原则：生产线布置须预留柔性发展空间。

g. Flexibility: Flexible development space should be reserved for the production line layout.

h. 硬件防错原则：生产线硬件设计与布局要预防错误，减少生产上的损失。

h. Hardware error prevention: Prevent errors from the hardware design and layout of the production line to reduce production losses.

②生产线布置的注意事项。

② Precautions for production line layout.

为减少空间浪费和降低作业人员巡回作业的强度，PC生产流水线工位呈环形布置。在PC生产线的周边，根据生产需要设置工器具存放区、半成品堆放区。

In order to reduce the waste of space and the work intensity of operators, the stations of PC production line are arranged in an annular shape. A tool storage area and a semi-finished products stacking area are arranged around the PC production line according to the need of production.

各生产线分开合理布置，中间区域可作为工作人员往来通道、巡视观摩通道。

Each production line is arranged separately and reasonably, and the middle area can be used as the staff channel and the inspection channel.

如果采用两条 PC 构件生产线和一条钢筋生产线的配置，则采用两条 PC 构件流水生产线左右分开，钢筋生产线居中布置的形式，以减少钢筋成品、半成品搬运距离。

If two PC member production lines and a reinforcement production line are designed, the two PC member production lines shall be separated left and right, and the reinforcement production line in the middle to reduce the transportation distance of finished and semi-finished products.

PC 构件生产线两侧均设排水沟与电缆沟，

有必要时设置暖气沟。电气管线分离，排水沟与电缆沟不得共用，废水与管线不得同用一条暗沟。

Drain ditches and cable trenches are provided on both sides of the PC member production line, and heating ditches if necessary. Electrical pipelines, drain ditches and cable trenches shall be separated, and wastewater and pipelines shall not be use the same blind ditch.

应根据生产设备高度，确定车间桁吊高度。为提高车间桁吊利用率，各桁吊必须能够在整个流水线内贯穿通行。桁吊必需在钢结构屋盖覆盖前安装完毕。

The height of truss cranes in the workshop is determined according to the height of the production equipment. In order to improve the utilization rate of truss cranes in the workshop, each truss crane must be able to pass through the whole assembly line. The truss crane must be installed before the steel structure roof is capped.

8

屋面及防水工程施工

Construction of Roof and Waterproofing Works

学习目标:

Learning objectives:

了解常用防水材料的类型、性能及使用方法；掌握卷材防水、涂料防水和细石混凝土防水屋面的施工要点及质量控制措施；了解地下工程防水方案；掌握卷材防水、涂料防水、水泥砂浆防水和自防水混凝土的构造及施工要点；掌握防水工程施工中质量通病的防治措施。

Learn about the types, properties, and applications of common waterproof materials; master the construction key points and quality control measures for roofs with membrane waterproofing, coating waterproofing, and fine aggregate concrete waterproofing; learn about the waterproofing scheme of underground structures; master the structures and construction key points of membrane waterproofing, coating waterproofing, cement mortar waterproofing, and self-waterproof concrete; master the prevention and control measures for common quality defects in the construction of waterproofing works.

技能抽查要求：

Skill checking requirements:

能按国家规范要求，正确使用常用检测工具对卷材防水屋面、刚性防水屋面的施工质量进行检查验收，能正确填写工程质量验收记录表。

Be able to inspect and accept the construction quality of roofs with membrane waterproofing and rigid waterproofing by correctly using common testing tools according to the national specification requirements, and be able to correctly fill in the record forms for construction quality acceptance.

建筑八大员岗位资格考试要求：

Qualification examination requirements for eight major posts of construction engineering:

掌握屋面、地下室及卫生间防水施工工艺；掌握防水工程施工的质量及安全要求。

Master the construction technology of roof, basement and toilet waterproofing; master the construction quality and safety requirements of waterproofing works.

8.1 防水工程概述

8.1 Overview of Waterproofing Works

8.1.1 防水工程的作用

8.1.1 Functions of Waterproofing Works

防水工程分为屋面防水工程和地下防水工程。防水工程质量的优劣，不仅关系到建（构）筑物的使用寿命长短，且直接关系到其使用功能的好坏。因此，评估建筑防水工程质量时，除了考虑设计的合理性、防水材料的选择外，更要注意其施工工艺及施工质量。

Waterproofing works are classified into roof waterproofing and waterproofing for underground structures. The quality of waterproofing works is not only related to the service life of buildings(structures), but also directly affects the service functions. Therefore, in addition to the rationality of design and the correct selection of waterproof materials, the focus should also be laid on the construction technology and quality of waterproofing works.

8.1.2 防水材料

8.1.2 Waterproof Materials

防水卷材：包括沥青防水卷材、高聚物改性沥青防水卷材、合成高分子防水卷材等。

Waterproof membrane: Including asphalt waterproof membrane, polymer modified asphalt waterproof membrane, synthetic polymer waterproof membrane, etc.

防水涂料：包括沥青防水涂料、高聚物改性沥青防水涂料、合成高分子防水涂料等。

Waterproof coating: Including asphalt waterproof coating, polymer modified asphalt waterproof coating, synthetic polymer waterproof coating, etc.

防水密封材料：包括不定型材料（如密封膏等）、定型材料（如止水带等）。

Waterproof sealant materials: Including unshaped materials(sealant, etc.), shaped materials(waterstop, etc.).

防水混凝土：包括普通防水混凝土、补偿收缩防水混凝土、预应力防水混凝土、掺外加剂防水混凝土以及钢纤维或塑料纤维防水混凝土等。

Waterproof concrete: Including ordinary waterproof concrete, shrinkage-compensating waterproof concrete, prestressed waterproof concrete, waterproof concrete mixed with admixtures, steel fiber or plastic fiber waterproof concrete.

防水砂浆：包括水泥砂浆（刚性多层抹面）、掺外加剂水泥砂浆以及聚合物水泥砂浆等。

Waterproof mortar: Including cement mortar(rigid multi-layer plastering), cement mortar mixed with admixtures and polymer cement mortar.

其他防水材料：包括各类粉状憎水材料，如建筑拒水粉、复合建筑防水粉等；还有各类渗透剂防水材料。

Others: Including various kinds of powdery hydrophobic materials, such as powdery water repellent agent for buildings, composite waterproof powder for buildings; and waterproof materials in various penetrants.

8.2 屋面防水工程

8.2 Roof Waterproofing Works

屋面防水工程是房屋建筑的一项重要工程。根据建筑物的类别、重要程度、使用功能要求及防水层耐用年限等，可将屋面防水分为四个等级，并按不同等级进行设防。屋面防水工程的防水等级和设防要求见表8-1。

Roof waterproofing is an important part of building construction. According to categories, importance levels, functional requirements of buildings, and durability of waterproof layers, roof waterproofing is rated in four grades with different waterproofing requirements as shown in Table 8-1.

表 8-1　屋面防水工程的防水等级和设防要求

Table 8-1 Waterproofing Rating and Requirements for Roof Waterproofing Works

项目 Item	屋面防水等级 Roof waterproofing rating			
	I	II	III	IV
建筑物类别 Category of buildings	特别重要或对防水有特殊要求的建筑 Buildings particularly important or with special requirements for waterproofing	重要建筑和高层建筑 Important buildings and high-rise buildings	一般建筑 General buildings	非永久性的建筑 Non-permanent buildings
使用年限 Service life	25年 25 years	15年 15 years	10年 10 years	5年 5 years
防水层选用材料 Materials selected for waterproof layer	合成高分子防水卷材、高聚物改性沥青防水卷材、金属板材、合成高分子防水涂料、细石混凝土等 Synthetic polymer waterproof membrane, polymer modified asphalt waterproof membrane, metal sheet, synthetic polymer waterproof coating, fine aggregate concrete, etc.	前述材料＋高聚物改性沥青防水涂料、平瓦、油毡瓦等 Aforesaid materials + polymer modified asphalt waterproof coating, plain roofing tile, asphalt roofing shingle, etc.	前述材料＋三毡四油沥青防水卷材等 Aforesaid materials + three-pass asphalt felt and four-pass asphalt waterproof membrane, etc.	二毡三油沥青防水卷材、高聚物改性沥青防水涂料等 Two-pass asphalt felt and three-pass asphalt waterproof membrane, polymer modified asphalt waterproof coating, etc.
设防要求 Waterproofing requirements	三道或三道以上设防 Three or more passes of waterproofing	二道设防 Two passes of waterproofing	一道设防 One pass of waterproofing	一道设防 One pass of waterproofing

屋面防水工程一般包括屋面卷材防水、涂膜防水、刚性防水、瓦屋面防水、屋面接缝密封防水。

Roof waterproofing works generally include roof membrane waterproofing, coating waterproofing, rigid waterproofing, tiled roof waterproofing and roof joint-sealing waterproofing.

屋面的保温层和防水层严禁在下雨天、下雪和刮五级以上大风的情况下施工，温度过低也不宜进行施工。屋面工程完工后，应对屋面细部构造、接缝、保护层等进行外观检验，并用淋水或蓄水的方式进行防水检验，防水层不得有渗漏或积水现象。

It is prohibited to pave the thermal insulation layer and waterproof layer for roofs on rainy days, snowy days, and under strong winds above scale 5, and construction is also unsuitable at too low temperatures. The roof works finished shall be visually inspected to check detailed structures, joints, protective layers, etc., and shall be waterproof inspected by water spraying or impounding. Any leakage or water ponding is unacceptable for waterproof layers.

8.2.1 卷材防水屋面

8.2.1 Roof with Membrane Waterproofing

卷材防水屋面是用胶结材料粘贴卷材进行防水的屋面。这种屋面具有重量轻、防水性能好的优点，其防水层的柔韧性好，能适应一定程度的结构振动和胀缩变形。所用卷材有沥青防水卷材、高聚物改性沥青防水卷材和合成高分子防水卷材等三大系列，目前主要采用高聚物改性沥青防水卷材和合成高分子防水卷材，如图 8-1 所示。

Membranes are paved by adhibiting with cementing materials for roof waterproofing. A roof with this kind of waterproofing is advantageous in lightweight and good waterproof performance. Its waterproof layer has good flexibility and adapts to a certain degree of structural vibration, and expansion and contraction deformation. Among the three series available, i.e., the asphalt waterproof membrane, polymer modified asphalt waterproof membrane, and synthetic polymer waterproof membrane, the latter two are mainly used at present, as shown in Figure 8-1.

（1）卷材防水屋面的构造

(1) Structure of roof with membrane waterproofing

防水卷材屋面的构造如图 8-2 所示。

The structure of roof with membrane waterproofing is shown in Figure 8-2.

图8-1　高聚物改性沥青防水卷材和合成高分子防水卷材

Figure 8-1 Polymer Modified Asphalt Waterproof Membrane and Synthetic Polymer Waterproof Membrane

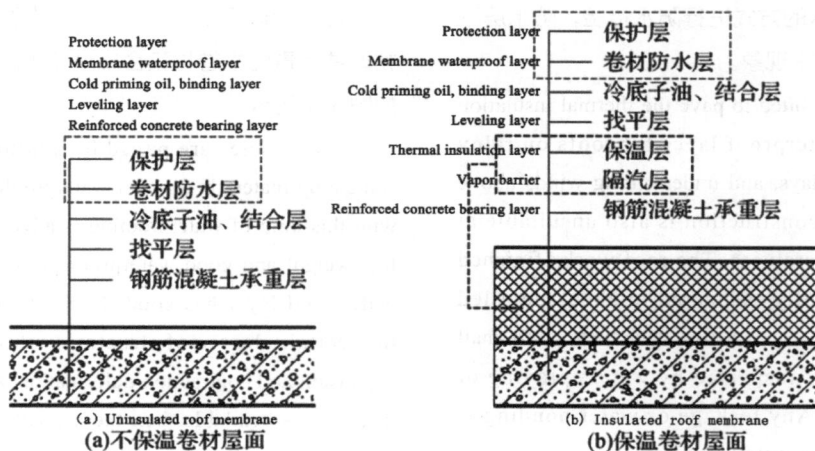

Protection layer
Membrane waterproof layer
Cold priming oil, binding layer
Leveling layer
Reinforced concrete bearing layer

保护层
卷材防水层
冷底子油、结合层
找平层
钢筋混凝土承重层

Protection layer
Membrane waterproof layer
Cold priming oil, binding layer
Leveling layer
Thermal insulation layer
Vapor barrier
Reinforced concrete bearing layer

保护层
卷材防水层
冷底子油、结合层
找平层
保温层
隔汽层
钢筋混凝土承重层

(a) Uninsulated roof membrane
(a)不保温卷材屋面

(b) Insulated roof membrane
(b)保温卷材屋面

图8-2　卷材防水屋面的构造

Figure 8-2　Structure of Roof with Membrane Waterproofing

（2）卷材防水层施工

(2) Construction of membrane waterproof layer

卷材防水屋面的施工工艺为：屋面基层处理→隔汽层施工→保温层施工→找平层施工→刷基层处理剂→铺贴卷材附加层→铺贴卷材防水层→保护层施工。

The construction process of a roof with membrane waterproofing is as follows: Base layer treatment on the roof → paving vapor barrier → paving thermal insulation layer → paving leveling layer → brushing base treatment agent → pasting additional layer of membrane → pasting membrane waterproof layer → paving protection layer.

①基层施工。

① Base layer paving.

基层应有足够的强度和刚度，承受荷载时不致产生显著变形。基层一般采用水泥砂浆、细石混凝土或沥青砂浆找平，做到平整、坚实、清洁、无凹凸及尖锐颗粒。其平整度为：用2 m长的直尺检查，基层与直尺间的最大空隙不应超过5 mm，空隙仅允许平缓变化且每2 m长度内不得多于一处。铺设屋面隔汽层和防水层之前，基层必须清扫干净。

The base layer shall be paved with sufficient strength and rigidity to withstand loads without visible deformation. The base layer is generally leveled with cement mortar, fine aggregate concrete or asphalt mortar, to achieve a flat, solid, clean surface free of unevenness and sharp particles. The evenness shall be inspected with a 2 m long ruler. The maximum gap between the base layer and the ruler shall not exceed 5 mm; only the gap with gentle change is acceptable, and there shall be not more than one gap per 2 meter. The base layer must be cleaned before paving the vapor barrier and waterproof layer on the roof.

②隔汽层施工。

② Vapor barrier paving.

隔汽层可采用气密性好的卷材或防水涂料。一般是在结构层（或找平层）上涂刷冷底子油一道和热沥青两道，或铺设一毡二油。

The vapor barrier may be paved with membrane or waterproof coating with good air tightness. As a general practice, one pass of cold priming oil and two passes of hot asphalt are applied, or one pass of asphalt felt and two passes of asphalt are applied on the structural layer(or leveling layer).

隔汽层必须是整体连续的。在屋面与垂直面衔接的地方，隔汽层还应延伸到保温层顶部并高出 150 mm，以便与防水层相接。采用油毡隔汽层时，油毡的搭接宽度不得小于 70 mm。采用沥青基防水涂料时，其耐热度应比室内或室外的最高温度高出 20 ~ 25℃。

The vapor barrier must be applied continuously. The vapor barrier shall also be extended to the top of and beyond the thermal insulation layer for 150 mm where the roof is connected to a vertical face to allow for connection to the waterproof layer. For the vapor barrier paved with asphalt felt, the lap width shall not be less than 70 mm. When an asphalt-based waterproof coating is used, its heat resistance shall be 20-25℃ higher than the maximum indoor or outdoor temperature.

隔汽层一般采用卷材空铺（也就是边上用胶黏剂与结构层黏接），或用防水涂料刷一层。

The vapor barrier is generally paved with membranes by the border adhibiting method(that is, only the edges are applied with adhesive for bonding with the structural layer), or a pass of waterproof coating is brushed.

③保温层施工。

③ Thermal insulation layer paving.

a. 施工准备。

a. Preparations.

材料准备：材料进场应查验其生产厂家提供的产品合格证、检测报告。材料外表或包装物有明显标志，标明材料生产厂家、材料名称、生产日期、执行标准、产品有效期等。

Materials: The materials shall be mobilized along with the product certificates and test reports provided by the manufacturers. The materials received on the site shall be clearly marked on the external surfaces or packages, indicating the material manufacturer, designation, production date, applicable standard, product lifetime, etc.

主要机具：砂浆搅拌机、混凝土搅拌机、运料手推车、木抹子、水平刮杠、水平尺等。

Main machines and tools: Mortar mixer, concrete mixer, transport trolley, wood float, darby slicker, level ruler, etc.

b. 工艺流程。

b. Process flow.

基层清理→管根封堵→涂刷隔汽层→标定标高、坡度→保温层施工→找坡层施工→验收。

Base layer cleaning → pipe root plugging → vapor barrier paving → level and slope marking → thermal insulation layer paving → sloping layer making → acceptance.

c. 施工要点。

c. Key points of construction.

铺设保温材料的基层应平整、干燥和干净。

The base layer paved with a thermal insulation layer shall be flat, clean and dry.

涂刷隔汽层是为了防止结构层或室内的潮气进入保温层，可用掺 0.2% ~ 0.3% 乳化剂的水溶液，也可用沥青溶液（冷底子油），基层处理剂应涂刷均匀，无露底，无堆积。涂刷时，应用刷子用力涂，使处理剂尽量刷进基层表面的毛细孔中，这样才能起防潮作用。

A vapor barrier is paved to prevent moisture in the structural layer or indoors from invading the thermal insulation layer. The water solution mixed with 0.2%-0.3% emulsifier or asphalt solution(cold priming oil) may be used. The base treatment agent shall be evenly applied without exposure and build-up. The brush shall be pressed to apply the treatment agent to make it penetrate into the capillary pores on the surface of the base layer as far as possible, so as to play a moisture-proof role.

在与室内空间有关联的天沟、檐沟处，均应铺设保温层；天沟、檐沟、檐口与屋面交接处，屋面保温层的铺设应延伸到墙内，其伸入的长度不应小于墙厚的 1/2。

The thermal insulation layers shall be paved at roof gutters and eave gutters associated with indoor space. At the junction of roof gutters, eave gutters and cornices with the roof, the roof thermal insulation layer shall be extended into the wall, for a length not less than 1/2 of the wall thickness.

保温层施工时应按设计坡度方案标定出标高和坡度。

The level and slope shall be marked according to the design slope scheme for paving the thermal insulation layer.

④找平层施工。

④ Leveling layer paving.

a. 施工准备。

a. Preparations.

材料准备：所用材料必须进场验收，并按要求对各类材料进行复试。其质量、技术性能必须符合设计要求和施工及验收规范的规定。

Materials: The materials received on the site must be inspected and shall be re-tested as required to ensure their quality and technical performance must meet the design requirements and the provisions of the construction and acceptance specifications.

主要机具：砂浆搅拌机、混凝土搅拌机、运料手推车、铁锹、铁抹子、木抹子、水平刮杠、水平尺。

Main machines and tools: Mortar mixer, concrete mixer, transport trolley, shovel, iron trowel, wood trowel, darby slicker, and level ruler.

b. 工艺流程。

b. Process flow.

基层清理→管根封堵→标定标高、坡度→浇水湿润或喷涂沥青稀料→找平层施工→刮平、抹平、压实→养护→填缝→验收。

Base layer cleaning → pipe root plugging → level and slope marking → watering or spraying asphalt flux → leveling layer making → screeding, floating and compaction → curing → caulking → acceptance.

c. 施工要点。

c. Key points of construction.

基层清理：将结构层、保温层上表面的松散杂物清扫干净，将凸出基层表面的灰渣等黏结杂物铲平，不得影响找平层的有效厚度。

Base layer cleaning: Clean loose foreign matter on the surface of the structure layer and thermal insulation layer, and remove the bonded foreign matter including projecting mortar residues to level the surface of the base layer without affecting the effective thickness of the leveling layer.

穿过屋面的管道、烟囱根部封堵：大面积做找平层前，应先将伸出屋面的预埋管件、烟囱、女儿墙、檐沟、伸缩缝根部处理好。

Plugging at the root of pipes and chimney penetrating the roof: The embedded pipes, chimneys, parapets, eaves gutters and expansion joints protruding from the roof shall be properly treated at the root before proceeding with the leveling layer of a large area.

找平层施工：按设计坡度方案标定出标高和坡度。贴点标高、冲筋并设置分格缝，也可以在施工找平层后切割。

Leveling layer paving: First, make the level and slope markings according to the design slope scheme; then, place dabbed mortar, level reference, and dividing joints, or cutting may also be carried out on the finished leveling layer.

浇水湿润：抹找平层前，根据找平层类型，应适当浇水湿润，但不可洒水过量，避免积水。

Wetting by watering: The base layer shall be properly wetted by watering according to the type of the leveling layer to be plastered, but excessive watering is not acceptable to avoid water ponding.

铺水泥砂浆：按分格块装灰、铺平，用刮杠靠冲筋条刮平，找坡后用木抹子搓平，铁抹子压光。待浮浆沉实后、人踏上去有脚印但不下陷时，再用铁抹子压第二遍即可。

Paving cement mortar: Place and level mortar in divided cells, screed it with a darby slicker against the level reference, float it with a wood trowel after slope making and trowel it with an iron trowel; after the laitance is settled so that stepping on it leaves footprints without sinking, trowel it again with the iron trowel.

养护：找平层抹平、压实 24 小时后可浇水、覆盖养护，一般养护期为 7 天，经干燥后铺设防水层。

Curing: Cure the leveling layer by watering after floating and compaction 24 hours. Usually, curing lasts for 7 days. A waterproof layer is paved on the dried leveling layer.

填缝：一般可以采用玛蹄脂弹性材料嵌缝，与找平层应齐平，不得有明显的凸起和凹陷。

Caulking: Generally, mastic elastic materials are used for caulking, to be flush with the leveling layer without obvious bulges and depressions.

d. 注意事项。

d. Precautions.

屋面及檐口、檐沟、天沟找平层的排水坡度，必须符合设计要求。平屋面采用结构找坡不应小于 3%，采用材料找坡宜为 2%，天沟、檐沟纵向找坡不应小于 1%，沟底落水差不得超过 200 mm。在与突出屋面结构的连接处以及在房屋的转角处，均应做成圆弧或钝角。其圆弧

半径应符合要求，高聚物改性沥青防水卷材为 50 mm，合成高分子防水卷材为 20 mm。

The drainage slope of the roof, cornice, eaves gutter and roof gutter leveling layer must meet the design requirements. A flat roof shall be sloped at not less than 3% for structural sloping and should be at 2% for sloping with materials, a roof gutter and eaves gutter shall be longitudinally sloped at not less than 1%, and the drop at the gutter bottom shall not exceed 200 mm. Arcs or obtuse angles shall be made at junctions with structures projecting from the roof and at building corners. Subject to the specified requirements, the arc radius shall be 50 mm for the polymer modified asphalt waterproof membrane and 20 mm for the synthetic polymer waterproof membrane.

为防止由于温差及混凝土构件收缩而使防水屋面开裂，找平层应留分格缝，缝宽宜为 5 ～ 20 mm。找平层分隔缝纵横间距不宜大于 6 m。

The leveling layer shall be provided with dividing joints to prevent the waterproof roof from cracking due to temperature differences and shrinkage of concrete members, and the joint width should be 5-20 mm. The dividing joints of the leveling layer should be spaced not greater than 6 m transversely and longitudinally.

采用水泥砂浆或细石混凝土找平层做基层时，其厚度和技术要求应符合表 8-2 的规定。

The base layer made of cement mortar or fine aggregate concrete leveling layer shall meet the thickness and technical requirements shown in Table 8-2.

表 8-2 找平层厚度和技术要求

Table 8-2 Thickness and Technical Requirements of Leveling Layer

类别 Items	基层种类 Base layer type	厚度/mm Thickness/mm	技术要求 Technical requirements
水泥砂浆找平层 Cement mortar leveling layer	整体混凝土 Monolithic concrete	15～20	体积比1：2.5～1：3（水泥：砂） Volume ratio of 1：2.5-1：3 (cement：sand)
	整体或板状材料保温层 Integral or plate material thermal insulation layer	20～25	
	装配式混凝土板、松散材料保温层 Prefabricated concrete slab, loose material thermal insulation layer	20～30	
细石混凝土找平层 Fine aggregate concrete leveling layer	松散材料保温层 Loose material thermal insulation layer	30～35	混凝土强度等级不低于C20 Concrete strength not less than C20.
沥青砂浆找平层 Asphalt mortar leveling layer	整体混凝土 Monolithic concrete	15～20	质量比1：8（沥青：砂） Mass ratio 1：8 (asphalt：sand)
	装配式混凝土板、整体或板状材料保温层 Prefabricated concrete slab, integral or plate material thermal insulation layer	20～25	

⑤卷材防水层施工。

⑤ Membrane waterproof layer adhibiting.

a. 卷材的铺贴顺序与方向。

a. Adhibiting sequence and direction of membrane.

时间：待屋面上其他工作完工，基层处理剂干燥后。

Time: After other construction tasks on the roof are finished and the base treatment agent is dried.

顺序：先高后低、先远后近。

Sequence: Top-to-bottom and far-to-near.

方向：屋面坡度＜3%，平行于屋脊方向施工（节省材料，便于施工）；坡度3%～15%，可平行或垂直于屋脊方向施工；坡度＞15%或屋面受震动，垂直于屋脊方向（防止沥青流淌）施工。

Direction: In the case of a roof slope less than 3%, parallel to the ridge strike(saving materials and facilitating construction); in the case of a slope of 3%-15%, parallel or perpendicular to the ridge strike; in the case of a slope greater than 15% or the roof subject to vibration, perpendicular to the ridge strike(preventing asphalt from flowing).

b. 搭接要求。

b. Lapping requirements.

上下两层及相邻两卷材的搭接缝均应错开，上下两层卷材长边搭接缝应错开，错开的距离不得小于幅宽的1/3、同一层相邻两幅卷材短边搭接缝也应错开，错开的距离不得小于500 mm。

The lap joints of adjacent membranes in two layers shall be staggered, and the lap joints at the long side of the two layers shall be staggered, for a distance not less than 1/3 of the width, and the lap joints at the short side of two adjacent membranes on the same layer shall also be staggered, for a distance not less than 500 mm.

要求：卷材平行于屋脊，搭接缝应顺水流方向；卷材垂直于屋脊，搭接缝应顺主导风向。叠层铺设的各层卷材，在天沟与屋面的连接处，应采用叉接法搭接，搭接缝应错开，接缝宜留在屋面或天沟侧面，不宜留在沟底。

Requirements: For the membranes parallel to the ridge, the lap joint shall be placed in the water flow direction; for the membranes perpendicular to the ridge, the lap joint shall be placed in the prevailing wind direction. For each layer of membrane paved in a laminated structure, the connection to the roof gutter and the roof shall be lapped in a cross manner, and the lap joints shall be staggered and should be placed on the side of the roof or the roof gutter, instead of the gutter bottom.

c. 施工方法。

c. Construction methods.

常用的施工方法有：冷黏法、热熔法、自黏法等。

Common construction methods include cold adhibiting, hot fusion, self-adhibiting method, etc.

冷黏法施工是利用毛刷将胶黏剂涂刷在基层或卷材上，然后直接铺贴卷材，使卷材与基层、卷材与卷材黏结的方法。施工时，胶黏剂涂刷应均匀、不露底、不堆积。铺贴卷材时应平整顺直，搭接尺寸准确，接缝应涂胶黏剂，辊压黏结牢固，不得扭曲，溢出的胶黏剂随即刮平封口；也可采用热熔法接缝。接缝口应用密封材料封严，宽度不应小于 10 mm。

The cold adhibiting method is a method that apply adhesive on the base layer or the membrane with a brush, and then directly pave the membrane for bonding with the underlying base layer and membrane. Adhesive shall be applied evenly without exposure of the underlying layer or build-up. The membranes shall be paved smoothly and straightly, lapped in accurate dimensions and jointed firmly by applying adhesive and rolling

without distortion. The excessive adhesive shall be scraped immediately for sealing. This method may be coordinated with the hot fusion method. The joints shall be tightly sealed with sealing materials in a width of not less than 10 mm.

热熔法施工是指利用火焰加热器熔化热熔型防水卷材底层的热熔胶再进行粘贴的方法。施工时，在卷材表面热熔后，以卷材表面熔融至光亮黑色为度，应立即滚铺卷材，使之平展，并且黏结牢固。加热卷材时应均匀，不得过分加热或烧穿卷材。对厚度小于 3 mm 的高聚物改性沥青防水卷材严禁采用热熔法施工。

The hot fusion method is a method that melt the hot-melt adhesive on the bottom surface of the hot-melt waterproof membrane with a fired heater. The membrane surface is hot melted until the surface turns in bright black, and shall be immediately unrolled flat and firmly adhibited. This kind of membrane shall be evenly heated, and excessive heating or burn-through is unacceptable. For the polymer modified asphalt waterproof membrane with a thickness of less than 3 mm, the hot fusion method shall be strictly prohibited for construction.

自黏法施工是指采用带有自黏胶的防水卷材，不用热施工，也不需涂胶结材料，而进行黏结的方法。铺贴前，基层表面应均匀涂刷基层处理剂，待干燥后及时铺贴卷材。铺贴时，应先将自黏胶底面隔离纸完全撕净，排除卷材下面的空气，并辊压黏结牢固，不得空鼓。接缝部位宜采用热风焊枪加热后随即黏贴牢固，溢出的自黏胶随即刮平封口。接缝口用不小于 10 mm 宽的密封材料封严。

The self-adhibiting method is a method that use waterproof membranes with self-adhesive, without hot fusion and application of adhesive. The base surface shall be evenly applied with the base treatment agent and the membrane shall be applied timely when the agent is dried. The release paper on the bottom surface of the self-adhesive shall be completely torn off before paving, and the membrane shall be pressed to remove air and shall be rolled for firm bonding without hollowing. The joint part should be heated by a hot air welding gun and then firmly adhibited, and the spilled self-adhesive shall be scraped immediately for sealing. The joints shall be sealed with sealing materials in width of not less than 10 mm.

⑥保护层施工。

⑥ Protection layer paving.

卷材屋面应有保护层，以减少雨水、冰雹冲刷或其他外力造成的卷材机械性损伤，并可折射阳光、降低温度，减缓卷材老化速度，从而增加防水层的使用寿命。

The roof membrane shall be provided with a protective layer to reduce the mechanical damage to the membrane under the scouring effect of rain and hail or other external forces, and to refract sunlight, reduce temperature and slow down the aging of the membrane, thus increasing the service life of the waterproof layer.

保护层施工应在防水层经过验收合格，并将其表面清扫干净后进行。

The protective layer shall be paved after the waterproof layer is accepted and the receiving surface is cleaned.

8.2.2 涂膜防水屋面
8.2.2 Coated Waterproof Roof

涂膜防水屋面是在屋面基层上涂刷防水涂料，经固化后形成一层有一定厚度和弹性的整体涂膜，从而达到防水目的的一种防水屋面形式，其典型的构造层次如图 8-3 所示。这种屋面具有施工操作简便、无污染、冷操作、无接缝、能适应复杂基层、防水性能好、温度适应性强、容易修补等特点。适用于防水等级为 I 级、II 级的屋面。

The waterproof coating is applied to the base layer on the roof and is cured into an integral coating with a certain thickness and elasticity, so as to achieve the waterproof purpose. The typical structures are shown in Figure 8-3. This type of waterproof roof is characterized by convenient operation, environmental friendliness, cold operation, jointless structure, adaptability to complex base layer conditions, good waterproof performance, strong temperature adaptability and easy repair. It is suitable for roof waterproofing

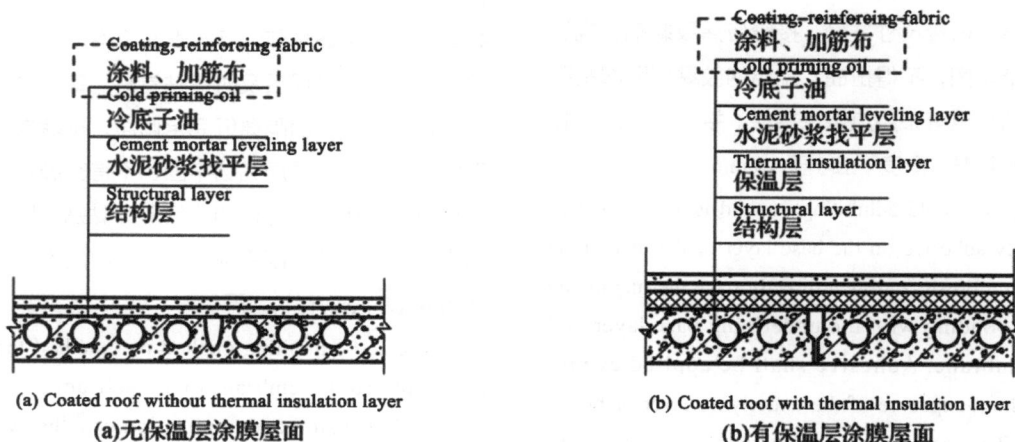

(a) Coated roof without thermal insulation layer
(a)无保温层涂膜屋面

(b) Coated roof with thermal insulation layer
(b)有保温层涂膜屋面

图8-3 涂膜防水屋面构造

Figure 8-3 Structure of Coated Waterproof Roof

rated I and II.

（1）材料要求

(1) Material requirements

根据防水涂料成膜物质的主要成分，适用于做涂膜防水层的涂料可分为：高聚物改性沥青防水涂料和合成高分子防水涂料两类。根据防水涂料形成液态的方式，可分为溶剂型、反应型和乳液型三类，见表8-3。

According to the main components of the film-forming materials of waterproof coatings, the coatings suitable for waterproof layers are divided into two categories: Polymer modified asphalt waterproof coatings and synthetic polymer waterproof coatings. According to the liquid formation mode, the waterproof coatings are divided into solvent type, reaction type and water emulsion type, as shown in Table 8-3.

表8-3　涂膜防水材料

Table 8-3 Coated Waterproof Materials

类别 Item		材料名称 Material name
高聚物改性沥青防水涂料 Polymer modified asphalt waterproof coating	溶剂型 Solvent type	再生橡胶沥青涂料、氯丁橡胶沥青涂料等 Recycled rubber asphalt coating, neoprene asphalt coating, etc.
	乳液型 Emulsion type	丁苯乳胶沥青涂料、氯丁胶乳沥青涂料、PVC（聚氯乙烯）煤焦油涂料等 Styrene-butadiene rubber latex asphalt coating, neoprene latex asphalt coating, PVC(polyvinyl chloride) coal tar coating, etc.
合成高分子防水涂料 Synthetic polymer waterproof coating	乳液型 Emulsion type	硅橡胶涂料、丙烯酸酯涂料、AAS（丙烯腈-丙烯酸丁脂-苯乙烯共聚物）隔热涂料等 Silicone rubber coating, acrylate coating, AAS(acrylonitrile, acrylate, styrene) thermal insulation coating, etc.
	反应型 Reaction type	聚氨酯防水涂料、环氧树脂防水涂料等 Polyurethane waterproof coating, epoxy resin waterproof coating, etc.

（2）施工工艺

(2) Construction process

①施工准备。

① Construction preparation.

主要机具设备：搅拌器、吹尘器、铺布机具、大鬃毛刷（板长24～40 cm）、油刷、大小橡皮刮板、磅秤等。

Main machines and tools: Agitator, dust blower, paver, large-sized bristle brush(plate length of 24-40 cm), oil brush, large- and small-sized rubber scraper, weighing scale, etc.

②工艺流程。

② Process flow.

基层表面清理、修整→喷涂基层处理剂（底涂料）→特殊部位附加增强处理→涂布防水涂料→保护层施工。

Cleaning and trimming base layer surface → spraying base treatment agent(primer) → additional reinforcement treatment of special parts → application of waterproof coating → paving protective layer.

a. 喷涂基层处理剂。

a. Applying base treatment agent.

基层处理剂应与上部涂料的材料性能相容，与常用防水涂料的稀释液充分混合进行刷涂或喷涂。喷涂均匀，覆盖完全，干燥后方可进行涂膜防水层施工。

The base treatment agent shall be compatible

with the material of the overlying coating in properties, and is mixed fully with the diluent commonly used for waterproof coatings before uniform application by brushing or spraying for full coverage, and the waterproof layer shall not be coated until the base treatment agent is dried.

b. 特殊部位附加增强处理。

b. Additional reinforcement at special parts.

在管道根部、阴阳角等其他类似部位，应做不少于一布二涂的附加层；在天沟、檐沟与屋面交接处以及找平层分格处均应空铺宽度不小于 200 ～ 300 mm 的附加层，构造做法应符合设计要求。

An additional layer with not less than one pass of fiber and two passes of coating shall be made at the root of pipes, internal and external corners and other similar parts; an additional layer in a width of not less than 200-300 mm shall be paved by the border adhibiting method at the joints of the roof gutter, eaves gutter with the roof and the dividing joints for leveling layer. The structure and construction method shall meet the design requirements.

c. 涂布防水涂料。

c. Application of waterproof coating.

防水涂料可采用手工抹压、涂刷和喷涂分层施工。涂膜防水必须由两层以上涂层组成，每层应刷 2 ～ 3 遍，且应根据防水涂料的品种，分层分遍涂布，不能一次涂成，并待先涂的涂层干燥成膜后，方可涂后一遍涂料，其总厚度必须达到设计要求。

Waterproof coating may be applied by manual troweling, brushing and spraying in layers. The waterproof coating must be composed of more than two layers, with 2-3 passes of application for each. The waterproof coating shall be applied in layers and passes according to the requirements of varied paint types, instead of application in one shot. The

next pass of coating shall not be applied until the receiving layer is dried into a film, and its total thickness must meet the design requirements.

涂料的涂布顺序为：先高跨后低跨，先远后近，先立面后平面。同一屋面上先涂布排水较集中的水落口、天沟、檐口等节点部位，再进行大面积涂布。涂层应厚薄均匀、表面平整，不得有露底、漏涂和堆积现象。两涂层施工间隔时间不宜过长，否则易形成分层现象。涂层中夹铺增强材料时，宜边涂边铺胎体。胎体增强材料长边搭接宽度不得小于 50 mm，短边搭接宽度不小于 70 mm。当屋面坡度小于 15% 时，可平行屋脊铺设；屋面坡度大于 15%，应垂直屋脊铺设；采用两层胎体增强材料时，上下层不得互相垂直铺设；搭接缝应错开，其间距不应小于幅宽的 1/3。找平层分格缝处应增设胎体增强材料的空铺附加层，其宽度以 200 ～ 300 mm 为宜。涂膜防水层收头应用防水涂料多遍涂刷或用密封材料封严。在涂膜未干前，不得在防水层上进行其他施工作业，涂膜防水屋面上不得直接堆放物品。涂膜防水屋面的隔汽层设置原则上与卷材防水屋面相同。

The coating sequence is based on the following principles: High span before low span, far-to-near, elevation-to-plan. On the same roof, the drain outlet, roof gutter, cornice and other nodes with concentrated drainage shall be coated first before large-area application. The coating shall be uniform in thickness and smooth in surface, without exposed underlying base, missing coating and build-up. The application interval between two coatings should not be too long; otherwise, delamination may occur. The reinforcing materials sandwiched between coatings should be laid while the coatings are applied. The lap width at the long side of the reinforcing matrix shall not be less than 50 mm, and that at the short side shall not be less than 70 mm. When the roof pitch is less than

15%, it may be laid in parallel with the ridge; when the roof pitch is greater than 15%, it shall be laid perpendicular to the ridge. When the reinforcing matrix is laid in two layers, the upper and lower layers shall not be laid perpendicularly to each other, and the lap joints shall be staggered at a spacing not less than 1/3 of the width. An additional layer of the reinforcing matrix shall be paved by the border adhibiting method at the dividing joints of the leveling layer, and in a width of 200-300 mm. The coated waterproof layer shall be closed by application with waterproof coating for multiple passes or with sealing materials. Other construction operations shall not be carried out on the waterproof layer until the coating film is dried, and the coated waterproof roof shall not be used as a stacking place. In principle, the same vapor barrier is made for the coated waterproof roof as that for the roof with membrane waterproofing.

d. 保护层施工。

d. Protection layer paving.

为了防止涂料过快老化，涂膜防水层应设置保护层。在涂刷最后一道涂料时，如采用纸石、云母作为保护层，可边涂刷边均匀撒布，不得露底，待涂料干燥后，将多余的撒布材料清除。当采用浅色涂料作为保护层时，应在涂膜固化后进行。

In order to prevent the coating from aging too fast, the waterproof coating shall be covered with a protection layer. If paper spar and mica are used to build the protection layer when the last pass of coating is applied, the stone materials may be evenly spread while the coating is applied, without exposing the underlying layer. After the coating is dried, the excess materials are removed. In the case that light-colored paint is used to create the protection layer, it shall be applied after the coating is cured.

8.2.3 复合防水屋面

8.2.3 Composite Waterproof Roof

复合防水屋面用不同的防水材料，充分利用各种材料的优势，形成屋面防水的多道防线，以提高屋面整体防水的功能为目的。换句话说，用两种以上的防水材料组合成的屋面防水层就是复合防水屋面。如防水涂料与防水卷材就可组成复合防水层，由于涂膜防水层具有黏结强度高，可修补防水层基层裂缝缺陷、防水层无接缝、整体性好的特点；卷材与涂料复合使用时，涂膜防水层宜设置在卷材防水层的下面。卷材防水层强度高、耐穿刺、厚薄均匀、使用寿命长，宜设置在涂膜防水层的上面。

Different waterproof materials are used to build multiple waterproof layers, bringing into play their respective advantages, to improve the overall roof waterproof function, that is, a composite waterproof roof is composed of waterproof layers made of more than two waterproof materials. For example, the composite waterproof layers made up of waterproof coating and waterproof membrane features a jointless structure and good integrity, since the coated waterproof in a high bonding strength can repair the cracking defects of the waterproof base layer; in the membrane-coating combination, it is preferred to pave the membrane over the coating, since the membrane features high strength, puncture resistance, uniform thickness, and long service life.

8.2.4 刚性防水屋面

8.2.4 Rigid Waterproof Roof

刚性防水屋面常采用普通细石混凝土防水层、补偿收缩混凝土防水层、块体刚性防水层、预应力混凝土防水层、钢纤维混凝土防水层；前两种应用较为广泛。适用于防水等级为Ⅰ～Ⅲ级的屋面防水，不能用于设有松散材料保温层的屋面、受较大震动或冲击的屋面和坡度大于 15% 的屋面，如图 8-4 所示。

Rigid waterproof roofs are commonly embodied as the ordinary fine aggregate concrete waterproof layer, shrinkage-compensating concrete waterproof layer, block rigid waterproof layer, prestressed concrete waterproof layer and steel fiber reinforced concrete waterproof layer, among which the former two are widely used. The roof waterproofing rated I-Ⅲ shall not be used for roofs with an insulation layer made of loose materials, roofs subject to heavy vibration or impact, and roofs with a pitch greater than 15%, as shown in Figure 8-4 for details.

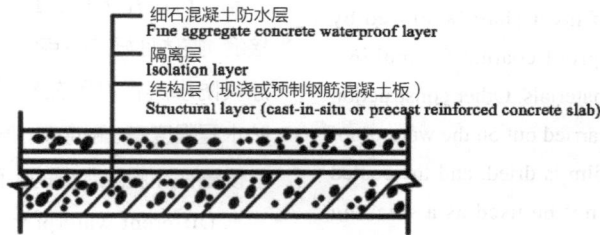

图8-4　细石混凝土防水屋面构造
Figure 8-4　Structure of Fine Aggregate Concrete Waterproof Roof

8.3　地下防水工程

8.3 Underground Waterproofing Works

地下防水工程是防止地下水对地下构筑物或建筑物基础的长期浸透，保证地下构筑物或地下室使用功能正常发挥的一项重要工程。由于地下工程常年受到地表水、潜水、上层滞水、毛细管水等的作用，所以，对地下工程防水的处理比屋面防水要求更高，防水技术难度更大。如何正确选择合理有效的防水方案是地下防水工程要解决的首要问题。

Underground waterproofing is an important part to prevent groundwater from long-term infiltration into underground structures or building foundations and ensure the normal functions of underground structures or basements. Since the underground works are permanently affected by surface water, phreatic water, perched water, capillary water, etc., their waterproof treatment is more demanding and more technological difficult than that of roofs. Therefore, how to choose a reasonable and effective waterproofing scheme becomes a primary issue in underground waterproofing engineering.

8.3.1 防水方案及防水措施

8.3.1 Waterproofing Scheme and Measures

（1）防水方案

(1) Waterproofing schemes

地下工程的防水方案，应遵循"防、排、截、堵结合,刚柔相济、因地制宜、综合治理"的原则,根据使用要求、自然环境条件及结构形式等因素确定。

The waterproofing scheme of underground works shall follow the principle of "coordination of prevention, drainage, interception and blocking, rigid-flexible combination, adjustment to local conditions and comprehensive treatment", and shall be developed according to the functional requirements, natural environmental conditions and structural forms.

常用的防水方案有以下三类:

There are three commonly used waterproofing schemes:

①结构自防水。

① Structural self-waterproofing.

依靠防水混凝土本身的抗渗性和密实性来进行防水,结构本身既是承重围护结构,又是防水层。该方案具有施工简便、工期较短、改善劳动条件、节省工程造价等优点,是解决地下防水问题的有效途径,从而被广泛采用。

The structure built with impermeable and compact waterproof concrete playing a role in waterproofing serves not only as a load-bearing enclosure but also as a waterproof layer. Thanks to the advantages of convenient construction, short construction period, improved working conditions and cost reduction, etc., this option offers an effective way to solve underground waterproofing, so it is widely used.

②设防水层。

② Waterproof layer paving.

在结构物的外侧增加防水层,以达到防水的目的。常用的防水层有水泥砂浆、卷材、沥青胶结料和金属防水层,可根据不同的工程对象、防水要求及施工条件选用。

A waterproof layer is added over the structure to achieve the purpose of waterproofing. The commonly used waterproof layers are made of cement mortar, membrane, asphalt and metal, among which selection may be made according to different engineering objects, waterproofing requirements and construction conditions.

③渗排水防水。

③ Seepage and drainage facilities.

利用盲沟、渗排水层等措施来排除附近的水源以达到防水目的。适用于形状复杂、受高温影响、地下水为上层滞水且防水要求较高的地下建筑。

Such measures as blind drains and seepage and drainage layers are taken to drain the nearby water sources, so as to achieve waterproofing. This scheme is suitable for underground buildings with complex shapes, exposed to high temperatures and perched water, and with high waterproof requirements.

(2)防水措施

(2) Waterproofing measures

钢筋混凝土结构的地下工程,应采用防水混凝土,并根据防水等级的要求采用防水措施。

The reinforced concrete structure of underground works shall be built with waterproof concrete, and waterproof measures shall be taken according to the requirements of corresponding waterproof rates.

8.3.2 结构主体防水工程施工

8.3.2 Waterproofing Works Construction of Main Structure

防水混凝土结构是指因本身的密实性而具有一定防水能力的整体式混凝土或钢筋混凝土结构。它兼有承重、围护和抗渗的功能,还可满足一定的耐冻融及耐侵蚀要求。

Waterproof concrete structures are a kind

of monolithic concrete or reinforced concrete structures with a certain waterproof capacity due to their compactness performance. With the functions of load bearing, enclosure and impermeability, they can also meet certain requirements of freeze-thaw resistance and erosion resistance.

（1）防水混凝土的种类

(1) Types of waterproof concrete

防水混凝土一般分为普通防水混凝土、外加剂防水混凝土和膨胀水泥防水混凝土三种。

Waterproof concrete is generally classified into ordinary waterproof concrete, waterproof concrete mixed with admixtures, and expansive-cement waterproof concrete.

（2）防水混凝土施工

(2) Construction of waterproof concrete

防水混凝土结构工程的质量，与合理的设计、材料的性质、配合成分以及施工质量有关。因此，对施工中的各主要环节，如混凝土搅拌、运输、浇筑、振捣、养护等，均应严格遵循施工及验收规范和操作规程的各项规定进行。

The quality of waterproof concrete structures depends not only on reasonable design, properties of materials and mix proportion, but also on the construction process. Therefore, all main steps throughout the construction, such as concrete mixing, transport, pouring, vibration tamping and curing, shall be carried out in strict accordance with the construction and acceptance specifications and operating procedures.

①模板安装。

① Formwork erection.

防水混凝土所用模板，除满足一般要求外，应特别注意模板拼缝严密，支撑牢固。在浇筑防水混凝土前，应将模板内部清理干净。如若两侧模板需用对拉螺栓固定时，应在螺栓或套管中间加止水环，螺栓加堵头，如图8-5、图8-6、图8-7所示。

In addition to meeting the general requirements, special attention shall be paid to the formwork used for waterproof concrete that shall be tightly spliced and securely supported. The formwork shall be cleaned inside before pouring waterproof concrete. If the formwork on both sides needs to be fixed with split bolts, waterstop rings shall be added in the middle of the bolts or sleeves, and the bolts shall be capped, as shown in Figure 8-5, Figure 8-6 and Figure 8-7.

②钢筋施工。

② Reinforcement placement.

钢筋不得用钢丝或铁钉固定在模板上，必

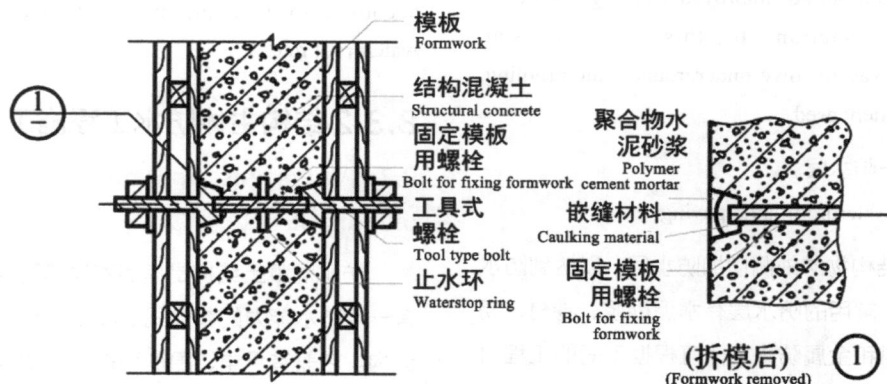

图8-5　工具式螺栓的防水做法示意图

Figure 8-5　Schematic Diagram of Tool Type Waterproofing Bolt

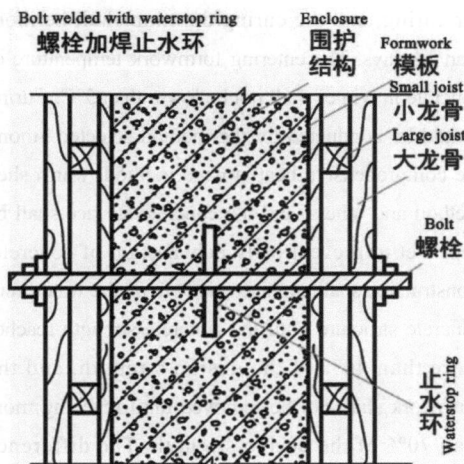

图8-6 螺栓加焊止水环

Figure 8-6 Bolt Welded with Waterstop Ring

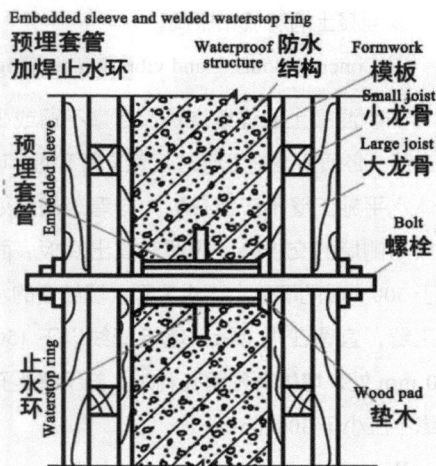

图8-7 预埋套管支撑

Figure 8-7 Embedded Sleeve Support

须采用相同配合比的细石混凝土或砂浆作为垫块，并确保钢筋保护层厚度符合规定，不得有负误差。如结构内设置的钢筋需用铁丝绑扎时，不得接触模板。

Instead of fixing onto the formwork with steel wires or iron nails, steel bars must be placed on the spacer made of fine aggregate concrete or mortar with the same mix proportion, and the concrete cover shall be made in the thickness as required, without a negative error. The steel bars placed in the structure shall be bound with iron wires if necessary, not in contact with the formwork.

③混凝土搅拌。

③ Concrete mixing.

选定配合比时，其试配要求的抗渗水压应较其设计值提高 0.2 MPa，准确计算并称量每种用料，然后投入混凝土搅拌机。

The anti-seepage water pressure required for trial mixing with the selected mix proportion shall be increased by 0.2 MPa compared with the design value. After consumption is accurately calculated for materials, they shall be weighed and then fed into the concrete mixer.

防水混凝土必须采用机械搅拌，搅拌时间不应小于 120 s。掺加外加剂时，应根据外加剂的技术要求确定搅拌时间。

Mechanical mixing must be adopted for waterproof concrete, for a duration not less than 120 s. Admixtures are added and mixed for a duration depending on their technical requirements.

④混凝土运输。

④ Concrete transport.

运输过程中应采取措施防止混凝土拌合物产生离析，以及塌落度和含气量的损失，同时要防止漏浆。

Concrete shall be transported with the measures taken to prevent segregation of concrete mixture, loss of slump and air content, and leakage.

防水混凝土拌合物在常温下应于 0.5 h 以内运至现场，运送距离较远或气温较高时，可掺入缓凝型减水剂，缓凝时间宜为 6 ~ 8 h。

The waterproof concrete mixture shall be transported to the site within 0.5 h at normal temperatures. Retarding water reducer may be added, with a retarding duration recommended of 6~8 h, to deal with a long transport distance or high temperatures.

⑤混凝土的浇筑和振捣。

⑤ Concrete pouring and vibration tamping.

防水混凝土应连续浇筑，尽量不留或少留施工缝。必须留设施工缝时，宜注意下列事项：墙体水平施工缝不应留在剪力与弯矩最大处或底板与侧墙的交接处，应留在高出底板表面不小于 300 mm 的墙体上；拱（板）墙结合的水平施工缝，宜留在拱（板）墙接缝线以下 150 ~ 300 mm 处；墙体有预留孔洞时，施工缝距孔洞边缘不应小于 300 mm。

Waterproof concrete shall be poured continuously, to avoid and minimize construction joints. Construction joints must be set, the following matters should be paid attention to: The horizontal construction joint of walls shall be set at not less than 300 mm above the surface of the base slab, instead of the places with the maximum shear force and bending moment or the joint between the base slab and the side wall; the horizontal construction joint between vault(slab) and wall should be set at 150-300 mm below the vault(slab)-wall seam line; the construction joint shall be set not less than 300 mm away from the edge of holes and openings provided in wall.

⑥混凝土的养护。

⑥ Concrete curing.

防水混凝土终凝后（一般浇筑后 4 ~ 6 h），即应开始覆盖浇水养护，养护时间应在 14 天以上，冬季施工混凝土入模温度不应低于 5℃，宜采用综合蓄热法、暖棚法等养护方法，并应保持混凝土表面湿润，防止混凝土早期脱水。防水混凝土结构须在混凝土强度达到设计强度 40% 以上时方可在其上面继续施工，达到设计强度 70% 以上时方可拆模。拆模时，混凝土表面温度与环境温度之差不得超过 15℃，以防混凝土表面出现裂缝。

The waterproof concrete finally set(generally 4-6 h after pouring) shall be covered and watered for curing, and the curing duration shall be more than 14 days. The entering formwork temperature of concrete in winter shall not be lower than 5℃. Curing should be conducted by the method selected among the comprehensive heat storage method, warm shed method and others, and the concrete surface shall be kept wet to prevent early dehydration of concrete. Construction shall not be carried out on the waterproof concrete structure until the concrete strength reaches more than 40% of the design strength, and the formwork shall not be removed until reaching more than 70% of the design strength. The difference between the concrete surface temperature and the ambient temperature shall not exceed 15℃ to prevent cracking on the concrete surface during formwork removal.

防水混凝土浇筑后严禁打洞，因此，所有的预留孔和预埋件在混凝土浇筑前必须埋设准确。对防水混凝土结构内的预埋铁件、穿墙管道等防水能力薄弱之处，应采取措施，仔细施工。

It is strictly prohibited to drill holes in the poured waterproof concrete. Therefore, all the openings and embedded parts must be provided and buried accurately before concrete pouring. Proper measures shall be taken for careful construction of the parts vulnerable to water leakage in the waterproof concrete structures, including the embedded iron parts, and through-wall pipes.

（3）水泥砂浆防水层的施工

(3) Construction of cement mortar waterproof layer

水泥砂浆抹面防水层可分为多层刚性做法防水层和掺外加剂的水泥砂浆防水层两种，其构造做法如图 8-8 所示。

The waterproof layer of cement mortar plastering is divided into two types, the multi-layer rigid waterproof layer and the waterproof layer made of cement mortar mixed with admixtures. Their structures are shown in Figure 8-8.

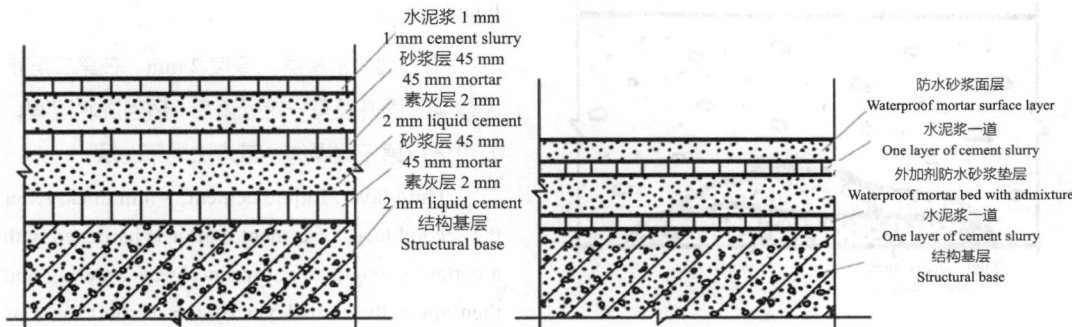

水泥浆 1 mm
1 mm cement slurry
砂浆层 45 mm
45 mm mortar
素灰层 2 mm
2 mm liquid cement
砂浆层 45 mm
45 mm mortar
素灰层 2 mm
2 mm liquid cement
结构基层
Structural base

防水砂浆面层
Waterproof mortar surface layer
水泥浆一道
One layer of cement slurry
外加剂防水砂浆垫层
Waterproof mortar bed with admixture
水泥浆一道
One layer of cement slurry
结构基层
Structural base

（a）多层刚性防水层

(a) Multi-layer rigid waterproof layer

（b）刚性外加剂防水层

(b) Rigid waterproof layer mixed with admixtures

图8-8　水泥砂浆抹面防水层构造做法

Figure 8-8　Structure of Waterproof Layer of Cement Mortar Plastering

防水层的做法分为外抹面防水（迎水面）和内抹面防水（背水面）。防水层的施工顺序一般是先抹顶板，再抹墙面，最后抹地面。

The waterproof layer may be applied by the external plastering(water face) and internal plastering(back face) methods, and generally in the following consequence: Plastering at the top slab first, then the wall surface, and finally the ground.

①基层处理。

① Base treatment.

刚性防水层的基层处理十分重要。基层处理包括清理、浇水、刷洗、补平，使基层表面保持潮湿、清洁、平整、坚实、粗糙。

Base treatment is very important for a rigid waterproof layer. Base treatment includes cleaning, watering, brushing and filling-in, to keep the base surface wet, clean, flat, solid and rough.

超过 1 cm 的棱角及凹凸不平处，应剔成慢坡形，并浇水清洗干净，用素灰和水泥砂浆分层找平，如图 8-9 所示。

Edges, corners and uneven surfaces exceeding 1 cm shall be chiseled into a gentle slope, watered and cleaned, and leveled with liquid cement and cement mortar in layers, as shown in Figure 8-9.

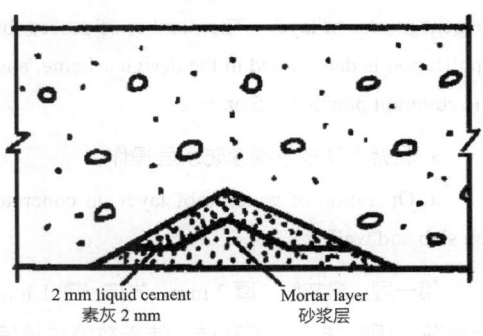

2 mm liquid cement
素灰 2 mm
Mortar layer
砂浆层

图8-9　混凝土基层凹凸不平的处理

Figure 8-9　Treatment of Uneven Concrete Base

处理混凝土表面的蜂窝及小孔洞时，应先将松散不牢的石子除掉，并浇水清洗干净，用素灰和水泥砂浆交替抹到与基层面相平，如图 8-10 所示。

Voids and pits on the concrete surface shall be treated as follows: Remove loose stones, clean by watering, and alternately plaster with liquid cement and cement mortar to be flush with the base surface, as shown in Figure 8-10.

②水泥砂浆防水层施工。

② Paving of cement mortar waterproof layer.

水泥砂浆防水层是用纯水泥和水泥砂浆分层交叉涂抹而成，防水层涂抹的遍数由设计

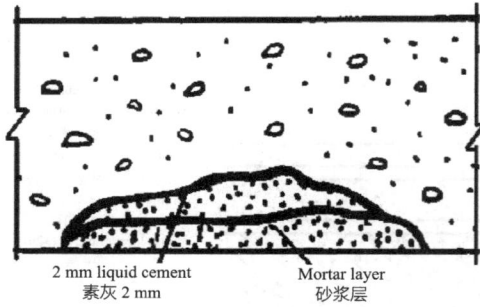

2 mm liquid cement
素灰 2 mm

Mortar layer
砂浆层

图8-10　混凝土基层蜂窝孔洞的处理

Figure 8-10 Treatment of Voids and Pits in Concrete Base

方案确定，较常采用的是 5 遍做法。

The cement mortar waterproof layer is formed by alternately applying liquid cement and cement mortar in layers. The number of passes for application is determined in the design scheme, and the common practice is 5 passes.

a. 混凝土顶板与墙面防水层操作。

a. Operation of waterproof layer on concrete top slab and wall.

第一层：素灰层，厚 2 mm。先抹一道 1 mm 厚素灰，用铁抹子往返刮抹，使素灰填实基层表面的孔隙；随即在已刮抹过素灰的基层表面再抹一道 1 mm 厚素灰找平层；抹完后，用湿毛刷在素灰表面按顺序涂刷一遍。

First layer: Liquid cement, 2 mm thick. First plaster a pass of 1 mm thick liquid cement back and forth with an iron trowel to fill the pores on the base surface; immediately plaster a pass of 1 mm thick liquid cement for leveling on the base surface plastered with the first pass of liquid cement; brush the base surface with two passes of liquid cement in order with a wet brush.

第二层：水泥砂浆层，厚度 4 ~ 5 mm。在素灰层初凝后进行，使水泥砂浆薄薄地嵌入水泥浆层厚度的 1/4 最为理想。

Second layer: Cement mortar, 4-5 mm thick. Embed cement mortar into the liquid cement initially set, ideally to 1/4 of the thickness of the liquid cement.

第三层：素灰层，厚度 2 mm。在第二层水泥砂浆凝固并具有一定强度后，适当浇水湿润，方可进行第三层操作，其方法同第一层。

Third layer: Liquid cement, 2 mm thick. After the second layer of cement mortar is hardened with a certain strength, wet it properly by watering and then apply the third layer by the method same as that of the first layer.

第四层：水泥砂浆层，厚度 4 ~ 5 mm。按照第二层的操作方法将水泥砂浆抹在第三层上，抹后在水泥砂浆凝固前水分蒸发的过程中，分次用铁抹子压实，一般以抹压 3 ~ 4 遍为宜，最后再压光。

Fourth layer: Cement mortar, 4-5 mm thick. Plaster cement mortar on the third layer according to the operation method of the second layer, compact the plastered mortar with an iron trowel several times(3-4 times recommended) during the water evaporation process before the cement mortar is solidified, and finally press and polish the layer.

第五层：第五层是在第四层水泥砂浆抹压两遍后，用毛刷均匀地将水泥砂浆刷在第四层表面，随第四层抹实。

Fifth layer: Apply cement mortar with a brush evenly on the surface of the fourth layer already trowelled twice, and then trowel it together with the fourth layer.

b. 砖墙面和拱顶防水层的操作。

b. Operation of waterproof layer on brick wall surface and vault.

刷水泥浆一道，是为第一层，厚度约为 1 mm，用毛刷往返涂刷均匀，涂刷后，可抹第二、三、四层等，其操作方法与混凝土基层防水相同。

The first layer is applied with liquid cement with a thickness of about 1 mm, which is evenly brushed back and forth with a brush; the second, third and fourth layers are then plastered by the method same as that of waterproofing on the concrete base.

c. 地面防水层的操作。

c. Operation of waterproof layer on floor.

地面防水层操作与顶板、墙面防水层操作不同的地方是：素灰层（一、三层）不采用刮抹的方法，而是把拌和好的素灰倒在地面上，用棕刷往返用力涂刷均匀，第二层和第四层是在素灰层初凝前后把拌和好的水泥砂浆层按厚度要求均匀铺在素灰层上，按顶板、墙面操作要求抹压，各层厚度也均与顶板、墙面防水层相同。地面防水层在施工时要防止践踏，应由里向外进行施工。

The operation of waterproof layers on floors differs from that on walls and roofs as follows: Instead of plastering, the liquid cement for the first and third layers is poured on the floor and evenly applied back and forth with a palm brush. For the second and fourth layers, the mixed cement motor is evenly spread on the liquid cement layers according to the thickness requirements at the time when the liquid cement is about to be initially set, and troweled according to the operation requirements of the wall surface and top slab. The thickness of each layer is also the same as that of the waterproof layer on the wall surface and top slab. The waterproof layer on floors shall be constructed by the inside-out method to prevent stepping on it.

8.3.3 卷材柔性防水层施工

8.3.3 Construction of Flexible Membrane Waterproofing Layer

柔性防水层多采用卷材，目前在地下工程的防水中通常选用高聚物改性沥青防水卷材和合成高分子防水卷材，柔性防水层的缺点是发生渗漏后修补较为困难。

The flexible waterproof layers are built with membranes, and the polymer modified asphalt waterproof membrane and synthetic polymer waterproof membrane are mainly used in the waterproofing works of underground works. The disadvantage of a flexible waterproof layer lies in the difficulties to repair in case of leakage.

卷材柔性防水层施工的铺贴方法，按其与地下防水结构施工的先后顺序分为外防外贴法和外防内贴法两种。

A flexible membrane waterproof layer is paved in two methods, the external adhibiting method and the internal adhibiting method both for outer waterproofing, according to the construction sequence relative to the underground waterproof structure.

（1）外防外贴法

(1) External adhibiting method for outer waterproofing

外防外贴法是将立面防水卷材直接铺设在防水结构的外墙外表面，如图 8-11 所示。

The elevation waterproofing membrane is directly paved on the outer surface of the exterior wall of the waterproof structure, as shown in Figure 8-11.

施工程序：浇筑垫层→砌永久性保护墙→砌 300 mm 高的临时保护墙→墙上粉刷水泥砂浆找平层→转角处铺贴附加防水层→铺贴底板防水层→浇筑底板和墙体混凝土→防水结构外墙水泥砂浆找平层施工→立面防水层施工→验收、保护层施工。

The construction procedure is as follows: Cushion pouring → permanent protection wall masonry → 300 mm high temporary protection wall masonry → plastering of cement mortar leveling layer on the wall → paving of additional waterproof layer at the corner → paving of base slab waterproof layer → concrete pouring of base slab and wall → cement mortar leveling layer on the exterior wall of waterproof structure → paving of elevation waterproof layer → acceptance and paving of protection layer.

（2）外防内贴法

(2) Internal adhibiting method for outer waterproofing

在地下建筑墙体施工前先砌筑保护墙，然后将卷材防水层铺贴在保护墙上，最后施工并浇筑地下建筑墙体，如图 8-12 所示。

Before the construction of the underground wall, the protection wall is built first, then the waterproof layer of membrane is paved on the protection wall, and finally the underground wall is constructed by pouring, as shown in Figure 8-12.

其施工程序是：先在垫层上砌筑永久性保护墙，然后在垫层及保护墙上抹 1：3 水泥砂浆，砂浆干后满涂冷底子油，沿保护墙与垫层铺贴卷材防水层。卷材防水层铺贴完成后，在立面防水层上涂刷最后一层沥青胶时，趁热黏上干

净的热砂或散麻丝，待冷却后，随即抹一层 10～20 mm 厚 1：3 水泥砂浆保护层。在平面上可铺设一层 30～50 mm 厚 1：3 水泥砂浆或细石混凝土保护层，最后进行防水结构的施工。

The construction procedure is as follows: First build a permanent protection wall masonry on the cushion, then plaster 1：3 cement mortar on the cushion and the protection wall, fully apply cold priming oil on the dried motar, and pave the membrane waterproof layer along the protection wall and the cushion; when the last asphalt is applied on the elevation waterproof layer after the membrane waterproof layer is paved, spread clean hot sand or loose hemp fiber on the hot asphalt; after cooling, immediately pave a layer of 10-20 mm thick 1：3 cement mortar protection layer; pave a layer of 30-50 mm thick 1：3 cement mortar or fine aggregate concrete protection layer can be laid on the horizontal waterproof layer; finally, the waterproof structure is built.

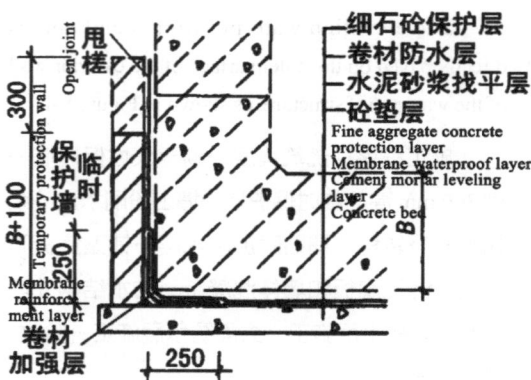

图8-11 外防外贴法（单位：mm）

Figure 8-11 External Adhibiting Method for Outer Waterproofing

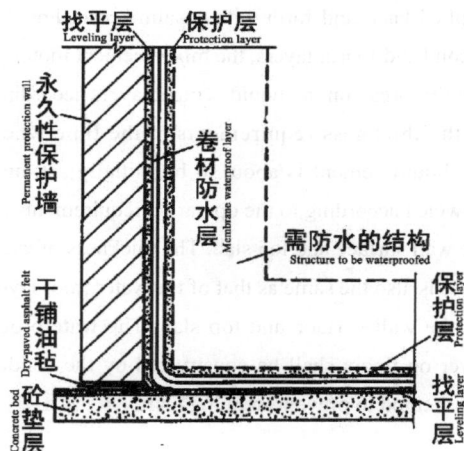

图8-12 外防内贴法

Figure 8-12 Internal Adhibiting Method for Outer Waterproofing

9

装饰工程

Decoration Works

学习目标：

Learning objectives:

掌握一般抹灰、装饰抹灰的施工工艺、施工要点、质量要求、质量验收标准及检测方法；掌握饰面工程、地面工程、吊顶工程、隔墙工程、涂料与刷浆工程、门窗工程的施工工艺、施工要点与施工质量验收标准及检测方法。

Master the construction technology, key construction points, quality requirements, quality acceptance standards and testing methods of general plastering and decorative plastering; master the construction technology, key construction points, construction quality acceptance standards and testing methods of finishing works, flooring works, ceiling works, partition wall works, coating and whitewashing works, as well as door and window works.

技能抽查要求：

Skill checking requirements:

能进行地板砖铺贴施工、地板砖铺贴施工质量检测；能进行墙面一般抹灰施工、墙面一般抹灰施工质量检测；能进行墙面釉面砖镶贴施工、墙面釉面砖镶贴施工质量检测。

Be able to carry out construction and corresponding quality inspections of floor tile paving, general wall plastering, and wall-glazed tile paving.

建筑八大员岗位资格考试要求：

Qualification examination requirements for eight major posts of construction engineering:

掌握抹灰工程的施工工艺和质量要求；掌握木地板的施工工艺和质量要求；掌握吊顶工程的施工工艺和质量要求；掌握隔墙工程的施工工艺和质量要求；掌握门窗工程的施工工艺和质量要求；掌握饰面板（砖）的施工工艺和质量要求；掌握涂饰工程的施工工艺和质量要求；掌握装饰装修工程施工的安全要求。

Master the construction technology and quality requirements of plastering works, wood floor works, ceiling works, partition wall works, door and window works, veneer(facing tile) works and coating works; master the safty requirements of decorating works.

建筑装饰工程是在土建基础上采用适当的材料和正确的构造，以科学的施工工艺及方法，保护建筑主体结构，满足人们的视觉要求和使用功能需求，提高建筑物档次的工程。其主要作用是：保护结构主体，延长使用寿命；美化建筑，增强艺术效果；优化环境，创造使用条件。建筑装饰工程是建筑施工的重要组成部分，其主要特点是项目繁多，工程量大，工期长，用工量大，造价高，装饰材料和施工技术更新快，施工管理复杂。它主要包括抹灰、饰面、楼地面、吊顶、隔墙、门窗、幕墙、涂料和裱糊等工程。

Architectural decoration works apply scientific construction technologies and methods with appropriate materials and proper structures on the basis of civil engineering to protect the main structure of buildings, meet people's visual requirements and functional needs, improve the grade of buildings. It is mainly to protect the main structure, prolong service life, beautify the building, improve the aesthetic appeal, optimize the environment and create use conditions. As an important part of building construction, architectural decoration works feature a large number of work items, large quantities of works, long construction periods, high demand for labor, high cost, fast update of decoration materials and construction technology, and complex construction management. It mainly includes plastering, finishing, flooring, ceiling, partition walls, doors and windows, curtain wall, coating and pasting.

建筑装饰工程的施工顺序对控制施工质量起着显著作用。室外抹灰和饰面工程的施工，一般应自上而下进行；室内装饰工程可采用自上而下、自下而上以及自中而下再自上而中三种施工顺序。

The construction sequence of architectural decoration works plays an obvious role in controlling the construction quality. Generally, the exterior plastering and finishing works are carried out from top to bottom; interior decoration works can be carried out in three sequences which are from top to bottom, from bottom to top, as well as from middle to bottom and then top to middle.

室内吊顶、隔墙的罩面板和花饰等工程，应待室内地（楼）面湿作业完工后施工。室内装饰工程的施工顺序，应符合下列规定：

Cover boards and ornaments for interior ceilings and partition walls shall be carried out after the completion of wet construction of the interior floor. The construction sequence of interior decoration works shall comply with the provisions below:

①抹灰、饰面、吊顶和隔断工程，应待隔墙、钢木门、窗框、暗装管道、电线管和电器预埋件、预制钢筋混凝土楼板灌缝完工后进行。

① Plastering, finishing, ceiling and partition works shall be carried out after the completion of partition walls, steel and wooden doors, window frames, concealed pipes, electric conduits and electrical embedded parts, and joint filling of precast reinforced concrete floor slabs.

②钢木门窗及其玻璃工程，根据地区气候条件和抹灰工程的要求，可在湿作业前进行；铝合金、塑料、涂色镀锌钢板门窗及其玻璃工程，宜在湿作业完工后进行，如需在湿作业前进行，必须采取相关保护措施。

② Steel, wooden doors and windows and relevant glass works can be carried out before wet construction according to the climate conditions and requirements of plastering works; aluminum alloy, plastic, colored doors and windows made of galvanized steel sheets and relevant glass works should be carried out after the completion of wet construction; if it is necessary to carry out before wet construction, relevant protection measures must be taken.

③有抹灰基层的饰面板工程、吊顶及轻型

花饰安装工程，应待抹灰工程完工后再进行。

③ Any veneer, ceiling and light ornaments installation works involved with plastering base courses shall be carried out after the plastering works are completed.

④涂料、刷浆工程以及吊顶、隔断、罩面板的安装，应在塑料地板、地毯、硬质纤维等地（楼）面的面层和明装电线施工前、管道设备试压后进行。木地（楼）板面层的最后一遍涂料应待裱糊工程完工后施涂。

④ Coating, whitewashing works, and installation of ceilings, partitions and cover boards shall be carried out before the construction of the surface layer of the floor, such as the plastic floor, carpet and hard fiber floor and the exposed installation of electric wire, but after the pressure test of pipeline equipment. The last pass for the surface layer of a wood floor shall be carried out after the completion of pasting works.

⑤裱糊工程，应待顶棚、墙面、门窗及建筑设备的涂料和刷浆工程完工后进行。

⑤ Pasting works shall be carried out after the completion of coating and whitewashing works of ceilings, walls, doors, windows and building equipment.

9.1 抹灰工程

9.1 Plastering Works

抹灰是将各种砂浆，如装饰性石屑浆、石子浆涂抹在建筑物的墙面、顶棚、地面等表面上，除了保护建筑物外，还可以作为饰面层起到装饰作用。

Plastering is to apply different kinds of mortar like decorative aggregate mortar and stone mortar to the walls, ceilings, floors and other surfaces of buildings to protect buildings and decorate finishing as a veneer.

抹灰工程按使用材料和装饰效果分为一般抹灰和装饰抹灰。一般抹灰适用于石灰砂浆、水泥砂浆、混合砂浆、聚合物水泥砂浆、膨胀珍珠岩水泥砂浆、麻刀灰、纸筋灰、石膏灰等抹灰工程。装饰抹灰的底层和中层与一般抹灰做法基本相同，其面层主要有水刷石、水磨石、斩假石等。

Plastering works are categorized into general plastering and decorative plastering based on the materials used and the decorative effects. General plastering is applicable to plastering of lime mortar, cement mortar, composite mortar, polymer cement mortar, expanded perlite cement mortar, hemp lime plaster, paper strip mixed lime plaster and gypsum plaster. The construction methods for the base and intermediate layers of decorative plastering are basically the same as those of general plastering, and the surface layers mainly include granitic plaster, terrazzo, artificial stone with, etc.

9.1.1 一般抹灰施工

9.1.1 General Plastering

（1）一般抹灰的构造

(1) Structure of general plastering

抹灰一般分三层，即底层、中层和面层（或罩面），如图9-1所示。

Plastering is generally divided into the bottom layer, intermediate layer and surface layer(or mask layer), as shown in Figure 9-1.

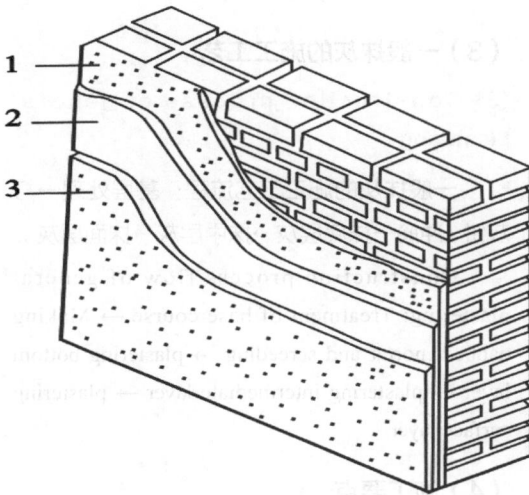

1.底层；2.中层；3.面层

1. bottom layer; 2. intermediate layer; 3. surface layer

图9-1　一般抹灰

Figure 9-1　General Plastering

底层主要起与基层黏结的作用，厚度一般为 5 ~ 9 mm，要求砂浆有较好的保水性，其稠度较中层和面层大，砂浆的组成材料要根据基层的种类不同而选用相应的配合比。底层砂浆的强度不能高于基层强度，以免抹灰砂浆在凝结过程中产生较强的收缩应力，破坏强度较低的基层，从而产生空鼓、裂缝、脱落等质量问题。中层起找平的作用，砂浆的种类基本与底层相同，只是稠度稍小，中层抹灰较厚时应分层，每层厚度应控制在 5 ~ 9 mm；面层起装饰作用，

要求涂抹光滑、洁净，因此要求用细砂，或用麻刀、纸筋灰浆。各层砂浆的强度要求应为"底层 > 中层 > 面层"，并不得将水泥砂浆抹在石灰砂浆或混合砂浆上，也不得把罩面石膏灰抹在水泥砂浆层上。

The bottom layers serve as binder layers to bond with the base layers, 5-9 mm thick generally. The bottom mortar shall have good water retention performance, with a consistency larger than intermediate mortar and surface mortar. A matching mixture ratio shall be selected for the material composition of mortar depending on the type of the base layers. The strength of the bottom mortar shall not be higher than the base mortar; otherwise, there would be strong shrinkage stress of plastering mortar during setting and damage to the base layers with low strength, thus causing quality problems such as hollowing, cracks and peeling. The intermediate layers is for leveling. The mortar used for the intermediate layers is basically the same as that for the bottom layers, only with smaller consistency. When the intermediate mortar is to be thickly applied, it shall be applied in layers with a thickness of 5-9 mm. The surface layers is for decoration, with a smooth and clean finish, applied with fine sand, hemp lime plaster or paper strip mixed lime plaster. The strength decreases progressively from the bottom mortar to the intermediate mortar and then to the surface mortar. Cement mortar shall not be plastered on lime mortar or mixed mortar, and finishing gypsum plaster shall not be plastered on cement mortar.

抹灰层的平均总厚度，不得大于下列规定：

The average total thickness of plastering layers shall follow the provisions below:

顶棚：板条、空心砖、现浇混凝土 15 mm，预制混凝土 18 mm，金属网 20 mm。

Ceiling: 15 mm for batten, hollow brick and cast-in-situ concrete, 18 mm for precast concrete, and 20 mm for metal mesh.

内墙：普通抹灰 18 ~ 20 mm，高级抹灰 25 mm。

Interior wall: 18-20 mm for common plastering and 25 mm for high-quality plastering.

外墙 20 mm，勒脚及突出墙面部分 25 mm。

Exterior wall: 20 mm for general parts and 25 mm for the plinth and protruding parts.

石墙 35 mm。

35 mm for stone wall.

当抹灰厚度≥ 35 mm 时，应采取加强措施。涂抹水泥砂浆每遍厚度宜为 5 ~ 7 mm；涂抹石灰砂浆和水泥混合砂浆，每遍厚度宜为 7 ~ 9 mm。面层抹灰经赶平压实后的厚度，麻刀石灰不得大于 3 mm；纸筋石灰、石膏灰不得大于 2 mm。

Strengthening measures shall be taken when the plastering thickness is ≥ 35 mm. The thickness of each pass of cement mortar should be 5-7 mm; the thickness of each pass of lime mortar and cement mixed mortar should be 7-9 mm. The thickness of the surface layer after trowelling and compacting shall not be greater than 3 mm for hemp lime plaster, and shall not be greater than 2 mm for paper strip mixed lime plaster and gypsum plaster.

（2）一般抹灰的分类
(2) Classification of general plastering

一般抹灰按质量要求分为普通抹灰和高级抹灰两个等级。

General plastering is classified into common plastering and high-quality plastering based on quality requirements.

普通抹灰为一道底层和一道面层或一道底层、一道中层和一道面层，要求表面光滑、洁净，接槎平整，分格缝清晰。

Common plastering layers consist of one bottom layer and one surface layer, or one bottom layer, one intermediate layer and one surface layer. The surface shall be smooth and clean and flat at the joint-based connections. The dividing joints shall be clear.

高级抹灰为一道底层、数道中层和一道面层，要求表面光滑、洁净，颜色均匀无抹纹，分格缝和灰线应清晰美观。

High-quality plastering layers consist of one bottom layer, several intermediate layers and one surface layer. The surface shall be smooth and clean, with uniform color and without plastering marks. The dividing joints and mortar control lines shall be clear and eye-pleasing.

（3）一般抹灰的施工工艺
(3) Construction process of general plastering

一般抹灰的施工工艺流程：基层处理→打灰饼、冲筋→抹底层灰→抹中层灰→抹面层灰。

Construction process flow of general plastering: Treatment of base course → Making dabbed mortar and screeding → plastering bottom layer → plastering intermediate layer → plastering surface layer.

（4）施工要点
(4) Key construction points

①基层处理。

① Base treatment.

抹灰前应对砖石、混凝土及木基层表面做处理。清除灰尘、污垢、油渍和碱膜等，并洒水湿润。表面凹凸明显的部位，应事先剔平或用 1：3 水泥砂浆补平，对于平整光滑的混凝土表面拆模时随即做凿毛处理或作拉毛处理（图 9-2），或用混凝土界面处理剂处理。

The surface of masonry, concrete and wood base is treated before plastering. Dust, dirt, oil stain and alkali film, etc. are removed and the surface is sprinkled with water. Uneven parts on the surface shall be leveled in advance or filled with 1：3 cement mortar for leveling. Flat and smooth concrete surfaces shall be roughened during formwork removal (as shown in Figure 9-2), or treated with a concrete interface agent.

(a)墙面拉毛处理
(a) surface roughening

(b)墙面挂网处理
(b) surface netting

图9-2 墙面拉毛及挂网
Figure 9-2 Surface Roughening and Netting

抹灰前应检查门、窗框位置是否正确，与墙连接是否牢固。连接处的缝隙应用水泥砂浆或水泥混合砂浆（加少量麻刀）分层嵌塞密实。

Before plastering, check whether the positions of doors and window frames are correct and whether they are firmly connected to the wall. The gap at the joint shall be filled tightly in layers with cement mortar or cement mixed mortar(with a little hemp fibres).

凡室内管道穿越的墙洞和楼板洞、凿剔墙后安装的管道、墙面的脚手孔洞均应用1∶3水泥砂浆填嵌密实。

All wall and floor openings through which interior pipelines pass, pipelines installed after chiseling the wall, and scaffold holes on the wall surface shall be filled with 1∶3 cement mortar.

不同基层材料（如砖石与木，混凝土结构）相接处应铺钉金属网并绷紧牢固，金属网与各结构的搭接宽度从相接处起每边不少于100 mm。

Metal mesh shall be nailed at the joints of different base materials(e.g. joints between masonry and wood, masonry and concrete structure) and tightened firmly, with overlapping width between metal mesh and each structure not less than 100 mm each edge from the joint.

②做灰饼、标筋。

② Dabbed mortar and screeding.

为控制抹灰层的厚度和墙面的平整度，在抹灰前应先检查基层表面的平整度，并用与抹灰层相同的砂浆设置50 mm×50 mm的标志或宽约100 mm的标筋，如图9-3所示。

To control the thickness of plastering courses and the flatness of wall surfaces, the flatness of the base course surface shall be checked before plastering, and a 50 mm×50 mm mark or a screed strip with a width of about 100 mm shall be provided with mortar of the same kind as that of the plastering course, as shown in Figure 9-3.

③做护角。

③ Corner protection.

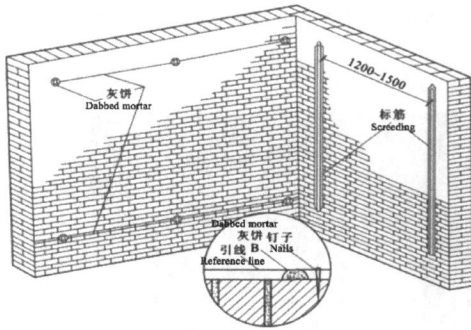

图9-3 做灰饼、标筋（单位：mm）

Figure 9-3 Dabbed Mortar and Screeding

抹灰工程施工前，对室内墙面、柱面和门洞的阳角，宜用 1 : 2 水泥砂浆做护角（图9-4），其高度不低于 2 m，每侧宽度不少于 50 mm。对外墙窗台、窗楣、雨篷、阳台、压顶和突出腰线等，上面应做成流水坡度，下面应做滴水线或滴水槽，滴水槽的深度和宽度均不应小于 10 mm。各滴水槽要整齐一致。

Before the construction of plastering works, the external corners of interior walls, columns and door openings should be protected with 1 : 2 cement mortar(as shown in Figure 9-4), with a height of not less than 2 m and a width of not less than 50 mm on each side. For window sill, window lintel, awning, balcony, coping and protruding string course of the exterior wall, drain slope shall be designed for the top surface, and water drip or drip mold with a depth and width of not less than 10 mm shall be designed for the bottom surface. They shall be neat and consistent.

④分层抹灰。

④ Layered plastering.

待标筋砂浆七至八成干后，就可以进行底层砂浆抹灰。

Plastering of the bottom layer can be carried out after the screeding mortar is 70% to 80% dry.

抹底层灰可用托灰板（大板）盛砂浆，用

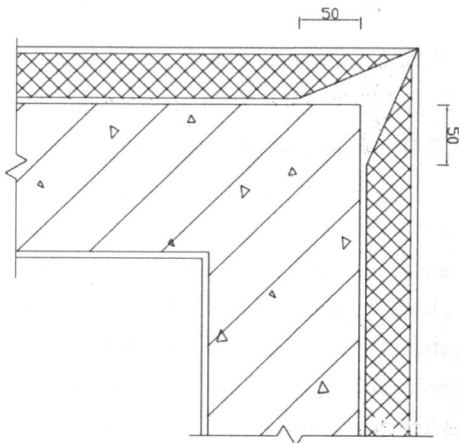

图9-4 做护角（单位：mm）

Figure 9-4 Corner Protection

力将砂浆推抹到墙面上，一般应从上而下进行，在两标筋之间的墙面砂浆抹满后，用长刮尺两头靠着标筋，从下而上进行刮灰，使抹上的底层灰与标筋面相平。再用木抹来回抹压，去高补低，最后再用铁抹压平一遍。

The bottom mortar can be held with a hawk to plaster the mortar on the wall surface from top to bottom. After the wall surface between the two screed strips is fully plastered, a long darby shall be placed against the screed strips on both ends to plaster from bottom to top so that the surface of the plastered bottom mortar is flush with the screed strip surface. Then, the surface is trowelled and compacted with a wooden trowel for leveling, and then is compacted and leveled with a steel trowel for the last pass.

中层砂浆抹灰应待水泥砂浆（或水泥混合砂浆）底层凝结后或石灰砂浆底层灰七八成干后，方可进行。

Plastering of the intermediate layer can be carried out after the cement mortar(or cement mixed mortar) of the bottom layer is set or the lime mortar on the bottom mortar is 70% or 80% dry.

中层砂浆抹灰时，应先在底层灰上洒水，待其收水后，即可将中层砂浆抹上去，一般应从上而下、自左向右涂抹，不用再做标志及标筋，整个墙面抹满后，用木抹来回搓抹，去高补低，再用铁抹压抹一遍，使抹灰层平整、厚度一致。

Intermediate mortar can be plastered after the bottom mortar is watered and basically dry. Generally, the mortar is plastered from top to bottom and left to right and no marks and screed strips are needed. After the whole wall surface is plastered, a wooden trowel is used for trowelling and plastering to level the surface, and then a steel trowel is used for trowelling and compacting for the last pass to keep the plastering surface flat and level.

面层灰应待中层灰凝固后才能开抹。先在

中层灰上洒水湿润，将面层砂浆（或灰浆）均匀地抹上去，一般应从上而下、自左向右涂抹整个墙面，抹满后，用铁抹分遍压抹，使面层灰平整、光滑，厚度一致。铁抹运行方向应注意：最后一遍抹压宜采用垂直方向，各分遍之间应互相垂直抹压。墙面上半部与墙面下半部面层灰接头处应压抹理顺，不留抹印。

Plastering of the surface course shall be carried out after the intermediate plaster is set. The surface mortar(or plaster) is evenly applied on the intermediate layer from top to bottom and left to right after the intermediate mortar is watered. After the surface is fully plastered, an iron plaster is used to trowel and compact in several passes to keep the surface mortar flat, smooth and level. For the last pass of trowelling, the steel trowel shall be moved in the vertical direction, and the plastering direction of the next pass shall be perpendicular to that of the previous pass. The mortar joint between the upper half and the lower half of the wall surface shall be trowelled and smoothed without plastering marks.

（5）一般抹灰的质量要求

(5) Quality requirements for general plastering

根据《建筑装饰装修工程质量验收规范》，一般抹灰工程质量验收应符合下列要求：

According to the Code for Acceptance of Construction Quality of Building Decoration, the quality acceptance of general plastering works shall follow the requirements below:

①抹灰工程应分层进行。当抹灰总厚度大于或等于 35 mm 时，应采取加强措施。不同材料基体交接处表面的抹灰，应采取防止开裂的加强措施。当采用加强网时，加强网与各基体的搭接宽度不应小于 100 mm。

① Plastering works shall be carried out in layers. When the total thickness of plastering is greater than or equal to 35 mm, reinforcement measures shall be taken. For plastering surfaces on the joints of different material bases, reinforcement measures shall be taken to avoid cracks. If a reinforcing mesh is used, the overlapping width of the reinforcing mesh and the material base shall not be less than 100 mm.

②抹灰层与基层之间及各抹灰层之间必须黏结牢固，抹灰层应无脱层、空鼓，面层应无爆灰和裂缝。

② The plastering courses must be firmly bonded with the base course and other plastering courses. The plastering course shall be free of delamination and hollowing, and the surface course shall be free of exploded pits and cracks.

一般抹灰工程的表面质量应符合下列规定：

The surface quality of general plastering works shall meet the following requirements:

①普通抹灰表面应光滑、洁净、接槎平整，分格缝应清晰。

① Common plastering surface shall be smooth and clean and flat at the joint-based connections, and the dividing joints shall be clear.

②高级抹灰表面应光滑、洁净、颜色均匀、无抹纹，分格缝和灰线应清晰美观。

② High-quality plastering surface shall be smooth and clean, with uniform color and without plastering marks. The dividing joints and mortar control lines shall be clear and eye-pleasing.

③抹灰层的总厚度应符合设计要求；水泥砂浆不得抹在石灰砂浆层上；罩面石膏灰不得抹在水泥砂浆层上。

③ The total thickness of the plastering layer shall meet the design requirements; cement mortar shall not be plastered on lime mortar, and finishing gypsum plaster shall not be plastered on cement mortar.

④抹灰分格缝的设置应符合设计要求；宽度和深度应均匀，表面应光滑，棱角应整齐。

④ The plastering dividing joints shall meet the design requirements; their width and depth shall be uniform, the surface shall be smooth, and the edges and corners shall be neat.

⑤有排水要求的部位应做滴水线（槽）。滴水线（槽）应整齐、顺直，滴水线应内高外低，滴水槽的宽度和深度均不应小于 10 mm。

⑤ Water drip(mold) shall be made at positions with drainage requirements. The water drip(mold) shall be neat and straight, the inner level of the water drip shall be higher than the outer level, and its width and depth shall not be less than 10 mm.

一般抹灰工程质量的允许偏差和检验方法应符合表 9-1 的规定。

The allowable deviation and inspection method for the quality of general plastering works shall comply with the provisions of Table 9-1.

表 9-1　一般抹灰的允许偏差检验方法
Table 9-1 Allowable Deviation and Inspection Method of General Plastering

项次 S/N	项目 Item	允许偏差/mm Allowable Deviation/mm		检验方法 Inspection Method
		普通抹灰 Common plastering	高级抹灰 High-quality plastering	
1	里面垂直度 Inner perpendicularity	4	3	用2 m垂直检测尺检查 Check with 2 m vertical inspection ruler
2	表面平整度 Surface roughness	4	3	用2 m靠尺和塞尺检查 Check with 2 m guiding rule and feeler gauge

项次 S/N	项目 Item	允许偏差/mm Allowable Deviation/mm		检验方法 Inspection Method
		普通抹灰 Common plastering	高级抹灰 High-quality plastering	
3	阴阳角方正度 Squareness of internal and external corners	4	3	用直角检测尺检查 Check with a right-angle inspection ruler
4	分格条（缝）直线度 Straightness of dividing strips(joints)	4	3	拉5 m线，不足5 m拉通线，用钢直尺检查 Pull a 5 m line or a full-length line (if less than 5 m), and check with a straight steel ruler
5	墙裙、勒脚上口直线度 Straightness of upper edge of dado and plinth	4	3	拉5 m线，不足5 m拉通线，用钢直尺检查 Pull a 5 m line or a full-length line(if less than 5 m), and check with a straight steel ruler

注：普通抹灰，本表第3项阴角方正度不可检查；顶棚抹灰，本表第2项表面平整度可不检查，但表面应平顺。

Notes: For common plastering, the squareness of the internal corner in item 3 of this table cannot be checked; for ceiling plastering, item 2 of this table may be omitted, but the surface shall be flat.

9.1.2 装饰抹灰施工

9.1.2 Decorative Plastering

装饰抹灰与一般抹灰的区别在于两者具有不同的装饰面层，其底层和中层的做法与一般抹灰基本相同，下面简单介绍水刷石、干粘石的施工，如图9-5、图9-6所示。

Except that the decorative surfaces of decorative plastering and general plastering are different, their construction methods for the bottom layer and the intermediate layer are basically the same. The construction methods for granitic plaster and drydash are briefly introduced below, as shown in Figure 9-5 and Figure 9-6.

图9-5 水刷石
Figure 9-5 Granitic Plaster

图9-6 干粘石
Figure 9-6 Drydash

（1）水刷石施工

(1) Granitic plaster works

水刷石饰面，是将水泥石子浆罩面中尚未干硬的水泥用水刷掉，使各色石子外露，形成具有"绒面感"的表面。水刷石是石粒类材料饰面的传统做法，这种饰面耐久性强，具有良好的装饰效果，造价较低，是传统的外墙装饰做法之一。

The granitic plaster is formed by washing the surface of the cement aggregate finish with water to remove the cement that has not yet dried and hardened, thus exposing the small stones of different colors and presenting a "suede effect" surface. Granitic plaster is a traditional method of stone finishing, which has strong durability, good decorative effect and low cost. It is one of the traditional decoration methods for exterior walls.

①水刷石施工工艺流程。

① Construction process of exposed granitic plaster.

抹灰中层验收→弹线、钉分格条→抹面层水泥石子浆→冲洗→起分格条、修整→养护。

Acceptance of plastering of intermediate layer → snapping lines and using dividing strips → plastering cement stone mortar as the surface layer → flushing → removing dividing strips and finishing → curing.

②施工要点。

② Key construction points.

弹线、分格。水泥石子浆大面积开抹前，为防止面层开裂，须在中层砂浆六至七成干时，应按设计要求弹线、分格，钉分格条时木分格条事先应在水中浸透。分格条两侧的八字形纯水泥浆，应抹成 45°。水刷石面层施工前，应根据中层抹灰的干燥程度确定是否需要浇水湿润。紧接着用铁抹子满刮水灰比为 0.37 ~ 0.4 的水泥浆（内掺 3% ~ 5% 水重的 108 胶）一道，

随即抹水泥石子浆面层。面层厚度视石子粒径而定，通常为石子粒径的 2.5 倍。水泥石子浆的厚度以 5 ~ 7 cm 为宜，用铁抹子一次抹平、压实。每块分格内抹灰顺序应自下而上，同平面的面层要求一次完成，不宜留施工缝，如必须留施工缝，应留在分格条位置上。

Line snapping and division. Before the large-area construction of cement stone mortar, when the intermediate mortar is 60% to 70% dry, snap lines and nail dividing strips according to the design requirements, so as to prevent the surface course from cracking. Soak the dividing strips in water in advance if wooden strips are used. The splay-shaped pure cement mortar for fixing the dividing strips on both sides shall be plastered at 45°. Before the construction of the granitic plaster, the intermediate mortar shall be watered based on its drying degree. Then, an iron trowel is used to fully apply a layer of cement mortar(108 adhesive mixed with 3%-5% water by weight) with a water-cement ratio of 0.3-0.4, and then the cement stone mortar is plastered as the surface layer. The thickness of the surface layer depends on the particle size of the stone and is usually 2.5 times the particle size of the stone. The cement stone mortar shall be trowelled and compacted with a steel trowel at one time, with a thickness of 5-7 cm. The plastering in each grid shall be from bottom to top. The surface course of the same plane shall be completed at one time, without construction joints. If any construction joints are to be reserved, they shall be located on the dividing strips.

修整。罩面灰收水后，用铁抹子溜一遍，将遗留的孔隙抹平。然后用软毛刷蘸水刷去表面灰浆，再拍平；阳角部位要往外刷，水刷石罩面应分遍拍平压实，石子应分布均匀、紧密。

Finishing. After the finish mortar is basically dry, an iron trowel is used to trowel the surface and smooth the remaining pores. Then, the mortar on the surface is removed by brushing with a wet soft brush and compacted. The external corner shall be brushed outward. The finishing surface of the granitic plaster shall be leveled and compacted in several passes, and the stone shall be evenly and closely distributed.

喷刷、冲洗。喷刷、冲洗是水刷石施工的重要工序，喷刷、冲洗不净会使水刷石表面色泽灰暗或明暗不一致。罩面灰浆初凝后，达到刷不掉石子的程度时，即可开始喷刷，喷刷时可以两人配合操作：一人用毛刷蘸水轻轻刷掉罩面灰浆，另一人用喷雾器，或用手压喷浆机紧跟着喷刷，先将分格四周喷湿，然后由上向下喷水，喷射要均匀，喷头至罩面距离 10 ~ 20 cm。不仅要将表面的水泥浆冲掉，还要将石渣间的水泥冲出来，使得石渣露出灰浆表面 1 ~ 2 mm，甚至露出粒径的 1/2，使之清晰可见，均匀密布。最后，用清水从上往下全部冲洗干净。

Brushing and flushing. Brushing and flushing are important processes for the granitic plaster works. If the surface is not brushed and flushed clean enough, the color of the granitic plaster will be dark or inconsistent in tone. After the initial setting of the finish mortar, and the stones are firmed enough to attach to the wall during brushing, brushing can be started by the cooperation of two people, with one gently removing the finish mortar with a wet brush, and the other spraying with a sprayer or a hand sprayer. First, the surrounding parts of the grid are sprayed with water and then the grid is sprayed with water from top to bottom evenly, with the nozzle 10-20 cm away from the finish surface. Not only the cement mortar on the surface but also the cement in the stones shall be flushed away so that the stone is exposed 1-2 mm above the mortar surface, or 1/2 of the particle size is exposed so that it is clearly visible and evenly distributed. In the end, the surface is flushed out with clean water from top to bottom.

起分格条。喷刷后，即可用抹子柄敲击分格条，用抹尖扎入木条上下活动，轻轻取出分格条。然后，修饰分格缝并描好颜色。水刷石是一项传统工艺，由于其操作技术要求较高，洗刷浪费水泥，墙面污染后不易清洗，故现今较少采用。

Removing the dividing strip. After brushing and flushing, the dividing strip is knocked with the trowel handle, and the trowel tip is inserted into the wooden strip and moved up and down, so as to gently take out the dividing strip. Then, the dividing joints are finished and colored. As a traditional process requiring technical skill, the granitic plaster is rarely used today for wasting cement and difficult to clean the polluted wall surface.

（2）干粘石施工

(2) Drydash works

干粘石是将干石子直接黏在砂浆层上的一种装饰抹灰做法。装饰效果与水刷石差不多，但湿作业量小，节约原材料，又能明显提高工作效率。

Drydash is a decorative plastering method to cast dry stones directly onto the mortar course. The decoration effect is similar to that of granitic plaster, but drydash requires less wet construction, saves raw materials, and significantly improves work efficiency.

①干粘石施工工艺流程。

① Construction process of drydash.

抹灰中层验收→弹线、黏分格条→抹黏结层砂浆→撒石粒、拍平→起分格条、修整。

Acceptance of plastering of intermediate layer → snapping lines and using dividing strips → plastering mortar for binder layer → spreading stones and compacting → removing dividing strips and finishing.

②施工要点。

② Key construction points.

抹黏结层。待中层水泥砂浆干至七成左右，洒水湿润后，黏分格条。待分格条黏牢后，在墙面刷水泥砂浆一遍，随即按格抹砂浆黏结层（1：3水泥砂浆、厚度4～6 mm，砂浆稠度≤8 cm），黏结层砂浆一定要抹平，不显抹纹，按分格大小，一次抹一块或数块，应避免在块中甩槎。

Plastering the binder layer. After the cement mortar of the intermediate layer is about 70% dry, spray water and paste dividing strips. After the dividing strip is firmly pasted, apply cement mortar on the wall surface for one pass, and then plaster binder mortar by grids (1 : 3 cement mortar, thickness 4-6 mm, mortar consistency ≤ 8 cm). Level the mortar of the binder layer without marks. Plaster one grid or several grids at one time and do not retain overlap joints in the grid.

甩石子。干粘石所选石子的粒径比水刷石要小些，一般为4～6 mm。黏结砂浆抹平后，应立即甩石子，先甩四周易干部位，然后甩中间，要做到大面均匀，边角和分格条两侧不漏黏，由上而下快速进行。石子使用前应用水冲洗净晾干，甩时用托盘盛装，托盘底部用窗纱钉成，以便筛净石子中的残留粉末。如发现饰面上石子有不匀或过稀现象，应用抹子或手直接补贴，否则会使墙面出现死坑或裂缝。

Casting stones. The particle size of the stones for daydash is smaller than that for the granitic plaster, generally 4-6 mm. Cast stones immediately after the binder mortar is leveled. Evenly cast from the surrounding parts that are easy to dry to the middle part to ensure overall uniformity. Do not miss out corners and both sides of dividing strips and cast quickly from top to bottom. Flush stones out with water and let dry before use. Hold the stones in a tray with a mesh bottom, so as to screen out the residual powder in them. Check the surface for uneven or missed parts and cast supplementary stones with a trowel

or by hand to avoid pits and crack.

压石子。当黏结砂浆表面均匀地黏上一层石子后，用抹子或辊子轻轻压一下，使石子嵌入砂浆的深度不小于1/2的石子粒径。拍压后的石子表面应平整坚实，拍压时用力不宜过大，否则容易翻浆糊面，出现抹子或滚子轴的印迹。阳角处应在角的两侧石子黏上后再同时操作，否则当一侧石子黏上后再黏另一侧时不易黏上，会出现明显的接槎黑边。干粘石也可用机械喷石代替手工甩石，施工时利用压缩空气和喷枪将石子均匀有力地喷射到黏结层上。喷头对准墙面，距墙300～400 mm，气压以0.6～0.8 MPa为宜。在黏结层硬化期间，应洒水养护，保持湿润。

Compacting stones. Use a trowel or a roller to slightly compact the stones evenly cast on the binder mortar so that they are embedded into the mortar for not less than 1/2 of their particle size. Gently pat the stones so that the surface is flat and firm, or the mortar may be squeezed to cover the stones and marks of the trowel or roller shaft may be left on the surface. For external corners, cast stones on both sides of the corner at the same time, or they may be difficult to attach to one side of the corner after they are cast on the other side and there will be an obvious black mark on the joint. Alternatively, this can be done in a mechanical way to cast stones on the binder layer evenly and strongly with compressed air and a casting gun. Hold the gun in front of the wall surface with the nozzle 300-400 mm away from the wall and an air pressure of 0.6-0.8 MPa. Spray water on the binder layer during the setting period for curing and keeping the surface moist.

起分格条与修整。干粘石墙面达到表面平整、石子饱满，即可将分格条取出，取分格条时应注意不要掉石子。如局部石子不饱满，可立即刷108胶水溶液，再甩石子补齐。将分格条取出后，随即用小溜子和素水泥浆将分格缝修补好。

Removing dividing strips and finishing. Remove the dividing strips when stones are evenly distributed and the wall surface is flat and level. Take care to prevent the stones from dropping out. Apply 108 adhesive on uneven parts(if any) in time and cast supplementary stones. After removing the dividing strip, repair the dividing joint with a narrow trowel and plain cement paste to keep it straight and clear.

干粘石施工操作简便，但是此种墙面久经风吹雨打易产生脱粒现象，现在已较少被采用。

As stones may drop out after being exposed to wind and rain for a long time, drydash works is rarely used today even with easy operation.

9.2　饰面工程

9.2 Finishing Works

饰面工程就是将人造的、天然的饰面材料镶贴于基层表面形成装饰层的施工过程。饰面的材料种类很多，一般可分为饰面砖和饰面板两大类。饰面砖如釉面砖、外墙面砖、陶瓷锦砖；饰面板如大理石、花岗岩等天然石材饰面板，预制水磨石、水刷石、人造大理石等人造石材饰面板，以及金属板和木质饰面板。

Finishing works shape a decorative layer on the base surface with artificial and natural finishing materials. There are many kinds of finishing materials, which can be generally categorized into facing tiles and facing veneers. Facing tiles include glazed tiles, exterior wall tiles and ceramic mosaic tiles; facing veneers include natural stone veneers such as marble and granite, artificial stone veneers such as precast terrazzo, granitic plaster and artificial marble, as well as metal and wood veneers.

本节主要介绍饰面砖的镶贴施工工艺和饰面板挂贴施工工艺。

This section mainly introduces the construction of facing tiles and facing veneers.

9.2.1 饰面砖施工

9.2.1. Construction of Facing Tiles

（1）饰面砖构造

(1) Structure of facing tiles

使用镶贴施工的饰面砖及基层的相关构造如图 9-7。

See Figure 9-7 for the relevant structures of facing tiles and base with the tiling technology.

（2）施工工艺流程

(2) Construction process

饰面砖镶贴施工的工艺流程：基层处理→抹找平层→选面砖→预排砖→分格弹线→浸砖→做标志块→镶贴→面砖勾缝与擦缝→养护及清理。

Process of tiling facing tiles: Base treatment → plastering leveling layer → selecting facing tiles → pre-laying the tiles → dividing and snapping lines → soaking tiles → pasting marker blocks → tiling → joint grouting and wipe-off → curing and cleaning.

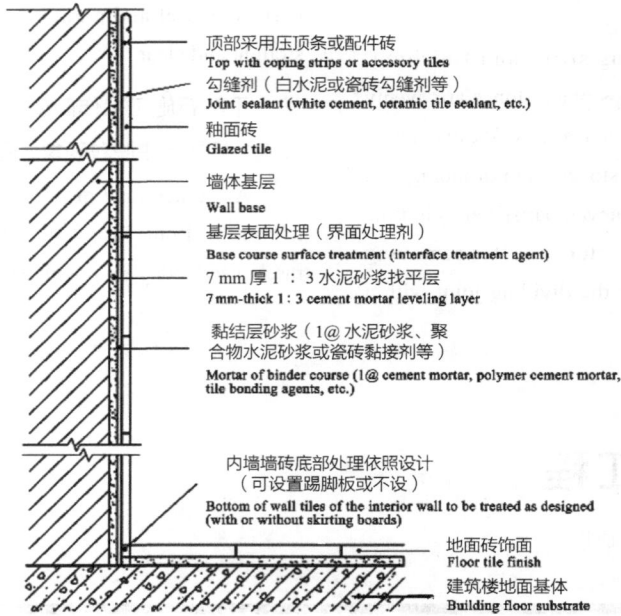

图9-7 饰面砖及基层构造

Figure 9-7 Structure of Facing Tiles and Base Course

（3）施工要点

(3) Key construction points

①基层处理。

① Base treatment.

先将凸出墙面的混凝土剔平，将表面尘土、污垢清扫干净，然后用 1 ：1 水泥细砂浆内掺水重 20% 的 107 胶，喷或用笤帚将砂浆甩到墙上，其甩点要均匀，终凝后浇水养护，直至水泥砂浆疙瘩全部黏到混凝土光面上，并有较高的强度（用手掰不动）为止。

First, remove the concrete protruding from the wall surface, and clear the dust and dirt off the surface. Evenly apply 1 ：1 cement fine mortar, mixed with 107 glue with 20% water by weight, on the wall surface by shooting or with a broom. Spray water after the final setting for curing until the cement mortar lump is completely adhered to the smooth finish surface of the concrete and has high strength(cannot be removed by hand).

②基层找平。

② Leveling of base.

按一般抹灰施工的方法进行，找平层要求平整、垂直、方正。

The leveling process is the same as that of the general plastering, and the leveling layer shall be flat, vertical and square.

③弹线分格。

③ Line snapping and dividing.

待基层灰六至七成干时，即可按图纸要求进行分段分格弹线，同时亦可进行面层贴标志块的工作，以控制面层出墙尺寸，使其保持垂直、平整，如图 9-8 所示。

Snap lines in sections and grids according to the drawing requirements after the base mortar is 60% to 70% dry and paste marker blocks at the same time, to control the protruding degree, keep vertical and flat of the surface layer, as shown in Figure 9-8.

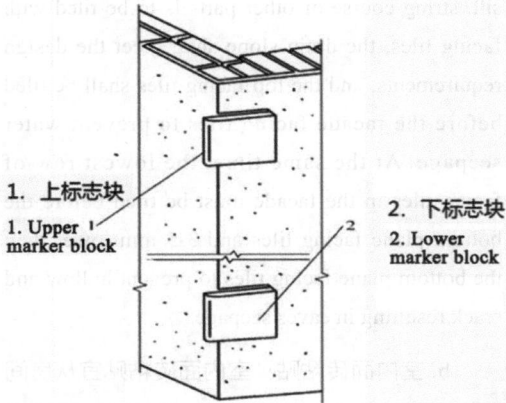

1. 上标志块
1. Upper marker block

2. 下标志块
2. Lower marker block

图9-8 标志块
Figure 9-8 Marker Block

④选砖、排砖。

④ Tile selection and laying.

选砖：选面砖是按照面砖颜色深浅、尺寸大小、表面平整度等进行筛选，保证面砖的饰面效果。

Tile selection: Select tiles according to the color, size, surface flatness, etc. of the tiles to ensure the finishing effect.

排砖：根据大样图及墙面尺寸进行横竖向排砖，以保证面砖缝隙均匀，符合设计图纸要求，注意大墙面、通天柱子和垛子要排整砖，以及在同一墙面上的横竖排列，均不得有一行以上的非整砖。非整砖行应排在次要部位，如窗间墙或阴角处等，但亦要注意一致和对称。如遇有突出的卡件，应用整砖套割吻合，不得用非整砖随意拼凑镶贴。

Tile laying: Lay tiles horizontally or vertically according to the detail drawing and wall size to ensure uniform space between facing tiles and meet the requirements of design drawings. Note that complete tiles shall be laid for large walls, full-length columns and buttresses, and more than one row of non-complete tiles is not allowed on the same wall surface in a vertical and horizontal row.

Arrange the non-complete tile row in secondary parts, such as walls between windows or internal corners, but ensure conformity and symmetry. In case of an extruding part, cut a complete tile to match the pattern, and do not randomly patch the part with a non-complete tile.

⑤浸砖。

⑤ Tile soaking.

釉面砖和外墙面砖镶贴前，首先要将面砖清扫干净，放入净水中浸泡 2 h 以上，取出待表面晾干或擦干净后方可使用。

Before tiling the glazed tiles and exterior wall tiles, clean and soak the facing tiles in clean water for more than 2 h and then take them out for drying in the air or wiping up before use.

⑥镶贴面砖。

⑥ Tiling.

a. 外墙镶贴：外墙镶贴应自上而下进行。高层建筑采取相关安全措施后，可分段进行。在每一分段或分块内的面砖，均为自下而上镶贴。从最下一层砖下皮的位置线先稳好靠尺，以此托住第一皮面砖。在面砖外皮上口拉水平通线，作为镶贴的标准。

a. Tiling exterior walls: Tile the exterior wall from top to bottom. High-rise buildings can be tiled in sections after relevant safty measures are taken. The facing tiles in each section or grid are tiled from bottom to top. Set the guiding rule stably against the position line of the lowest layer of tiles to support the first layer of facing tile. Pull a full-length line horizontally from the upper edge of the outer layer of tiles as the reference for tiling.

在面砖背面宜采用 1 ∶ 2 水泥砂浆镶贴，砂浆厚度为 6 ～ 10 mm，贴上后用灰铲柄轻轻敲打，使之附线，再用钢片开刀调整竖缝，并用小杠通过标准点调整砖面垂直度。

Tile the facing tiles by applying 6-10 mm

thick 1 : 2 cement mortar on the back of the tile and then tap the tiled tile gently with the handle of a plaster shovel to align it to the full-length line. Then, adjust the vertical joint with a steel knife, and adjust the verticality with a small bar based on reference points.

另外一种做法是，用 1 : 1 水泥砂浆加水重 20% 的 107 胶，在砖背面抹 3 ~ 4 mm 厚黏贴即可。但使用此种做法的前提是基层灰必须抹得平整，而且砂子必须用窗纱筛后使用。

Alternatively, apply 3-4 mm thick 1 : 1 cement mortar(mixed with 107 adhesive with 20% water by weight) on the back of the tile for tiling. But the base mortar must be plastered flat, and the sand must be screened with mesh.

另外，也可用胶粉来粘贴面砖，其厚度为 2 ~ 3 mm，使用此种做法必须保证基层灰抹得更平整。

Rubber powder can also be used for tiling, generally with a thickness of 2-3 mm, but the base mortar must be plastered even flatter.

如要求釉面砖拉缝镶贴时，面砖之间的水平缝宽度用米厘条控制，米厘条用贴砖用砂浆与中层灰临时镶贴，米厘条贴在已镶贴好的面砖上口，为保证其平整，可临时加垫小木楔。

If glazed tiles are to be tiled by setting joints, use a floating rule to control the width of the horizontal joint between facing tiles, and temporarily paste the floating rule with tiling mortar on the intermediate plaster. Paste the floating rule on the upper edge of the tiled facing tile and add a wooden wedge to ensure smoothness if necessary.

女儿墙压顶、窗台、腰线等部位平面也要镶贴面砖时，除流水坡度符合设计要求外，应采取顶面面砖压立面面砖的做法，预防渗水。同时还应采取立面中最低一排面砖必须压底平面面砖，并低出底平面面砖 3 ~ 5 mm 的做法，防止屋檐面空裂。

When the plane of parapet coping, window sill, string course or other parts is to be tiled with facing tiles, the drain slope shall meet the design requirements, and the top facing tiles shall be tiled before the facade facing tiles to prevent water seepage. At the same time, the lowest row of facing tiles in the facade must be tiled before the bottom plane facing tiles and 3-5 mm lower than the bottom plane facing tiles to prevent hollow and crack resulting in eaves seepage.

b. 室内面砖粘贴：室内面砖粘贴宜从房间阳角开始，并由上而下进行。按地面水平线嵌上一根八字尺或直靠尺，用水平尺校正，作为第一行面砖水平方向的依据。粘贴时，墙面砖的下口坐在八字尺或靠尺上，这样可防止面砖因自重而向下滑移，以确保其横平竖直。

b. Tiling interior facing tiles: Interior facing tiles should be tiled from the external corner of the room and from top to bottom. A Luban ruler or a straight guiding ruler shall be embedded according to the horizontal line on the ground, and calibrated with a level ruler, so as to serve as the reference for the first row of facing tiles in the horizontal direction. During tiling, the lower edge of the wall tiles shall be placed on the Luban ruler or guiding ruler, so as to prevent the tiles from sliding down by gravity and ensure that they are plumb and level.

在面砖背面刮满 1 : 2 水泥砂浆刀灰，厚度为 6 ~ 10 mm，按所弹尺寸线，将面砖坐在八字尺或靠尺上，贴于墙面并用力按压，使其略高于标志块，贴上后用灰铲柄轻轻敲打，使面砖紧密黏在墙面上，再用钢片开刀调整竖缝，并用小杠通过标准点调整平面和垂直度。对于高出标志块的应轻轻敲击，使其平齐，若低于标志块时应取下面砖，重新抹满刀灰再粘贴，不得在砖口处塞灰，否则会产生空鼓。

Fully apply 1 : 2 cement mortar on the back of the facing tile, with a thickness of 6-10 mm.

According to the snapped line, place the facing tile on the Luban ruler or guiding ruler and attach it to the wall surface to make it slightly higher than the marker block. After tiling, tap the tile gently with the handle of a plaster shovel to make it tightly adhere to the wall surface, then adjust the vertical joint with a steel knife, and adjust the plane and verticality with a small bar based on reference points. Taking the marker block as a reference surface, gently tap tiles protruding from the reference surface and remove tiles below the reference surface for applying mortar and tiling again. Do not fill mortar at the brick edge of the tile, so as to avoid hollowing.

然后，依次按上述方法往上粘贴，粘贴时应尽量注意与相邻面砖的平整及竖直方向的垂直和水平方向的平整，如因面砖的规格尺寸或几何形状不等时，应在粘贴时随时调整，使其缝隙宽窄一致。当贴到最上一行时，要求上口成一直线。上口如没有压条（镶边）应用一面圆的面砖。阳角的大面一侧用圆的面砖，这一排的最上面一块应用两面圆的面砖。

Then tile as per the above method one by one. Try to ensure flatness at the joint of adjacent tiles, verticality in the vertical direction and smoothness in the horizontal direction. If specifications, dimensions or geometric shapes of facing tiles vary, adjust them at any time during tiling to ensure the same joint width. The upper edge of the tiles in the top row shall form a straight line. A round facing tile shall be used if there are no trim strips(edging) at the upper edge of the tile. A round facing tile shall be used for the larger side of the external corner, and two round facing tiles shall be used for the top tile of this row.

⑦面砖勾缝与擦缝。

⑦ Joint grouting and wipe-off.

面砖铺贴拉缝时，用 1 ： 1 水泥砂浆勾缝，先勾水平缝再勾竖缝，勾好后要求凹进面砖外表面 2 ~ 3 mm。若水平竖缝为干挤缝，或小于 3 mm 者，应用白水泥配颜料进行擦缝处理。面砖缝子勾完后，用布或棉丝蘸稀稀盐酸擦洗干净。

When tiling facing tiles by setting joints, 1 : 1 cement mortar shall be used for joint grouting. The horizontal joints shall be pointed before the vertical joints. The joint shall be below the outer surface of the facing tile for 2-3 mm. If the horizontal and vertical joints are dry extruded joints or less than 3 mm, white cement with pigment shall be used to grout the joints. After pointing the tile joints, wipe them with cloth or cotton wool dipped in diluted hydrochloric acid.

9.2.2 饰面板工程

9.2.2 Facing Veneers

饰面板的施工方法，通常有挂贴法、粘贴法和干挂法几种，本节主要介绍干挂法施工，如图 9-9 所示。

The construction methods of veneers usually include the hanging and attaching method, the pasting method and the dry hanging method. This section mainly introduces the dry hanging method, as shown in Figure 9-9.

干挂法施工，即在饰面板材上直接打孔或开槽，用各种形式的连接件与结构基体用膨胀螺栓或其他架设金属连接而不需要灌注砂浆或细石混凝土。

In the dry hanging method, holes or slots are directly made on the veneers, and various connectors are connected with the structural base with expansion bolts or connected with other erection metals without pouring mortar or fine stone concrete.

饰面板与墙体之间留出 40 ~ 50 mm 的空腔。干挂法铺贴多适用于 30 m 以下的钢筋混凝土结构基体，不适用于砖墙和加气混凝土墙。其主要优点有：

图9-9　干挂法施工

Figure 9-9　Construction by Dry Hanging Method

A cavity of 40-50 mm is reserved between the veneer and the wall. The dry hanging method is applicable to reinforced concrete bases below 30 m but not applicable to brick walls and aerated concrete walls. Its main advantages are listed below:

在历经强风和地震时，或许会产生适量的变位，但不致出现裂缝和脱落。

Under the action of strong wind and earthquake, appropriate displacement is allowed without cracks and falling off.

冬季照常施工，不受季节限制。

Construction can be carried out in winter and is not limited by seasons.

没有湿作业，既改善了施工环境，也避免了浅色板材透底污染的问题以及空鼓、脱落等问题的产生。

This method features no wet construction, better construction environment and is free of problems such as penetration pollution of light-colored plates, and hollowing and falling off.

可以采用大规格的饰面石材铺贴，从而提高了施工效率。

Large-sized facing stones can be used for finishing by this method, thus improving construction efficiency.

可自上而下拆换、维修，无损于板材和连接件，使饰面工程拆改翻修方便。

The facing materials can be removed, replaced and repaired from top to bottom without damaging the boards and connectors, so as to facilitate the removal, modification and repair of the finishing works.

（1）饰面板干挂法施工的构造

(1) Construction structure of veneers by dry hanging method

饰面板干挂法施工工艺主要采用扣件固定法，如图 9-10 和图 9-11 所示。

The dry hanging construction is dominated by fastener fixing, as shown in Figure 9-10 and Figure 9-11.

不锈钢斜脚直角钩
Stainless steel inclined leg right-angle hook

φ6 斜洞　φ6 inclined hole
45° 斜　45° inclined

φ10 stainless steel
φ10 不锈钢

如系砖墙，凡膨胀螺栓处均
加 C20 细石混凝土块
C20 fine aggregate
concrete blocks are added
for all expansion bolts in
case of a brick wall

膨胀螺栓横向间距 ＝
板长 L＋板缝
Transverse spacing of expansion bolts ＝
boardlength L ＋ board joint

φ5 不锈钢销钉
φ5 stainless steel pin
φ8 不锈钢螺栓
φ8 stainless steel bolt

不锈钢直挂件
Straight stainless
steel hanger
嵌缝处理
Joint sealing

Board
thickness
of 80 or
thickness
as
designed

板厚
80 或按
具体设

板高 H
Board height H

板高 H
Board height H

不锈钢
角挂件
Stainless steel
angled hanger

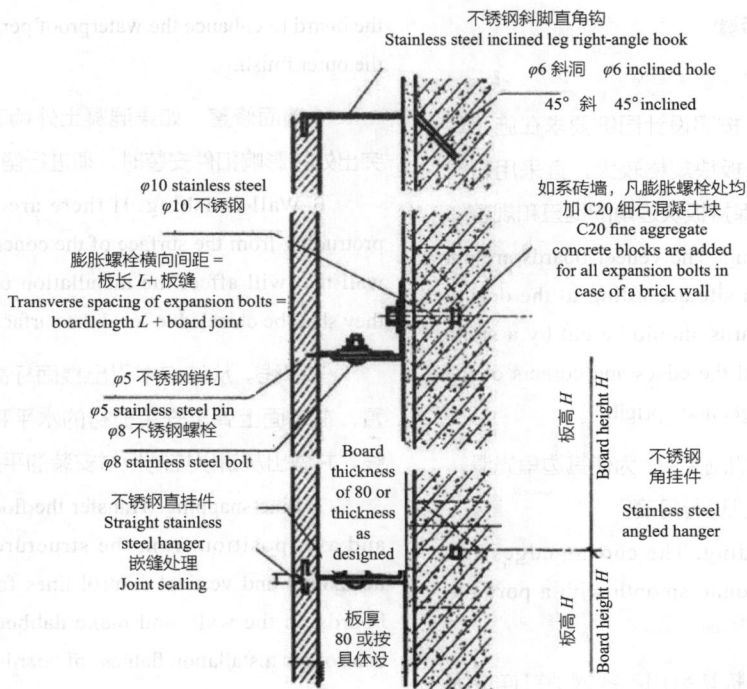

图9-10　干挂石材施工构造

Figure 9-10 Construction Structure of Dry Hanging Method

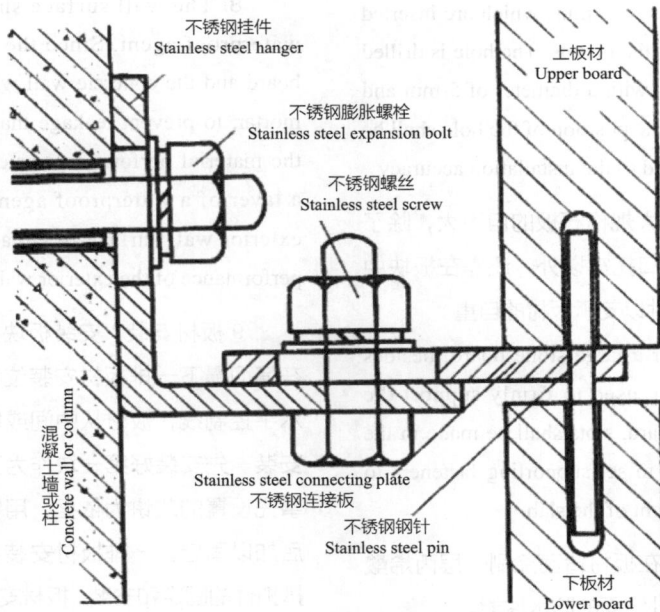

不锈钢挂件
Stainless steel hanger

上板材
Upper board

不锈钢膨胀螺栓
Stainless steel expansion bolt

不锈钢螺丝
Stainless steel screw

混凝土墙或柱
Concrete wall or column

Stainless steel connecting plate
不锈钢连接板

不锈钢钢针
Stainless steel pin

下板材
Lower board

图9-11　干挂石材节点大样图

Figure 9-11 Node Detail of Dry Hanging Method

（2）安装施工步骤

(2) Installation and construction steps

①板材切割。按照设计图纸要求在施工现场进行切制，由于板块规格较大，宜采用石材切割机切制，注意保持板块边角的挺直和规整。

① Board cutting. The veneer boards are cut on the construction site according to the design drawing. Large boards should be cut by a stone cutting machine, and the edges and corners of the boards shall be straight and upright.

②磨边。板材切割完后，为使其边角光滑，可采用手提式磨光机进行打磨。

② Edge grinding. The cutting edges and corners shall be ground smooth with a portable grinding machine.

③钻孔。相邻板块采用不锈钢销钉连接固定，销钉插在板材侧面孔内，孔径 5 mm，深度 12 mm，用电钻打孔。由于它关系到板材的安装精度，要求钻孔位置准确。

③ Drilling. Adjacent boards are connected and fixed by stainless steel pins, which are inserted into the side holes of the boards. The hole is drilled with an electric drill, with a diameter of 5 mm and a depth of 12 mm. The position of the hole shall be accurate as it is related to the installation accuracy.

④开槽。由于大规格石板的自重大，除了由钢扣件将板块下口托牢以外，还需在板块中部开槽设置承托扣件以支承板材的自重。

④ Slotting. For a large stone board, besides steel fasteners being used to firmly support the lower edge of the board, slots shall be made in the middle of the board to set supporting fasteners to support the dead weight of the slab.

⑤涂防水剂。在板材背面涂刷一层丙烯酸防水涂料，以增强外饰面的防水性能。

⑤ Applying waterproof agent. One coat of acrylic waterproofing paint is applied to the back of the board to enhance the waterproof performance of the outer finish.

⑥墙面修整。如果混凝土外墙表面有局部突出处并影响扣件安装时，须进行凿平修整。

⑥ Wall finishing. If there are local parts protruding from the surface of the concrete exterior wall that will affect the installation of fasteners, they shall be chiseled to level the surface.

⑦弹线。从结构中引出楼面标高和轴线位置，在墙面上弹出安装板材的水平和垂直控制线，并做出灰饼以控制板材安装的平整度。

⑦ Line snapping. Transfer the floor elevation and axis position from the structure, snap the horizontal and vertical control lines for installing boards on the wall, and make dabbed mortar to control the installation flatness of boards.

⑧墙面涂刷防水剂。由于板材与混凝土墙身之间不填充砂浆，为了防止因材料性能或施工质量可能造成的渗漏，在外墙面上涂刷一层防水剂，以加强外墙的防水性能。

⑧ The wall surface shall be coated with waterproof agent. Since the space between the board and the concrete wall will not be filled with mortar, to prevent leakage that may be caused by the material performance or construction quality, a layer of a waterproof agent is applied on the exterior wall surface to enhance the waterproof performance of the exterior wall.

⑨板材安装。安装板块是自下而上进行，在墙面最下一排板材安装位置的上下口拉两条水平控制线，板材从中间或墙面阳角开始就位安装。先安装好第一块作为基准，其平整度以事先设置的灰饼为依据，用铅锤吊直，经校准后加以固定。一排板材安装完毕，再进行上一排扣件的固定和安装。板材安装要求四角平整，纵横对缝。

⑨ Board installation. The boards are installed

from bottom to top. Two horizontal control lines are pulled along the upper and lower edges at the installation position for the lowest row of boards on the wall, and the boards are set in place for installation from the middle part or the external corner of the wall. Install the first board as a reference, use the dabbed mortar set up in advance as the basis of flatness, check the verticality with a plumb bob and fix the board after alignment. After a row of boards is installed, the fastenings of the previous row shall be fixed and installed. The four corners of the installed board shall be plumb and level and the vertical and horizontal joints shall be aligned.

⑩板材固定。钢扣件和墙身用膨胀螺栓固定，扣件为一块钻有螺栓安装孔和销钉孔的平钢板，根据墙面与板材之间的安装距离，在现场用手提式折压机将其加工成角型钢。扣件上的孔洞均呈椭圆形，以便安装时调节位置。

⑩ Board fixing. The steel fastener and the wall body are fixed with expansion bolts. The fastener is a flat steel plate drilled with bolt mounting holes and pin holes. According to the installation distance between the wall surface and the board, it is processed into angle steel by a portable flanging press on site. The holes in the fasteners are oval to facilitate adjustment of the position during installation.

⑪板材接缝的防水处理。石板饰面接缝处的防水处理采用密封硅胶嵌缝。嵌缝之前先在缝隙内嵌入柔性条状泡沫聚乙烯材料作为衬底，以控制接缝的密封深度和加强密封胶的黏接力。

⑪ Waterproofing of board joints. Joints of stone veneer shall be waterproofed by caulking seal silicone. Before caulking, polyethylene flexible stripped foam is embedded in the joint as a substrate to control the sealing depth of the joint and the adhesive strength of the sealant.